The Physics of Non-Ideal Plasma

The Physics of Non-Ideal Plasma

Vladimir E. Fortov
Igor T. Iakubov

Russian Academy of Sciences

World Scientific
Singapore • New Jersey • London • Hong Kong

Published by

World Scientific Publishing Co. Pte. Ltd.

P O Box 128, Farrer Road, Singapore 912805

USA office: Suite 1B, 1060 Main Street, River Edge, NJ 07661

UK office: 57 Shelton Street, Covent Garden, London WC2H 9HE

British Library Cataloguing-in-Publication Data
A catalogue record for this book is available from the British Library.

THE PHYSICS OF NON-IDEAL PLASMA

ISBN 981-02-3305-1

Printed in Singapore.

Preface

This book is devoted to the physical properties of dense plasma compressed so strongly that the effects of interparticle interaction are substantial, that is, non-ideal. These non-traditional plasma studies have increased in scale over the last 10–15 years, owing to the development of modern technical projects and facilities whose operation is based on a high local concentration of energy in dense media.

Non-ideal plasma is essential from the standpoint of operation of pulsed thermonuclear reactors with inertial confinement of hot plasma, powerful magnetic-flux and magnetohydrodynamic generators, power-generating plants and rocket engines with gas-phase nuclear reactors, plasmochemical and microwave reactors, plasma generators, and powerful sources of optical and X-ray radiation. Non-ideal plasma occurs as a result of stimulation of matter by strong shock, detonation and electric-explosion waves, concentrated laser radiation, and electron and ion fluxes, under conditions of powerful chemical and nuclear explosions upon pulsed evaporation of the liners of pinches and magnetocumulative generators. Non-ideal plasma occurs during hypersonic motion of bodies in dense planetary atmospheres, as a result of high-velocity impact, and in numerous situations characterized by extreme pressures and temperatures. The physics of electrode, contact and electric-explosion processes, under conditions of vacuum breakdown, are closely related to non-ideal plasma, which is essential to the operation of powerful plasma accelerators, microwave generators and plasma switches. Modern progress in understanding the structure and evolution of the giant planets in the solar system, as well as astrophysical objects, is largely based on the ideas and results of the physics of highly compressed plasma.

Along with the pragmatic interest in high pressure plasma, purely scientific interest is gaining momentum, because it is in this exotic state that the major part of matter in the universe finds itself. Besides, the non-ideality of plasma defines the behavior of matter in a wide range of the phase diagram, from solid and liquid to neutral gas, the phase boundaries of melting and boiling, and the metal-dielectric transition region.

These facts provide a permanent and stable stimulus to intense theoretical and experimental studies, which have recently produced a host of interesting and, what is most important, reliable data on the thermodynamic, optical, electrophysical and transport properties of dense plasma. This concrete information is contained in a wide flow of original publications. This takes place against the background of an increasing number of specialists, both researchers and engineers, who make use of non-ideal plasma to solve most diverse scientific and applied problems.

The present authors have attempted to systematize, generalize, and present, from a single viewpoint, the theoretical and experimental material relating to this new field of science. The contents give a good idea of the scope of the book. They have tried to expand the discussion as much as possible to include cases when non-ideality shows most clearly in the plasma state of matter. Therefore, the interesting problems pertaining to the dense plasma of metals and semiconductors, electrolyte plasma, "cryogenic" highly non-equilibrium plasma, and a detailed discussion of plasma applications have been omitted.

The physics of non-ideal plasma appears to present a very inconvenient subject for pure theory. This is because the strong interparticle interaction impedes the use of the conventional methods of theoretical physics. Therefore, the recent progress in understanding the properties of compressed plasma was only made possible by the emergence of experimental data obtained through non-conventional generation and diagnostic techniques. In this case, the experimental results provide a basis for model theories, as well as for defining the range of applicability of asymptotic approximations. Therefore, the authors tried to maintain the natural proportions between theory and experiment while giving primary consideration to physical results. In their opinion, this is what distinguishes this work from the available (and rather few) review publications in the field.

The authors hope that this book will prove useful to broad sections of specialists by giving them access to original studies and to help them get their bearings amid the present-day problems of dense plasma. The book is written such that the knowledge of standard university courses is sufficient for productive reading.

The physics of non-ideal plasma is developing very rapidly with more and more applications coming to light. Naturally, the material contained in this book will likewise be expanded and complemented. The authors would like to thank the readers in advance for their critical comments and suggestions.

The authors are deeply grateful to L. M. Biberman, E. P. Velikhov, A. M. Prokhorov and A. E. Scheindlin for encouragement and fruitful discussions. Of great value were stimulating discussions and creative contacts with the late Ya. B. Zel'dovich and V. M. Ievlev. The authors are also grateful to all their colleagues who helped to perform the numerous experiments and calculations which form the basis of this book. They are grateful to Henry Bronshtein who made the English translation.

Contents

One

Non-Ideal Plasma. Basic Concepts

1.1. Interparticle Interactions. Criteria of Non-Ideality

At low densities, a low-temperature, partly ionized plasma can be regarded as a mixture of ideal gases of electrons, atoms and ions. The particles travel at thermal velocities, mainly along straight paths, and collide with each other only occasionally. In other words, the free path times prove greater than those of interparticle interaction. With an increase in density, mean distances between the particles decrease and the particles start spending even more time interacting with each other, that is, in the fields of surrounding particles. Under these conditions, the mean energy of interparticle interaction increases. When this energy becomes comparable with the mean kinetic energy of thermal motion, the plasma becomes non-ideal. The properties of such plasma cease to be described by simple relationships of the theory of ideal gases and plasma, and become very unusual.

If the plasma is fully ionized, its state is defined by Coulomb interactions, whose specific feature is their long-range action. Therefore, in a rarefied plasma, the particles move in weak but self-consistent fields developed by all the particles. The interaction energy increases with compression. However, it starts receiving an ever growing contribution from strong pair interactions. Finally, under conditions of strong non-ideality, the role of these interactions becomes predominant. However, the pair interactions start to overlap, and the many-particle aspect of interaction takes a new form.

The peculiarities of interparticle interactions are discussed below. The parameters of non-ideality are introduced and assessed. This enables one to classify the states of non-ideal plasma and single out the density and temperature regions in which some or other effects of interparticle interaction become most pronounced.

The conditions of strong non-ideality call for a high concentration of energy in the plasma. Therefore, under natural conditions, the plasma non-ideality is usually most evident in phenomena which have always attracted researchers' attention. In the recent decade, the interest in non-ideal plasma assumed a practical significance and stimulated a number of new scientific and technical ideas. Therefore, the last section of this chapter deals with non-ideal plasma in natural phenomena and technology.

1.1.1. *Interparticle Interactions*

Interatomic interactions, Coulomb interactions between charges, and interactions between neutral and charged particles are realized in a plasma consisting of electrons, atoms and ions. If the plasma is fully ionized, its state is defined by Coulomb interactions whose specific character resides in their long-range interaction. In a partly ionized plasma, charge-neutral interactions become most pronounced and, at high densities, interatomic (in a molecular plasma, intermolecular) interactions.

1.1.2. *Coulomb Interaction*

The criterion of ideality of a plasma as regards the interaction between charged particles may be provided by the smallness of the quantity γ:

$$\gamma = Ze^2 N_e^{1/3} \beta \ll 1 , \tag{1.1}$$

where N_e is the electron concentration equal to the ion concentration because of the condition of local electrical neutrality, and $\beta = T^{-1}$ is a quantity reciprocal to temperature T.

The quantity γ is the ratio between the characteristic potential energy of Coulomb interaction $e^2/\bar{r}$ over mean interparticle distance $\bar{r}$ and the mean thermal energy T. The quantity $\bar{r}$ is related to the plasma density by the simple relation

$$(4\pi/3) N_i \bar{r}^3 = 1 .$$

Given high temperatures, multiple ionization can be attained in a plasma incorporating complex atoms and ions. The parameter of Coulomb interaction is then written as

$$\gamma = Z_1 Z_2 e^2 \beta / \bar{r} , \tag{1.2}$$

where Z_1 and Z_2 are the charges of interacting particles, and $\bar{r}$ is the mean distance between them.

Note that γ is the parameter of non-ideality of a classical Coulomb system. The electrons and ions in a plasma form a classical system if they are seldom found to be at distances comparable with the thermal wavelength $\lambda_e = \hbar/\sqrt{2mT}$. Since the drawing together of charged particles is only possible up to a distance of Ze^2/T, the condition of classicality can be written as

$$\lambda_e \ll Ze^2\beta\,. \tag{1.3}$$

Therefore, a plasma is classical at relatively low temperatures, $T \leq \text{Ry} = e^4m/\hbar^2$. In a more heated plasma, one must allow for interference quantum effects due to the uncertainty principle.

Another manifestation of quantum effects is degeneracy and, primarily, the degeneracy of electrons having the greatest thermal wavelength. The possibility of employing classical statistics is defined by the smallness of the number of electrons in an elementary volume with a linear size of λ_e. In other words, the $N_e\lambda_e^3 \ll 1$ inequality should be met. The condition of applicability of classical statistics can be transcribed, using the relationship between temperature and the Fermi energy ε_F, into

$$\xi = \beta\varepsilon_\text{F} \ll 1\,, \qquad \varepsilon_\text{F} = (3\pi^2N_e)^{2/3}\hbar^2/m\,. \tag{1.4}$$

Therefore, the isothermal compression of low-temperature (at $T \leq \text{Ry} = 1.5 \cdot 10^5$ K) plasma leads to an increase in Coulomb interaction energy which, after the parameter γ becomes greater than unity, exceeds the kinetic energy of particle motion. This hampers considerably the theoretical description of non-ideal plasma while making it impossible to use perturbation theory. It also forces one to employ qualitative physical models. Further compression of plasma causes an increase in non-ideality, though to a certain limit only. The thing is that a degeneracy of electrons occurs at $N_e\lambda_e^3 \approx 1$ with an increase in density. The Fermi energy (1.4), which increases with the plasma density, becomes the kinetic energy scale. The non-ideality criterion assumes the form of

$$\gamma_\text{q} = e^2N_e^{1/3}/\varepsilon_\text{F} \geq 1\,. \tag{1.5}$$

Therefore, a degenerate electron plasma becomes even more ideal with compression, and the parameter γ_q decreases because $\varepsilon_\text{F} \sim N_e^{2/3}$. It should be emphasized that at higher values of density, only electrons represent a quasi-homogeneous ideal Fermi gas. The ion component is non-ideal. Depending on the degree of its non-ideality, one may talk about ionic liquid, cellular structure and other model representations of the ion subsystem.

It is well-known that even in a rarefied plasma, when $\gamma \to 1$, one cannot directly employ the formulas of the ideal gas theory to determine the thermodynamic and transport properties of plasma. Quantities such as the second virial coefficient or free path length prove to be tending to infinity. This is due to the specific character of the Coulomb potential, that is, its slow decrease at large distances and infinite increase at small distances.

As a result of the long-range effect of the potential Ze^2/r, multi-particle interactions at great distances $r \gg \bar{r}$ in a plasma prove to be substantial. Consider now the charge density distribution in the neighborhood of an arbitrary particle. Such a particle repels like charges and attracts charges of the opposite sign of multiplicity of Z. The potential of the resultant self-consistent field created by the selected (test) particle and its environment is equal to

$$\varphi = (Ze/r)\exp(-r/r_{\mathrm{D}})\,, \tag{1.6}$$

where r is the distance from the test particle and r_{D} is the Debye screening distance:

$$r_{\mathrm{D}} = \left(4\pi e^2 \sum_k \beta Z_k^2 N_k\right)^{-1/2} . \tag{1.7}$$

Therefore, the Coulomb field of the test particle is screened over a distance of the order of r_{D}. This, in fact, leads to the convergence of the basic quantities, such as the free path length and virial coefficient. Estimate the typical value of the Debye screening distance. Let $T = 1$ eV, $N_e = 10^{15}$ cm^{-3}, then $r_{\mathrm{D}}. = 2 \cdot 10^{-5}$ cm. While doing so, approximately 50 charged particles of either sign will be found in a sphere of the radius r_{D}. The presence in the Debye sphere of a sufficiently large number of particles is the essential condition of validity of the estimates which led to Eqs. (1.6) and (1.7).

Let us now estimate the intensity of interparticle interaction in a Debye plasma. Note that Eq. (1.6) represents a superposition of the Coulomb potential created by the test particle and potential created by all the remaining particles of the Debye sphere. By deducting from Eq. (1.6) the test particle potential e/r and directing $r \to 0$, we obtain, at the location of test particle, the potential created by charged particles of the Debye sphere, $\varphi_{\mathrm{D}} = -Ze/r_{\mathrm{D}}$. The ideality criterion will then be written as

$$\Gamma = Ze^2/r_{\mathrm{D}}T \ll 1\,. \tag{1.8}$$

The parameter Γ is referred to as plasma parameter.

It is readily seen that the criterion (1.8) can be expressed in terms of $N_{\rm D} = (4\pi/3)N_e r_{\rm D}^3$, that is, the number of electrons in the Debye sphere, $N_{\rm D} \gg 1$. Therefore, the plasma ideality criterion, that is, the smallness of the energy of Coulomb interaction as compared with kinetic energy, coincides with the condition of applicability of the Debye approximation, namely, the number of charged particles in the Debye sphere must be great. At $T = 1$ eV, an electron concentration much lower than 10^{18} cm^{-3} corresponds to an ideal plasma with $\Gamma \leq 0.1$.

Under conditions when the electron component is degenerate, $N_e\lambda_e^3 \gg 1$. The distance of screening by degenerate electrons is defined by the Thomas-Fermi length,

$$r_{\rm TF} = (\pi/3N_e)\sqrt{\hbar^2/4me^2}\,.$$

In a two-component electron-ion system, in which the ions remain classical, the screening distance of the test charge is defined by the expression

$$r^{-2} = (r_{\rm TF}^e)^{-2} + (r_{\rm D}^i)^{-2} \approx (r_{\rm D}^i)^{-2}\,.$$

Hence it follows that the screening is done by the ions.

As the density increases, the Debye screening distance decreases and may become less than the interparticle distance $\bar{r}$. Under these conditions, r_D loses the meaning of screening distance. The screening is effected over the length of the order of $\bar{r}$. For example, such is the situation in liquid metal, where the charge is screened over distances of the order of the radius of Wigner-Seitz cell.

Another characteristic singularity of Coulomb potential is its behavior at small distances. It renders a purely classical Coulomb system unstable. The "electron-ion" binary correlation function $\exp(e^2/r)$ strongly tends to infinity. However, it is at small distances (of the order of wavelength λ_e) that quantum effects become appreciable. One can speak of short-range quantum repulsion and link its emergence to the effect of the uncertainty principle, which does not permit any strong drawing together of particles with preset relative momentum. This eliminates the tendency to infinity and leads to quantum corrections.

The range of existence of non-ideal classical and degenerate plasma are given in Fig. 1.1, which represents the $\{N_e, T\}$ diagram. They are defined by the $\Gamma = 1$, $N_e\lambda_e^3 = 1$ and $\gamma_{\rm q} = 1$ curves. The diagram also shows the range of applicability of limiting approximations describing the states of a weakly non-ideal plasma, namely, the Debye and Thomas-Fermi approximations.

As we can see, only states with extremely high pressures and temperatures located at the periphery of the phase diagram of matter are accessible to

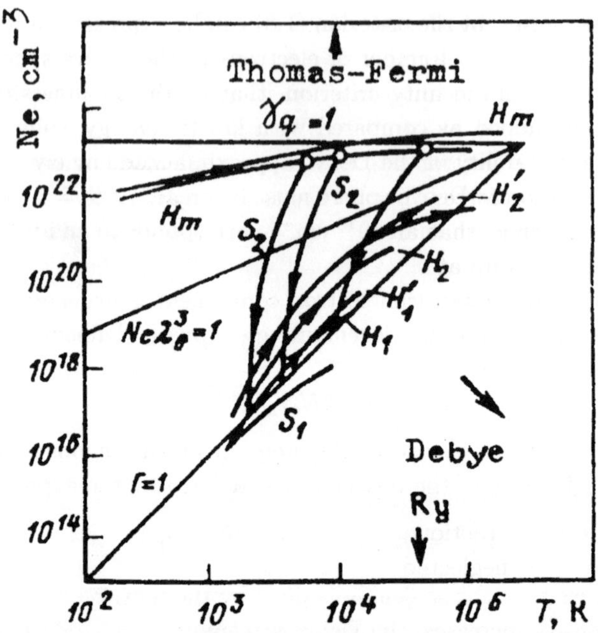

Fig. 1.1. Range of existence of non-ideal classical and degenerate low-temperature plasma.

consistent theoretical analysis. The range of existence of non-ideal plasma in Fig. 1.1 appears to be located within the triangle defined by the $\varepsilon_F = e^2 N_e^{1/3}$ and $\Gamma = 1$ curves, with the upper portion of this range relating to degenerate plasma and the lower portion to Boltzmann plasma, while the maximum possible values of the non-ideality parameter are finite and do not exceed several units. The Coulomb interaction grows weak at temperatures appreciably lower than those of the ionization potential, where other kinds of non-ideality become pronounced.

1.1.3. *Electron/Atom and Ion/Atom Interaction*

Given low electron concentrations in a low-temperature plasma, the interactions between charged particles can be ignored. Here, the interactions between charged and neutral particles may prove to be important in view of a high density for the latter. In this case, the electron and ion components of a weakly ionized plasma cannot be described in an ideal-gas approximation. The non-ideality due to charge/neutral interaction is appreciable primarily for the properties caused by the presence of charged particles, such as electrical conductivity and thermoelectromotive force.

In order to assess the importance of charge/neutral interaction, calculate the potential produced by atoms (molecules) at the electron (ion) location. Contributions to the electron/atom interaction are made by the exchange, electric and polarization forces. Because of their long-range effect, it is primarily the polarization forces that are appreciable at low densities. An atom polarized by electron sets up at the electron location a potential $\varphi(r) = -\alpha e/2r^4$, where r is the distance between them and α, the atomic polarizability. While doing so, $r > r_a$, where r_a is the atomic radius. The total potential set up by all atoms at the ion location is

$$\varphi = 4\pi \int_{r_a}^{\infty} \varphi(r) N_a(r) r^2 dr\,, \tag{1.9}$$

where $N_a(r)$ is the atomic concentration which depends on the distance from the ion. If the interaction is not yet strong, this relationship (ion/atom correlation) may be neglected. Then

$$\varphi = 2\pi e N_a \alpha / r_a\,.$$

The ideality criterion can be written as

$$\gamma_{ai} = 2\pi\alpha e^2 N_a \beta / r_a \ll 1\,, \tag{1.10}$$

where r_a is the radius of polarization interaction "cut-off" and is close to the atomic size.

The non-ideality caused by charge/neutral interaction may occur in highly polarizable gases such as metal vapors. For cesium, $\alpha = 400a_0^3$, $r_a = 4a_0$, where a_0 is the Bohr radius. In a plasma of cesium vapors at $T = 2,000$ K, $\gamma_{ai} \leq 0.1$ as long as $N_a \geq 10^{19}$ cm^{-3}.

The electron/atom interaction potential has the same polarization asymptotics $\varphi(r)$. However, this potential may not always be determined unambiguously at small distances. Full information on the electron/atom interaction is contained in the scattering phases. A set of those phases enables one to calculate the mean interaction energy using the Beth-Uhlenbeck formula (cf., [1]). This leads to rather bulky expressions. However, the electron/atom interaction at low temperatures can be described by a single parameter, namely, the electron-atom scattering length L (as long as $N_a|L|^3 \ll 1$). Thus, in solving a number of problems, the $V(r)$ potential can be replaced with a δ-like potential:

$$V(r) = 2\pi\hbar^2 L\delta(r)/m\,. \tag{1.11}$$

In this approximation, one can readily calculate the electron/atom interaction energy:

$$U = \int N_a(\mathbf{r}')V(\mathbf{r}-\mathbf{r}')|\psi(\mathbf{r}')|^2 d\mathbf{r}d\mathbf{r}' ,$$

where $\psi(r)$ is the electron wave function.

If the interaction is weak, one may neglect the electron/atom correlation and, using Eq. (1.11), obtain

$$U = 2\pi\hbar^2 L N_a/m . \tag{1.12}$$

The criterion of plasma ideality relative to the electron/atom interaction takes the form

$$\gamma_{ae} = 2\pi|L|\hbar^2 N_a\beta/m \ll 1 . \tag{1.13}$$

The approximation (1.11) is valid if $\lambda_e \gg |L|$, that is, at low temperatures.

Thanks to their high intensity and long-range effect, the plasma interactions (Coulomb and polarization) can be strong under conditions when the interaction between neutral particles is still minor. This interaction is usually adequately described by the Lennard-Jones potential. Simple inequalities can be written using the van der Waals equation of state. Hence the ideality criteria take the form

$$N_a b \ll 1 , \qquad N_a a\beta \ll 1 , \tag{1.14}$$

where a and b are the parameters in the van der Waals equation expressed in terms of the critical temperature and density:

$$T_c = 8a/27b , \qquad N_c = 1/3b . \tag{1.15}$$

Interatomic interactions are substantial at densities close to the critical values and, in vapors, in the range of loss of thermodynamic stability.

The discussion of plasma non-ideality criteria can also be found in [2–4].

1.1.4. *Compound Particles in Plasma*

First, consider how atoms are formed in a plasma. Whilst doing so, we shall proceed from a two-component classical plasma consisting of only electrons and ions. The inequality (1.3) can be rewritten as

$$\beta\mathrm{Ry} \gg 1 , \qquad \mathrm{Ry} = me^4/2\hbar^2 . \tag{1.16}$$

It follows from this inequality that the energy of bound state of electron and ion, that is, the bond energy of atom, exceeds considerably the energy of thermal motion of free particles. This points to the thermodynamic efficiency of the presence of atoms in the plasma. Such partly ionized plasma is of greatest practical interest from the standpoint of the physics of non-ideal plasma. It is also referred to as the three-component plasma, in view of the atoms, electrons and ions that remain unbound.

The concentrations of atoms, electrons and ions are related by Saha's equation:

$$\frac{N_e N_i}{N_a} = \frac{2\Sigma^+}{\Sigma}\left(\frac{m}{2\pi\hbar^2\beta}\right)^{3/2} \exp(-I\beta)\,, \tag{1.17}$$

where I is the ionization energy of atom and Σ and Σ^+ are the internal statistical sums of ion and atom, respectively.

The quantity $x = N_{e\cdot}/(N_a + N_i)^{-1}$ is referred to as the degree of ionization.

It is important that the electron and ion bound in the atom are separated from each other by a distance of the order of the Bohr radius a_0, which is the smallest characteristic dimension in the system. Therefore, the electron and ion making up the atom are screened from each other such that, in a first approximation, the interaction of such pairs with the surrounding particles may be generally ignored. Indeed, note that the bond energy of the hydrogen atom, $H = p + e$, is equal to 13.6 eV whereas the bond energy of ion, $H^- = p + 2e$, is only 0.8 eV, while the possibility of the stability of $H^{-2} = p + 3e$ is only discussed in the literature.

After the atoms are isolated, the residual quantum effects in a classical plasma are described by corrections. The latter are due to direct (interference, as is usually said) and exchange interactions. While proportional to λ_e/r_D, they are small if

$$\lambda_e/r_D \sim \gamma^{3/2}(\beta \mathrm{Ry})^{-1/2} \ll 1\,.$$

Nevertheless, given a high density or sufficiently low temperatures, a plasma may become multi-component. The A_2 molecules, A_2^+ molecular ions and A^- negative ions are formed in the plasma. These are the result of interatomic interaction and charged/neutral particle interaction. The concentrations of these composite particles are calculated using equations of chemical equilibrium. For instance, for the reaction of negative ion formation, $A^- \Leftrightarrow A + e$, we obtain the equation

$$N^-/(N_e \cdot N_a) = (\Sigma^-/2\Sigma)(\hbar^2/2\pi mT)^{3/2} \exp(\beta E)\,. \tag{1.18}$$

Here, N^- is the concentration of negative ions, Σ and Σ^- are the internal statistical sums of the negative ion and the atom, and E is the energy of electron — the atomic affinity.

The isolation of quasi-particles is always accompanied by the renormalization of interaction responsible for their emergence. The renormalized interaction may be greatly weakened at distances corresponding to the quasi-particle dimensions. This is the case if, say, the A_2^+ ion mainly occurs due to the quantum component of interaction between A^+ and A. After the isolation of A_2^+, the ion/atom interaction greatly weakens at distances of the order of ion size. The point is that the quantum forces are saturated rapidly with an increase in the number of partners. This principle also forms the basis for the chemical valency theory. In this case, the bond energy of the cluster ion A_3^+ would be substantially less than that of A_2^+. If, on the contrary, the A_2^+ ion occurs due to classical polarization interaction, the renormalization weakens the interaction, only partly because the polarization forces are additive. In the latter case, the A_3^+ ion has a bond energy close to that of A_2^+ and, at high densities, A_m^+ clusters where $m \gg 1$ may form.

1.2. The Range of Existence of Non-ideal Plasma. The Classification of States

1.2.1. *Two-Component Plasma*

If a plasma is fully ionized, it is two-component, the electron concentration N_e being equal to the ion concentration N_i. It is characterized by two independent parameters, namely, the degeneracy parameter ξ (1.4) and the interaction parameters γ (1.1) (if $\xi \ll 1$) and γ_q (1.5) (if $\xi \gg 1$).

It is convenient to show on the $\{N_e, T\}$ plane, in logarithmic co-ordinates, the lines $\gamma = 1$, $\gamma_\mathrm{q} = 1$ and $\xi = 1$, which will divide the $\{N_e, T\}$ plane into a number of characteristic regions. This is shown in Fig. 1.2 in the case of hydrogen plasma [5]. The degeneracy parameter depends on the mass of interacting particles. For electrons, $\xi \sim \hbar^2 m^{-1} \beta N_e^{2/3}$, and for ions, $\xi_i \sim \hbar^2 M^{-1} \beta N_i^{2/3}$. The same is true of the parameter γ_q: for electrons, $\gamma_\mathrm{qe} = \bar{r}/a_0$, where $a_0 = \hbar^2/me^2$, and for ions, $\gamma_{\mathrm{q}i} = \bar{r}/(\hbar^2/Me^2)$.

Region I: $\xi < 1$, $\xi_i < 1$, $\gamma < 1$ — A classical plasma with weak interaction of the electrons and ions.

Region II: $\xi < 1$, $\xi_i < 1$, $\gamma > 1$ — A classical plasma with strong interaction of the electrons and ions.

Region III: $\xi_e > 1$, $\xi_i < 1$, $\gamma_\mathrm{qe} > 1$, $\gamma_i > 1$ — The electrons form a degenerate system with strong interaction while the ions form a classical system with strong interaction.

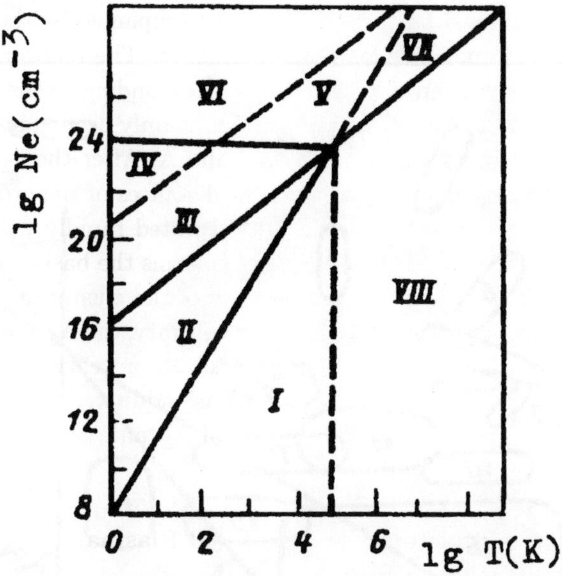

Fig. 1.2. Regions of parameters of a two-component plasma [5].

Region IV: $\xi_e > 1$, $\xi_i > 1$, $\gamma_{qe} > 1$, $\gamma_{qi} > 1$ — A quantum plasma with strong interaction of the electrons and ions.

Region V: $\xi_e > 1$, $\xi_i < 1$, $\gamma_{qe} < 1$, $\gamma_i > 1$ — The electrons form a degenerate system with weak interaction while the ions form a classical system with strong interaction.

Region VI: $\xi_e > 1$, $\xi_i > 1$, $\gamma_{qe} < 1$, $\gamma_{qi} > 1$ — The electrons are degenerate and interact weakly. The ions are degenerate and interact strongly.

Region VII: $\xi_e > 1$, $\xi_i < 1$, $\gamma_{qe} < 1$, $\gamma_i < 1$ — An electron/ion plasma with weak interaction. The electron component is degenerate.

It can be seen, from the foregoing analysis [5] of the different values of characteristic parameters, that the regions I and VII represent gaseous plasmas at various temperatures and densities; the regions V and VI correspond to a solid in which the electrons form a degenerate gas with weak interaction; and in the regions III and IV, the states which correspond to moderate values of γ for electrons (a typical value of γ in metals of 2–5) and rather high values of γ for ions are also condensed.

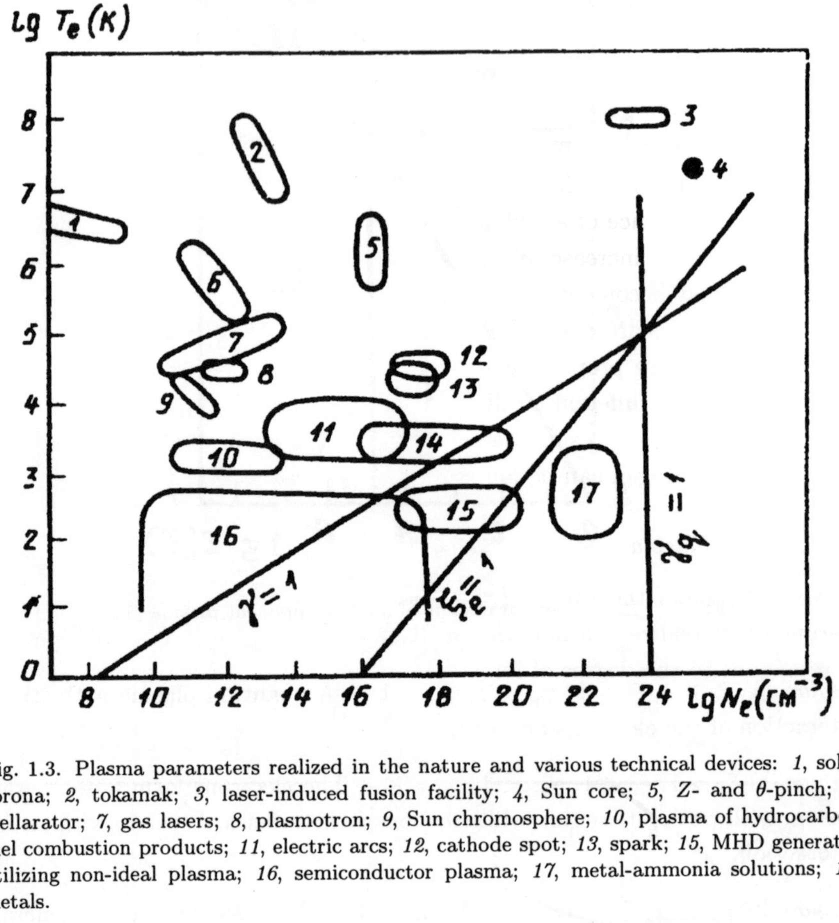

Fig. 1.3. Plasma parameters realized in the nature and various technical devices: *1*, solar corona; *2*, tokamak; *3*, laser-induced fusion facility; *4*, Sun core; *5*, Z- and θ-pinch; *6*, stellarator; *7*, gas lasers; *8*, plasmotron; *9*, Sun chromosphere; *10*, plasma of hydrocarbon fuel combustion products; *11*, electric arcs; *12*, cathode spot; *13*, spark; *15*, MHD generator utilizing non-ideal plasma; *16*, semiconductor plasma; *17*, metal-ammonia solutions; *18*, metals.

The region I represents a weakly non-ideal gas plasma well-known in physics of gas discharge and realized in numerous natural phenomena. The plasma parameters observed in the various systems are shown diagrammatically in Fig. 1.3 (see also [6, 7]).

In the regions of $\gamma < 1$, the perturbation theory can be employed to calculate the thermodynamic quantities. The particle interaction in those regions is low and the system may be regarded as a gas. In the regions of $\gamma > 1$, the perturbation theory is inapplicable; the particle interaction in such regions is appreciable and the system is similar to liquid. Such is region III where the liquid-metal states are located. The region of classical, strongly non-ideal plasma, region II, is restricted from below with respect to charge concentrations by the condition of strong interaction, $\gamma = 1$, and above by the electron degeneracy. Naturally, the regions of non-ideality are restricted from above with respect to temperature as well, because at high temperatures the kinetic energy prevails over the interaction energy.

The information provided by the $\{N_e, T\}$ diagram is incomplete. The regions of real existence of a stable two-component plasma are not indicated in Fig. 1.2. With an increase in T, the plasma recombines and becomes partly ionized and may become molecular. Finally, the matter condenses and crystallizes. These stages are realized differently for different substances. Their analysis would be more properly performed on the "density-temperature" plane. Such diagrams are different for different substances. They will be presented below for metals such as cesium and mercury, and for an inert gas. An analogous diagram for hydrogen can be found in [7].

1.2.2. *Metal Plasma*

Figure 1.4 shows the range of temperature and density where one can experimentally realize the non-ideal plasma of mercury and cesium. Curve 5 corresponds to the degree of ionization of $x = 0.5$. This curve conventionally

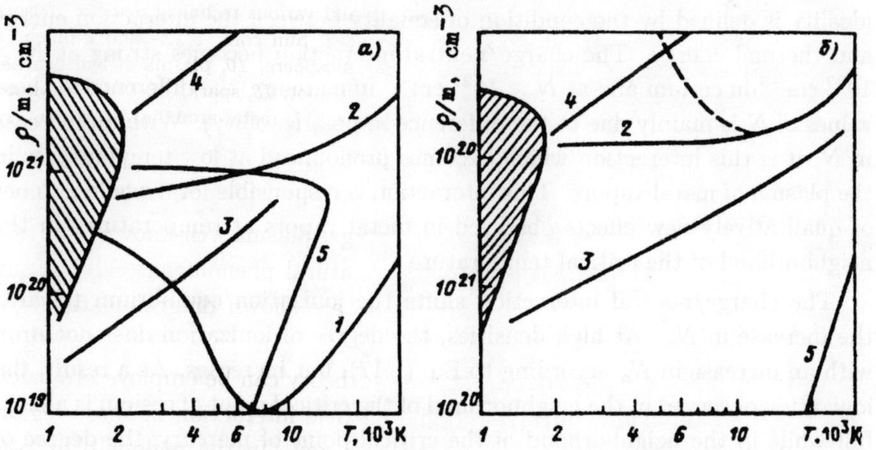

Fig. 1.4. Density-temperature diagram for cesium (a) and mercury (b): *1*, $\Gamma = 1$; *2*, $\gamma = 1$ (broken line $\gamma = 1$ if N_e, according to Saha's equation); *3*, $\gamma_{ai} = 1$; *4*, electron degeneracy; *5*, $x = 0.5$. The region of the coexistence of vapor and liquid is hatched.

separates the region of two-component fully ionized system from the region where ionization is only partial. A rise in temperature causes an increase in the configurational weight of the free states of electrons and ions. As a result, thermal ionization occurs. Accordingly, the high-temperature branch of curve 5 is constructed using Saha's equation (1.17). It is known, however, that with a strong interparticle interaction, the conventional Saha's equation does not describe the real ionization equilibrium. One of the most important effects is that, by becoming strong, the interaction leads to a decrease in the bond energy of atoms or, in other words, reduces the ionization potential. Given a very strong interaction, the bound states of electron and ion disappear altogether. It is sometimes said that pressure ionization takes place. The branch of curve 5 corresponding to high densities is constructed using the real equation of ionization equilibrium given in Fig. 1.4.

The region taken up by non-ideal plasma on the low-temperature side is defined by the curves of vapor and liquid existence on the high-density side. The region of classical non-ideality is defined by the electron degeneracy $N_e\lambda_e^3 = 1$ (in fact, $N\lambda_e^3 = 1$ because the system is already fully ionized). At $\rho \geq \rho_c$, the metallization of plasma occurs. The states above curve 4 can be regarded as liquid-metal ones. It should be noted, however, that the direct measurements in mercury by Kikoin *et al.* [8] have shown that a truly metal state is only attained at $\rho = 11\ g \cdot \text{cm}^{-3}$. (Fig. 1.4b)

On the low-density side, that is, on the ideal plasma side, the region of non-ideality is defined by the condition of equality between the interaction energy and thermal energy. The charge/neutral interaction becomes strong at $N \geq 10^{20}\ \text{cm}^{-3}$ in cesium and at $N \geq 10^{21}\ \text{cm}^{-3}$ in mercury (the difference in these values of N is mainly due to the difference in polarizability). With an increase in N, it is this interaction which becomes pronounced at low temperatures in the plasma of metal vapors. This interaction is responsible for a whole number of qualitatively new effects observed in metal vapors at temperatures in the neighborhood of the critical temperature.

The charge/neutral interaction shifts the ionization equilibrium towards the increase in N_e. At high densities, the degree of ionization does not drop with an increase in N_a according to Eq. (1.17), but increases. As a result, the ionization observed in the neighborhood of the critical point of cesium is almost full while in the neighborhood of the critical point of mercury, the degree of ionization x amounts to several tens per cent.

In view of this, in Fig. 1.4, $\Gamma = 1$ and $\gamma = 1$ curves are plotted, indicative of non-ideality as to interaction between charges. In order to illustrate the extent of the discrepancy between the ionization equilibrium in non-ideal plasma and ideal plasma equlibrium (1.17), Fig. 1.4b shows, with a broken line, a $\gamma = 1$ curve constructed using Saha's equation (1.17).

As is evident from Fig. 1.4, the region taken by non-Debye plasma is very wide. In the last decade, it was subjected to very intensive studies. Much less data are available on a strongly non-ideal plasma, in which $\gamma \geq 1$, that is, $\Gamma \geq 5$.

1.2.3. *Inert Gas Plasma*

As distinct from alkali metals and mercury, in the case of an inert gas plasma, one manages to realize an extremely wide range of parameters and continuously follows the variation of the physical properties of such plasma from the gas ($N_e \sim 10^{14}$ cm^{-3}) to solid state ($N_e \sim 10^{23}$ cm^{-3}) densities. In doing so, one deals with extremely high pressures, that is, from hundreds to a million bar at maximum temperatures of up to 10^5 K. Under these strongly supercritical conditions, a plasma with developed ionization is realized. The wave curves in Fig. 1.5 show, for xenon plasma, the regions of single and double ionizations, and points are indicated for which thermodynamic measurements were performed. In view of the high temperatures and peculiarities of electron shells, molecular and cluster formations are absent in such plasma while the Coulomb interaction is the prevailing type of interparticle interaction because of a small number of neutrals. The non-ideality parameter reaches, in this case, the maximum values of about 10. With the maximum plasma densities realized in the experiment $\rho = 4.5$–9.7 g/cm^3 (which exceeds several times the solid state density of xenon of $\rho \approx 3$ g/cm^3), the electron component of such plasma is partly degenerate and the parameter $N_e\lambda_e^3$ reaches values close to 0.5.

A characteristic feature in the description of inert gas plasma, when compressed to supercritical densities, is the need to allow for interatomic and interion repulsion. Indeed, at $\rho > \rho_c = 1.1$ g/cm^3, the characteristic interparticle distance in the plasma is comparable to the size of atoms and ions in the ground ($\sigma_{\mathrm{Ar}} = 3.4$ Å, $\sigma_{\mathrm{Xe}} = 4$ Å) and, the more so, in excited states. In these strongly supercritical (as to temperature) conditions, the polarization attraction is insignificant. However, the repulsion qualitatively described by virial expansion or, more precisely, by the "hard"-sphere model is more prominent. The dimensionless parameter characterizing this interaction, $N\sigma^3$, reaches in xenon a value of 0.2–0.3. Under these conditions, an appreciable shift and broadening of the energy levels of atoms and ions are observed due to strong interparticle interaction.

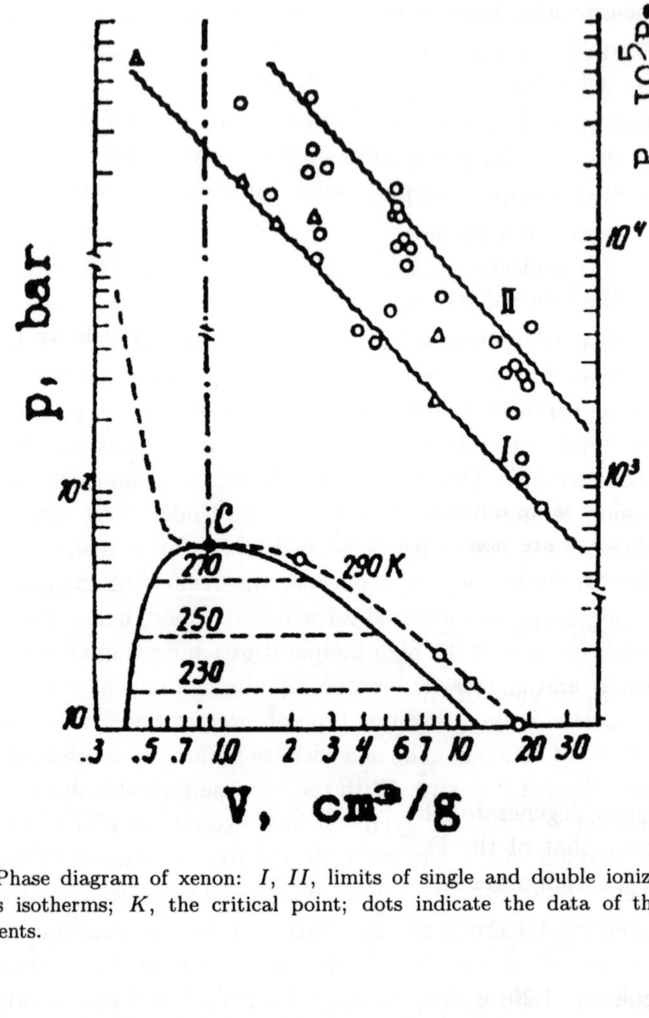

Fig. 1.5. Phase diagram of xenon: I, II, limits of single and double ionization; broken line shows isotherms; K, the critical point; dots indicate the data of thermodynamic measurements.

Finally, in a high-temperature ($T > 3 \cdot 10^4$ K) plasma, the amplitude of the Coulomb electron scattering $f = Ze^2/T$ appears comparable to the characteristic ion size. This leads to a non-Coulomb behavior of electron scattering in a high-temperature non-ideal plasma, which manifests itself primarily in electron transfer coefficients.

These conditions are attained during compression of gas in strong shock waves (in Fig. 1.1, these processes are shown at H_1, H_1', H_2 and H_2') and for conditions with more moderate parameters, during adiabatic compression (shown at S_1 in Fig. 1.1).

1.2.4. *Plasma with Multiply Charged Ions*

Powerful pulsed energy contributions to condensed matter give rise to a non-ideal superdense plasma with multiply charged ions. The density of such plasma is close to that of condensed state. Its pressure reaches several terapascals and the temperature is of the order of tens of electron-volts. The plasma consists of electrons and high-multiplicity ions, $Z \geq 10$. During expansion, it passes through a whole gamut of peculiar states, in which the degree of degeneracy of the electron component also varies.

Relatively uniform volumes of such plasma are produced under laboratory conditions during compression of porous metals (processes shown in Fig. 1.1) by powerful shock waves ($N_e\lambda_e^3 \approx 0.5 \div 2$, $\gamma_{\rm q} = 10$, $\Gamma = 2$). During adiabatic expansion of shock-compressed metals, a wide range of plasma states are realized from a highly compressed metallic liquid to a weakly ideal classical plasma. The respective states are shown at S_2 in Fig. 1.1 and overlap the entire range of existing non-ideal plasma.

If the electrons are non-degenerate, interaction of three different types is observed, namely, ion-ion, ion-electron, and inter-electron:

$$\gamma_{ZZ} = Z^2 e^2 N_i^{1/3} \beta \,, \qquad \gamma_{Ze} = Z e^2 N_i^{1/3} \beta \,, \qquad \gamma_{ee} = e^2 N_e^{1/3} \beta \,. \tag{1.19}$$

If $Z \gg 1$, then, in view of the condition of quasi-neutrality, one can write the inequalities

$$\gamma_{ZZ} \gg \gamma_{Ze} \gg \gamma_{ee} \,. \tag{1.20}$$

If the electrons degenerate, the values of their interaction energy must be compared with that of the Fermi energy (1.5). In doing so, the ions still remain very far from degeneracy. Therefore,

$$\gamma_{ZZ} = Z^2 e^2 N_i^{1/3} \beta \,, \qquad \gamma_{Ze} = Z e^2 N_i^{1/3} / \varepsilon_{\rm F} \,, \qquad \gamma_{ee} = e^2 N_e^{1/3} / \varepsilon_{\rm F}$$

and the hierarchy (1.20) is maintained. Hence the possibility of the existence of a number of interesting physical systems. These include a plasma formed by a quasi-crystal system of multiply charged ions ($\gamma_{ZZ} \gg 1$) and ideal electron gas ($\gamma_{ee} \ll 1$) with different intensity of interaction with ions.

In the diagram on the parameters of highly compressed matter (Fig. 1.6), a region is shaded in which the energy of interaction of a pair of electrons is low compared to their kinetic energy. However, the electrons interact with nuclei, as well as with their numerous partners that are present in the sphere effectively covered by the Coulomb forces. At $Z \gg 1$, these interactions are far from being minor. In the shaded region, the plasma is divided into Wigner

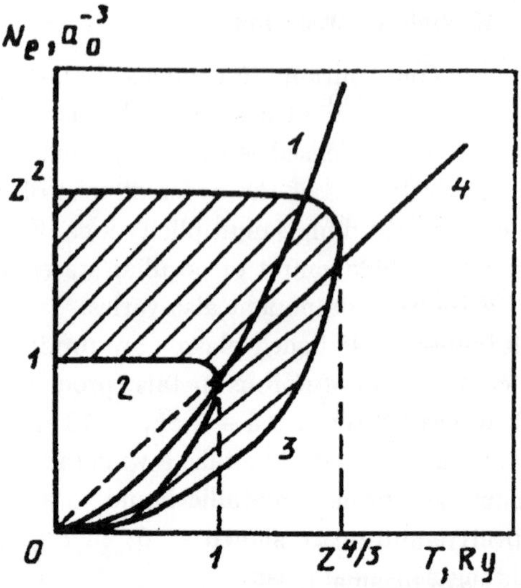

Fig. 1.6. Diagram of the parameters of highly compressed matter [9]: *1*, curve of electron degeneracy; *2*, $\gamma_{ee} = 1$; *3*, $\gamma_{Ze} = 1$; *4*, curve of equality of the exchange + quantum-mechanical effects to the correlation effects.

cells and may be described by the Thomas-Fermi method with corrections, or by the functional method of energy density (thermodynamic potential) [9].

1.3. Non-ideal Plasma in Nature. Scientific and Technical Applications

A high-density plasma with appreciable non-ideality effects is realized in numerous natural phenomena and technical devices. This includes the electron plasma in solid and liquid metals, semiconductors and electrolytes, superdense plasma of the matter of white dwarfs, the Sun, deep layers of the giant planets in the solar system, and astrophysical objects whose structure and evolution are defined by the plasma properties. In studying the giant planets of the solar system, one will run, both in the literal and figurative sense, into a non-ideal plasma formed upon hypersonic travel of space vehicles in the dense atmospheres of those planets.

Recently, a purely growing pragmatic interest has been observed in high-pressure plasma studies in view of the realization of a number of major energy-related projects and devices which depend for their action on a pulsed local concentration of energy in dense media. A non-ideal plasma appears to provide an advanced working medium in powerful continuous-operation and pulsed MHD generators, power-generating facilities [10] and rocket engines [11] with a gas-phase reactor, and in commercial-scale plasmochemical apparatuses. A non-ideal plasma is formed as a result of nuclear explosions [12], explosive evaporation of liners of pinches and magnetocumulative generators, the effect of powerful shock waves, laser radiation or electron beams on condensed matter, and in a number of other cases.

A special need for information on the physical characteristics of non-ideal plasma arises when realizing the idea of pulsed fusion, which is accomplished by way of laser, electron, ion or explosion compression of spherical targets [13], as well as when solving the most important problems related to high-velocity shock.

Of decisive importance, from the standpoint of physical analysis and calculation of the hydrodynamic consequences of such effects, are the data on the physical characteristics of plasma in a wide range of phase diagrams of matter. These range from a highly compressed condensed state to ideal degenerate and Boltzmann gases, including the high-temperature boiling curve and the neighborhood of the critical point.

In this section, we shall mention, very briefly, the most representative practical situations in which a non-ideal plasma is formed and utilized.

The most natural example of non-ideal plasma is provided by the plasma of conduction electrons in solid and liquid metals and semiconductors. Under normal conditions, we refer to a degenerate plasma ($\varepsilon_{\mathrm{F}} \gg T$) with an electron concentration of $N_e = 10^{22} - 2.5 \cdot 10^{23}$ cm^{-3}, in which a strong Coulomb interaction is realized because $2 \leq r_s \leq 5.6$. As a result of the effect of powerful shock waves on metals (cf., Chapter 3), such a plasma can be compressed by a factor of three-four to bring the maximum values of N_e to about $(5\text{–}6) \cdot 10^{23}$ cm^{-3} and the maximum temperatures to $5 \cdot 10^5$ K, thereby approaching the limit of degeneracy.

Liquid electrolytes and, in particular, ammoniacal solutions of alkali metals [14] represent a strongly non-ideal plasma in a very wide range of variation of the degeneracy and interaction parameters. This is attained by varying the fraction of metal dissolved in ammonia. Under these conditions, a powerful charge/neutral interaction is realized in the system along with the strong Coulomb interaction. These interactions have, for their results, unusual phase transitions and anomalously high electrical conductivities attained at moderate temperatures, even with small fractions of metal in the solution.

In intrinsic and impurity semiconductors, the number of electrons and holes is varied in a wide range by varying the temperature and concentration of impurities. Under conditions of intense light irradiation, electrons optically excited to the conduction band form a plasma. In a number of cases, the interparticle interaction in this plasma is so strong that it leads to a phase transition, that is, the formation of exciton liquid.

Of the great diversity of astrophysical objects in which a strongly interacting plasma is formed, we have only mentioned the Sun and the white dwarfs. According to Shatzman [15], the parameters of iron plasma in the center of the Sun are extremely high: $\rho \approx 120$ g/cm^3, $T \approx 13 \cdot 10^6$ K. The non-ideality parameter $\gamma \approx 41$, realized under these conditions, is apparently insufficient for plasma crystallization. Calculations using the method of molecular dynamics give the crystallization limit $\gamma_c \approx 144$. However, even at lower γ, partial clustering of the plasma is possible. This causes, in particular, a variation in its optical properties which determine the internal structure of the Sun.

A plasma with extreme parameters is realized in the late stages of the evolution of stars in the so-called white dwarfs [16], decaying stars with a mass of less than 1–1.2 mass of the Sun. In this case, the matter is in the state of equilibrium when the pressure of a quasi-uniform degenerate gas balances out the gravity forces compressing the star. At high densities, pyknonuclear reactions (upon compression above the density of $\log N \approx 34.2$) occur in such a plasma and, at sufficiently high temperatures, thermonuclear reactions. The effects of non-ideality for the white dwarf matter are appreciable; they increase the rates of nuclear reactions and define the structure, stability and evolution of those exotic objects.

Experimental investigations of the giant planets of the solar system using unmanned spacecraft will likewise involve problems associated with non-ideal plasma. The point is that the strong gravitational field of those planets forms a very dense atmosphere. For instance, the atmosphere of Jupiter (a mixture of helium and hydrogen) in the bottom layers reaches a pressure of about 1,000 atm. Under the conditions of hypersonic travel of spacecraft in such a dense atmosphere, a shock wave is formed in which the compression and irreversible heating of the plasma to tens of thousands of degrees at pressures of hundreds and thousands of atmospheres take place [17]. In order to provide effective protection of the spacecraft against the effects of such a plasma, as well as to ensure reliable radio communication, one needs reliable data on the thermodynamic, transport and radiation characteristics of strongly non-ideal shock-compressed plasma.

Note that, in accordance with the existing views, the internal structure and evolution of the giant planets in the solar system are defined by the properties of their matter being in the state of non-ideal plasma.

The most important technical applications of non-ideal plasma are its energy-related applications. This is because the development and realization of a whole series of advanced energy-related projects are associated with ionized high-density plasma. Along with thermonuclear systems of magnetic confinement of hot plasma, inertial controlled thermonuclear fusion is developed [18] as an alternative approach. With this approach, the thermonuclear reaction is accomplished as a "microexplosion" over a short (several nanoseconds) time, defined as the time of inertial expansion of hot reacting plasma. While doing so, the lowering of the energy threshold of initiation in systems of inertial fusion is accomplished by compressing the thermonuclear fuel of the target to a density approximately 1,000 times higher than that of a solid. Most diverse possibilities are considered as regards the compression and heating of a deuterium/tritium mixture in spherical micro-targets, namely, powerful laser radiation, streams of relativistic electrons and light and heavy ions, and macroscopic liner shock. In doing so, an energy of the order of 10^5–10^6 J must be delivered to a composite layered target of about 0.1 cm over a time of about 10^{-9} s. This results in the emergence of a complicated unsteady-state flow of dense non-ideal plasma with a pressure of up to 10^3 Mbar. In order to calculate the hydrodynamics of such flow and develop the optimum structures of thermonuclear micro-targets, one needs extensive data on the thermodynamic, optical and transport properties of non-ideal plasma of various compositions in an extremely wide range of pressures and temperatures.

A strongly non-ideal plasma characterized by a wide range of parameters also emerges in the case of other problems in calculating the interaction between powerful pulsed sources of radiation and matter. When the surface of a solid material is subjected to a high-power laser pulse, a region of inhomogeneous non-ideal plasma occurs, in which a wide range of states are realized. A relatively rarefied heated plasma ($N_e < 10^{21}$ cm^{-3}, $T \approx 1$ keV) moves towards the laser beam and absorbs its energy. This energy is transferred to the region of strongly non-ideal plasma ($N_e = 10^{21} \div 10^{23}$ cm^{-3}), where heat is converted to the energy of hydrodynamic motion. The strong recoil reaction causes the target material to compress and a region of superdense matter to form in the target ($N_e > 10^{23}$ cm^{-3}) (Fig. 1.7) [19].

Also directed toward the utilization of non-ideal plasma is another futuristic energy project — the gas-phase nuclear reactor [9, 20]. This is a cavity-type

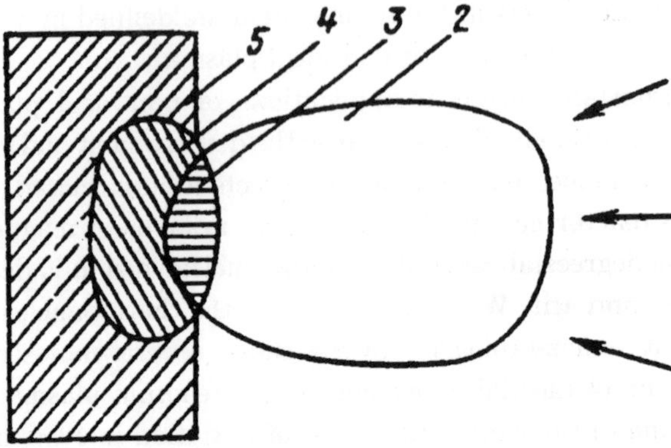

Fig. 1.7. Results of interaction between radiation and matter: *1*, laser pulse; *2*, corona flame ($\rho \leq 10^{-2}$ g/cm^3, $T \approx 1$ keV); *3*, non-ideal plasma of ablation products ($\rho = 0.01$ to 1 g/cm^3, $T \approx 100$ eV); *4*, superdense matter ($\rho = 10$ to 50 g/cm^3, $T = 10$ to 100 eV); *5*, cold matter of normal density.

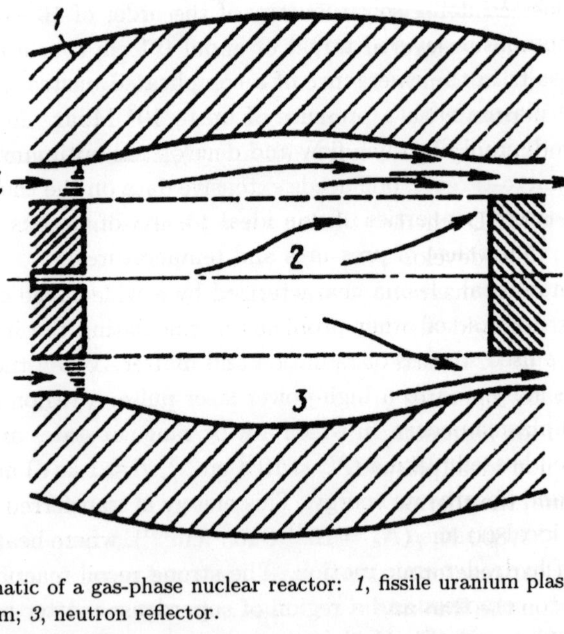

Fig. 1.8. Schematic of a gas-phase nuclear reactor: *1*, fissile uranium plasma; *2*, flow of the working medium; *3*, neutron reflector.

reactor (Fig. 1.8) with a high-pressure uranium plasma in the center. Flowing between uranium and the walls is a working medium heated by the thermal radiation of the uranium plasma. The mixing of the working medium and uranium is suppressed by a stabilizing magnetic field, profiling of the velocity field, and by other measures. Such an apparatus provides a basis for the development of nuclear power plants, compact space-borne power-generating facilities, engines and the like [9, 20]. The high temperatures and requirements of criticality result in the need to have a pressure of hundreds of atmospheres in the gas-phase reactor cavity. Under these conditions, at temperatures of tens of thousands degrees, uranium and the working media (mixtures of alkali metals, hydrogen and the like) are in the state of non-ideal plasma whose physical properties depend on strong interparticle interaction.

A high level of electrical conductivity with appreciable compressibility, characteristic of non-ideal plasma, renders such a plasma a suitable working medium for magnetohydrodynamic generators [21, 22]. The operating principle of magnetohydrodynamic (MHD) generator enables the travel of conducting medium in a transverse magnetic field. The power which can be extracted per unit active volume is, other things being equal, proportional to the electrical conductivity of the working medium: an increase in the electric conductivity enhances the plasma interaction with magnetic field.

Non-ideal plasma emerges under conditions of considerable compression, such that the electrical conductivity increases both due to a temperature rise and as a result of interparticle interactions. The conductivity may exceed the ideal plasma conductivity by several orders of magnitude at relatively moderate temperatures.

A closed-cycle MHD scheme utilizing a non-ideal cesium and sodium plasma is treated in [21]. The development of a compact magnetic-flux MHD generator with self-excitation of magnetic field is associated (see [23]) with the use of a shock-compressed non-ideal xenon plasma at a pressure of several gigapascals.

Biberman *et al.* [24] have proposed a MHD power plant scheme where the MHD generator operates in a closed cycle utilizing a non-ideal plasma of saturated cesium or potassium vapors. Use is made of the anomalously high conductivity of non-ideal plasma of saturated vapors, $\sigma \sim 100$ 1/Ohm.cm. In the case of cesium, the thermodynamic parameters of the working medium are at a level of $T \leq 1,800$ K, $p \leq 7$ MPa. Such an MHD generator can operate both with a nuclear reactor and in a fossil fuel-fired power plant scheme. The facility operates on the Rankine cycle. The cycle performed by the alkali metal (Fig. 1.9) includes heating and evaporation in the steam generator, expansion in the MHD unit, condensation in the condensing heat exchanger with heat transfer to the steam-water loop, and an increase in the pressure of the pump.

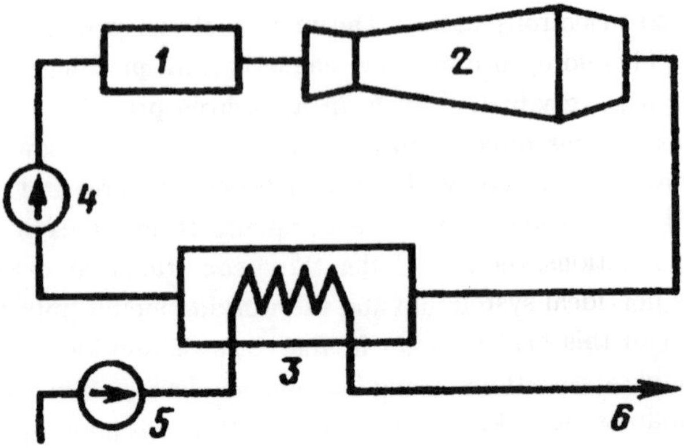

Fig. 1.9. MHD facility utilizing alkali metal vapors: *1*, steam generator; *2*, MHD unit (nozzle, MHD channel, diffuser); *3*, condensing heat exchanger; *4*, liquid metal pump; *5*, water pump; *6*, to the turbogenerator.

A non-ideal plasma of alkali metals, primarily of cesium, can also be regarded as the working medium for plasma thermoelements. This is favored by the high values of the thermal e.m.f. of cesium, $S = 10^3$ μV/deg, at moderate pressures and temperatures ($p \leq 100$ bar, $T \leq 2,000$ K). A model of such a thermoelement was considered by Alekseev *et al.* [25], who had been the first to measure the thermal e.m.f. of dense plasma. The thermal conversion element is a sleeve of niobium in which a beryllium oxide tube is inserted inside, from which cesium is poured. The tube is closed at the top with a metal plug serving as one of the electrodes. The temperature difference is developed by a resistance furnace placed in a protective casing under a desired pressure of inert gas. The thermoelement may have a power of up to 10 W/cm^2 and an efficiency of up to 20%. Apart from their ability to operate at high temperatures, such converters have an advantage over the semiconductor ones in that they can vary greatly their characteristics upon varying the applied pressure at constant temperature.

An example of strongly non-ideal plasma is provided by a plasma with condensed disperse phase (CDP). This is a low-temperature plasma containing disperse particles of matter in a liquid or solid state. While emitting electrons and getting positively charged, the particles can increase the electron concentration in the plasma. In the extreme case, particles emerge in a non-ionized gas to fully define its electrophysical properties. On the contrary, if the particles absorb electrons from the plasma, they are charged negatively and reduce the number of free electrons. The metal electronic work function amounts to about 2–4 eV. Therefore, in order to reach an equilibrium with the plasma, the particles acquire very high charges, $Z \approx 100$ for a particle with a radius of 0.1 μm. This results in strong interparticle interaction, $\gamma = Z^2e^2N_p^{1/3}/T \gg 1$. It turns out that in a CDP plasma under standard conditions ($T = 2,000$ K, $N_e = 10^8$–10^{14} cm^{-3}), and depending on concentrations, particle sizes and electronic work functions, the entire range of plasma states from Debye plasma to the strongly non-ideal system of charged particles is realized [26].

The interest in this field of plasma physics is sustained owing to a great number of possible applications. These include the electrophysics of rocket fuel combustion products, solid fuel-fired magnetohydrodynamic generators, technology of plasma sputtering and plasma treatment of materials, and physics of gaseous dust clouds in the atmosphere and of artificial formations in space. CDP particles may be introduced into a plasma by design: present-day technology permits mass produced particles with a size of less than or approximately equal to 1 μm. Submicron particles (droplets) may also occur as a result of volume condensation upon the efflux of plasma from nozzles, the plasma expansion in the channel of an MHD generator, the expansion of aggregates of matter into a void, in the products of ablation of solid surface subjected to the effect of high energy fluxes, and so on.

A non-ideal plasma results from powerful electric discharges in water. Its properties also define the dynamics of the motion of generated shock waves. In recent years, pulsed electric discharges in liquid have found extensive industrial application for forming, pressing and machining of parts, for crushing and grinding of materials, and for intensifying chemical production processes [27]. Electric explosion appears to hold promise under conditions of great oceanic depth in performing various mechanical operations. The operation of such facilities is based on the use of high-voltage discharge in liquid as the process of rapid conversion of the energy of capacitor bank to mechanical work. The discharge duration is 10^{-5}–10^{-4} s with an energy density of 10^{14}–10^{15} J/m^3, a temperature of 10^4–10^5 K, and a pressure in the discharge channel of up to 1 TPa. Under these conditions, the plasma may be substantially non-ideal. The plasma properties, especially its electrical conductivity (one of the principal parameters determining the efficiency of the method), affect both the process of formation and expansion of current-conducting channel, as well as the shock wave generation which is the principal mechanism of operation of electrohydraulic-pulsed facilities.

Two

Electrical Methods of Non-Ideal Plasma Generation

The electrical methods of non-ideal plasma generation include two principally different techniques, namely, the heating in resistance furnaces ampoules containing material under investigation and Joule heating of samples of material by electric current passed through the latter. Resistance furnace can be used to obtain homogeneous volumes of plasma without passing electric current through it and to record the plasma parameters to a relatively high degree of accuracy. On the other hand, the static character of the method restricts its capabilities to temperatures of up to 3,000 K. This is due to great difficulties associated with the choice of structural materials. The techniques utilizing Joule heating include high-pressure gas discharges in thick-walled capillaries, the heating of materials with current in an atmosphere of inert gas, discharges in liquids, and some other techniques. These techniques, which are mainly pulsed, permit plasma generation at considerably higher temperatures of up to 10^5 K. The principal difficulties of these methods are caused by the difficulty in attaining the homogeneity of plasma volumes and by various plasma instabilities.

2.1. Plasma Heating in Furnaces

A detailed description of stationary methods of plasma generation can be found in reviews by Alekseev and Iakubov [28, 29]. These methods are based on heating an ampoule containing material under investigation (measuring cell) in electric furnaces of various designs. The entire unit is placed in a chamber under a high pressure of inert gas. Shown by way of example in Fig. 2.1 is the schematic of a facility which was used [30] to investigate alkali metals at high temperatures and pressures. Argon from a cylinder *1* was fed via a system of

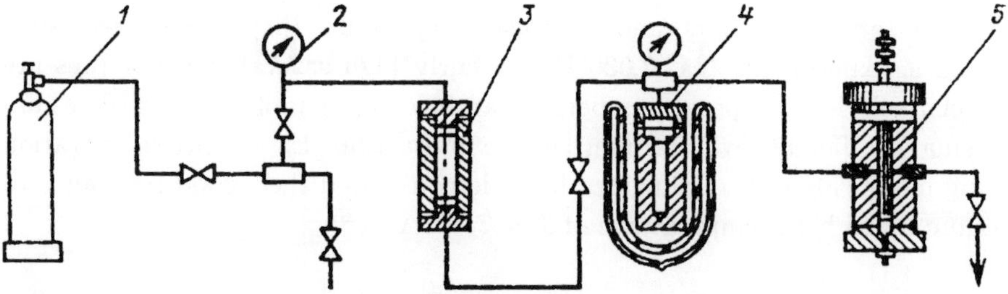

Fig. 2.1. Schematic of a facility for the generation and investigation of non-ideal plasma of metals in the region of high temperatures and pressures [30]: *1*, cylinder with argon; *2*, pressure gauges; *3*, argon cleaning system; *4*, thermocompressor; *5*, high-pressure chamber.

valves to a chamber *3* for cleaning. Cleaned argon was delivered to a nitrogen thermocompressor *4* with the pressure in the latter rising to 600 bar, and from the thermocompressor to a measuring chamber *5* accommodating a heater and a measuring cell. The pressure was monitored by gauges *2*. The temperature was determined using a thermocouple with its cold Junction placed in a melting ice bath. If the required pressure level exceeded 600 bar, the facility was modified: the gas was compressed to 5,000 bar using a booster [31].

High temperature values in the region of non-ideality present the principal difficulty of static methods, which is due to the difficulty in selecting structural materials for measuring cells. Especially serious is the extremely narrow choice of insulating materials. The best insulating properties are observed in insulators of pure aluminum, beryllium and thorium oxides. Speaking of metals, the best materials include plastic tungsten, tungsten and molybdenum alloys, and molybdenum. The construction of a reliable operating measuring cell calls for a high degree of experimental craftsmanship.

As regards the development of the static methods of non-ideal plasma research, one should note the importance of the pioneering work by Kikoin *et al.* [8] and Hensel and Frank [32]. In the 1970s, new designs of measuring cells were developed which helped considerably to improve the accuracy of the experimental data. The results of previous studies were often considerably refined and, in some cases, revised. At present, the experimental data obtained in different laboratories are usually found to be in adequate agreement with each other.

The most convenient object for investigations into the non-ideality effect is cesium, which has a low ionization potential (I = 3.89 eV), low critical pressure (about 110 bar), and a critical temperature that is quite accessible to static experiments (ca. 2,030 K). A fairly high saturated vapor pressure at relatively low temperatures permits an investigation of a non-ideal cesium plasma at different levels of interparticle interaction. Experience in operation using measuring cells of tungsten have demonstrated their good resistance to cesium up to high temperatures of $T \leq 2,500$ C [33].

2.1.1. *Measurement of Electrical Conductivity and Thermoelectromotive Force*

The requirements for measuring cells include a high corrosion resistance of their walls upon contact with plasma, airtightness of contacts between the insulating ceramics and metal electrodes, and the protection of material under investigation against contact with inert gas. Figure 2.2 illustrates a

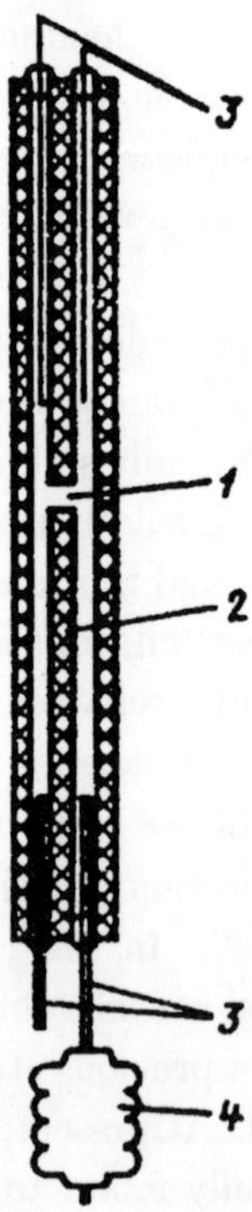

Fig. 2.2. Cell (closed) for measuring electrical conductivity (Alekseev *et al.* [34]): *1*, H-shaped measuring volume; *2*, beryllium oxide ceramics; *3*, electrodes; *4*, expanders of the material under investigation.

measuring cell of beryllium oxide with an H-shaped measuring volume in which the metal under investigation does not come into contact with the pressure-transmitting gas[34].

Two through holes with a jumper in the center were made in the cell body *2* of beryllium oxide. Current (solid) and potential (tubular) electrodes *3* were sealed at the ends. Using such a structure, one can measure the resistance in a volume whose uniformity is reliably guaranteed. An elastic expander, bellows *4*, was attached to one of the ends (in a closed ampoule, the material expands upon heating) while through the other end, liquid cesium was delivered in vacuum. After filling the cell and bellows with cesium, the measuring cell was sealed off.

The measuring current was passed through the bottom electrodes while the top electrodes were used to measure the voltage drop. Given the jumper dimensions, one can calculate the resistance. The error in measuring the cesium resistance at a temperature of up to 1,300 K amounts to 3% and in the region of metallization ("metal-dielectric" transition), to 10%. Meanwhile, in dense cesium vapors at a temperature above 2,000 K, this error may exceed 50%. The temperature is measured by W/Rh thermocouples.

Special precautions may be taken to eliminate the effect of leakage currents on the electrode surface. Borzhievsky *et al.* [35] measured the coefficient of electrical conductivity by the two-probe technique using a guard-ring electrode *3* made of sapphire (Fig. 2.3). Equal potentials of the electrode and guard-ring

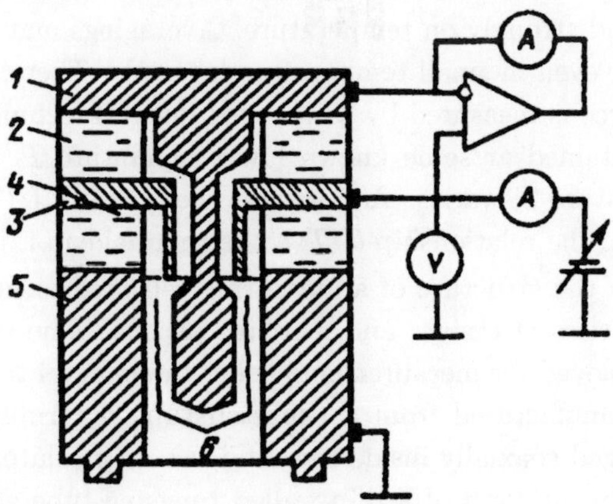

Fig. 2.3. Cell for measuring the coefficient of electrical conductivity equipped with a guard-ring electrode [35]: *1* and *5*, measuring electrodes; *2* and *4*, insulation rings; *3*, guard-ring; *6*, measuring volume.

electrode *3* are maintained. Therefore, the current being measured (between the electrodes *1* and *5*) could only flow in the radial direction in the plasma volume *6*. It was the objective of Borzhievsky *et al.* [35] to measure the coefficient of electrical conductivity of low-temperature cesium vapors, $T \leq 1,200$ K.

A thermoelectric effect shows up in matter in the presence of temperature gradient,

$$\mathbf{F} = \mathbf{j}/\sigma + S\nabla\mathbf{T}\,, \tag{2.1}$$

where S is the thermoelectric coefficient usually referred to as thermoelectromotive force, and $\mathbf{F}$ is the gradient of the electrochemical potential. In static experiments, the potential difference U is measured, which appears at the ends of an open circuit consisting of dissimilar conductors. If the circuit contains two contacts so that the end conductors are of like metal (tungsten) with a temperature gradient developed along the middle conductor (cesium), a potential difference appears at the circuit ends,

$$U = \int_{T_1}^{T_{20}} (S_{Cs} - S_W)dT\,,$$

where T_1 and T_2 define the temperature difference.

Since the temperatures are high, during the experiment it is very difficult to maintain with adequate accuracy small temperature differences between the junctions. One should bear in mind that under these conditions, the thermal e.m.f. may depend strongly on temperature. Averagings may also produce an appreciable error even in small temperature intervals. Therefore, the thermoelectromotive force is measured by a so-called integral technique. The "cold" junction is maintained at some known low temperature T_1 while the "hot" junction temperature T_2 varies. As a result, the quantity $U(T_2)$ is measured. By differentiating the relationship $U(T_2)$, the thermal e.m.f. is obtained.

Consider now the structure of a measuring cell used to measure the thermoelectric properties of cesium and mercury plasma. The cell illustrated in Fig. 2.4 was employed for measurements using the integral technique [34, 36]. This cell was manufactured from a tungsten tube *5* terminating in an expander *1*. Arranged coaxially inside the tube was an insulator *6* comprising a top electrode *4* in the form of a thin-walled tungsten tube with a thin tungsten bottom welded on its lower end and a W/Rh thermocouple *2* welded in the bottom. The thermocouple ends, insulated by beryllium oxide ceramics, passed out through the top end of the tube.

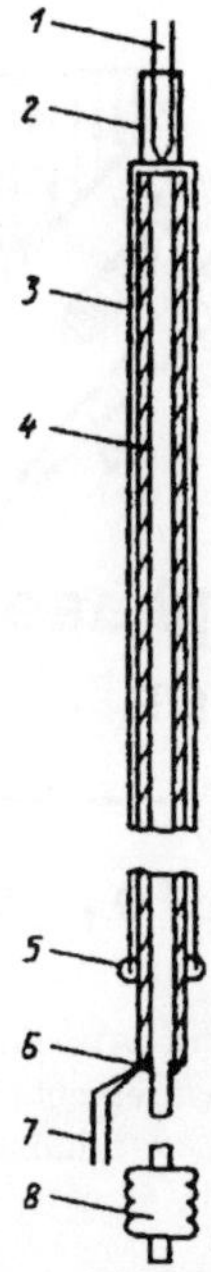

Fig. 2.4. Cell for measuring the thermal emf [34, 36]: *1*, bellows; *2*, thermocouples; *3* and *4*, tungsten electrodes; *5*, tungsten casing; *6*, beryllium oxide insulator; *7*, junction between the ceramics and the casing; *8*, junction between the bottom electrode and the insulator.

The measuring element was filled with mercury or cesium. The working volume was essentially a cylinder with a height of 0.5–2 mm and diameter of 3–5 mm defined by the insulator, lower end of the top electrode and the bottom of the outer tungsten ampoule serving the function of the second electrode *3*. The thermal e.m.f. being measured occurs owing to the temperature gradient developed along the cell.

Let us turn now to a discussion of the results of measurements of electrical conductivity and thermal e.m.f. The electrophysical properties of cesium appear to be the most studied at present. Figure 2.5 shows a region in which the electrical conductivity and thermoelectromotive force of cesium were studied using both static and dynamic methods [30, 35, 44].

The pioneering study of cesium was performed by Alekseev [30]. It was shown that the electrical conductivity in the region of high temperatures on the isobars had dropped by up to four orders of magnitude upon heating.

Figure 2.6 gives the results of more recent measurements of the electrical conductivity of rubidium [45], which demonstrate clearly how it decreases upon

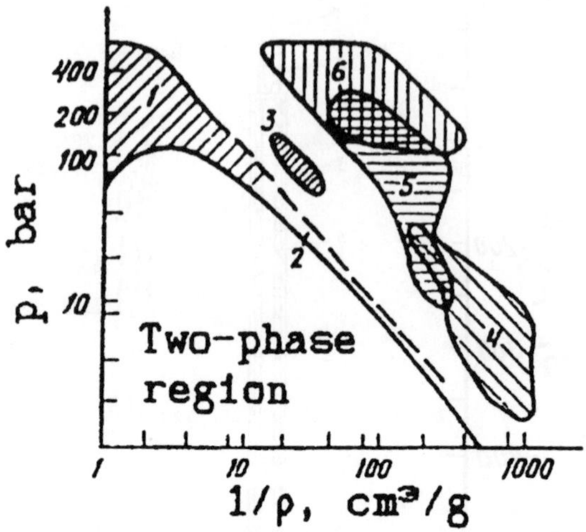

Fig. 2.5. Experimentally studied regions of p-V diagram of cesium. Static experiments: *1*, measurements of σ [30, 36–39] and measurements of S [25, 36, 40]; *2*, phase coexistence curve [37]; dynamic measurements of σ; *3*, adiabatic tube [41]; *4*, *5*, shock tube [42]; *6*, ohmic heating in an inert atmosphere [43].

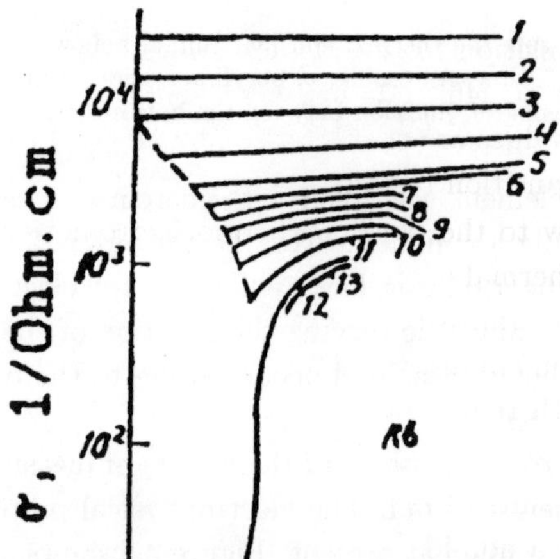

Fig. 2.6. The electrical conductivity of rubidium in the "metal-dielectric" transition region [45]: *1*, T = 300 K; *2*, 500 K; *3*, 900 K; *4*, 1,200 K; *5*, 1,400 K; *6*, 1,500 K; *7*, 1,550 K; *8*, 1,600 K; *9*, 1,650 K; *10*, 1,700 K; *11*, 1,750 K; *12*, 1,815 K; *13*, 1,830 K.

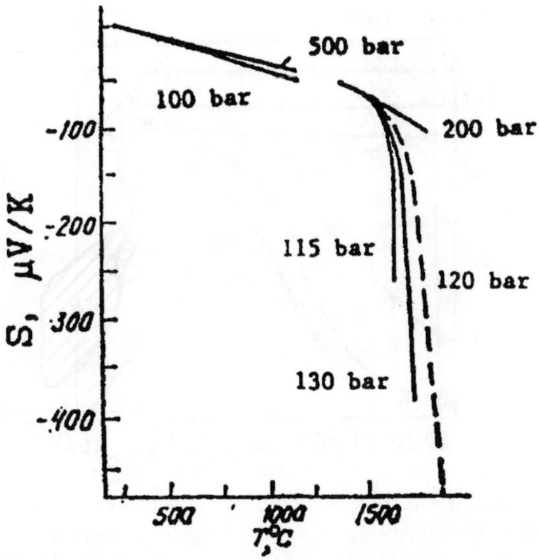

Fig. 2.7. Experimental values of the thermal e.m.f. of cesium: broken line, [36]; solid lines, [40].

heating and its transition to the non-ideal plasma state. Analogous data are cited by Pfeifer *et al.* [46] and Freyland and Hensel [47] for rubidium and potassium. These values are characterized by an error of ±3–4% in liquid and ±50–60% in the transition region.

Let us turn now to the data on thermoelectromotive force. The absolute magnitude of the thermal e.m.f. of cesium had grown rapidly with temperature in the region of the transition from metal to plasma state. One can see that the solid and broken curves, while featuring the same behavior of the thermal e.m.f. variation with temperature, diverge somewhat in the "metal non-ideal plasma" transition zone.

Consider the results of measuring the electrical conductivity and thermal e.m.f. of cesium in a non-ideal plasma of metal vapors. The results obtained in this region are far from being numerous. Here, however, qualitative results were also obtained which had proven to be of importance in selecting the theoretical models of non-ideal plasma. Figure 2.8 gives the isotherms of the electrical conductivity of cesium, as well as the electrical conductivity of cesium on the phase coexistence curve [37]. Figure 2.9 represents a chart combining the results obtained in different laboratories [36, 37, 48] for the 20 atm isobar and saturated vapors.

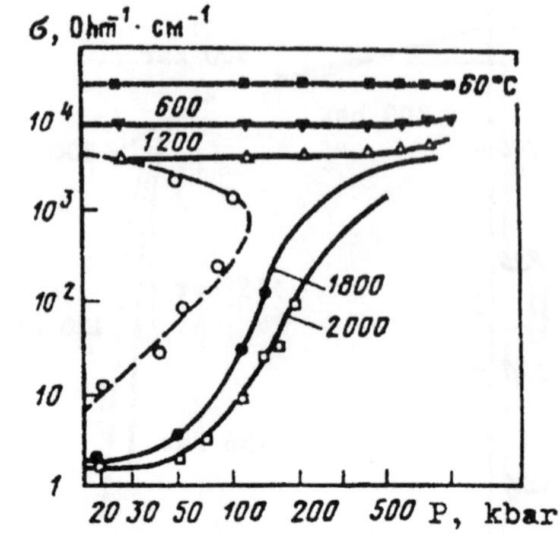

Fig. 2.8. Isotherms of cesium electric conductivity (solid) and electric conductivity on phase coexistence curve (dashed) [37].

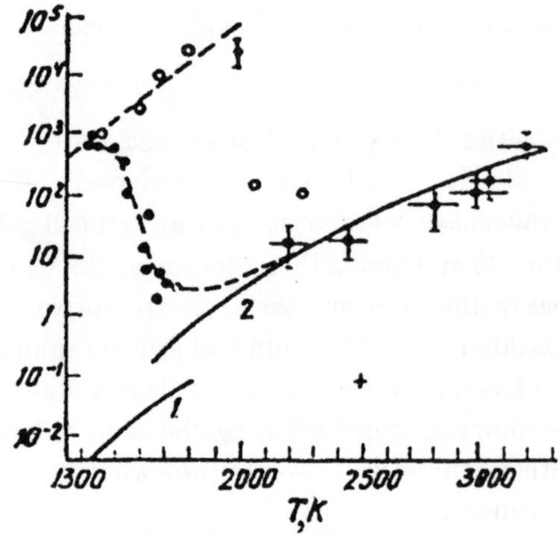

Fig. 2.9. The electrical conductivity of non-ideal plasma of cesium vapors. Measurement results: saturated vapor, ∘ [37]; on the 20 atm isobar, • [36], ×· [48], + [39], Δ [37]. Calculated results: saturated vapor by ideal gas approach, *1*; isobar 2 MPa with account for molecular ions [49].

Attention is drawn to the anomalously high level of electrical conductivity of saturated vapor. It exceeds by orders of magnitude the level of standard estimates for ideal plasma. Further, on isobaric heating, the electrical conductivity starting from these high values does not rise but, on the contrary,

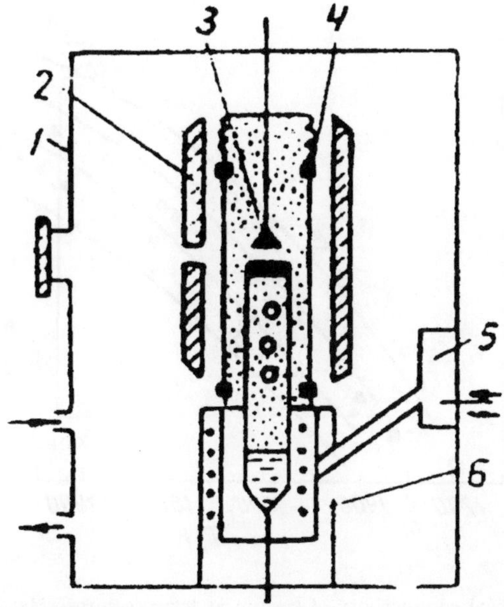

Fig. 2.10. Schematic of the facility for generation and studies of a low-pressure cesium plasma ($p \leq 1$ bar) in the high-temperature region [50]: *1*, vacuum chamber body; *2*, heaters; *3*, electrodes; *4*, protective ring; *5*, cesium delivery system; *6*, liquid cesium heating furnace.

drops sharply. It decreases on the 20 atm isobar to the minimum value of about $2 \cdot 10^{-2}$1/Ohm.cm and, on further heating, starts increasing in the usual manner as in a weakly non-ideal plasma.

The region of weakly non-ideal cesium plasma corresponds to low pressures of $p \leq 1$ atm, at which the measurements are performed in vacuum chambers *1* (Fig. 2.10), thus accommodating heaters *2*. Two insulated electrodes *3* with applied potential used to measure the current-voltage characteristics are instead in the heaters. From the latter, the resistance of the plasma gap between the electrodes is determined. In order to weaken the end effects, an insulated protective ring *4* is set on one of the electrodes.

The plasma of desired pressure is provided with the aid of a cesium delivery system *5*. For this purpose, a furnace *6* maintains a preset temperature of the liquid cesium surface. An optical pyrometer or thermocouple was used to determine the plasma temperature.

Figure 2.11 shows the results [50, 51] of measuring the electrical conductivity of cesium at low pressures. These values of electrical conductivity are characterized by an error of from 10–25%. At $p \leq 1$ atm, the plasma of cesium

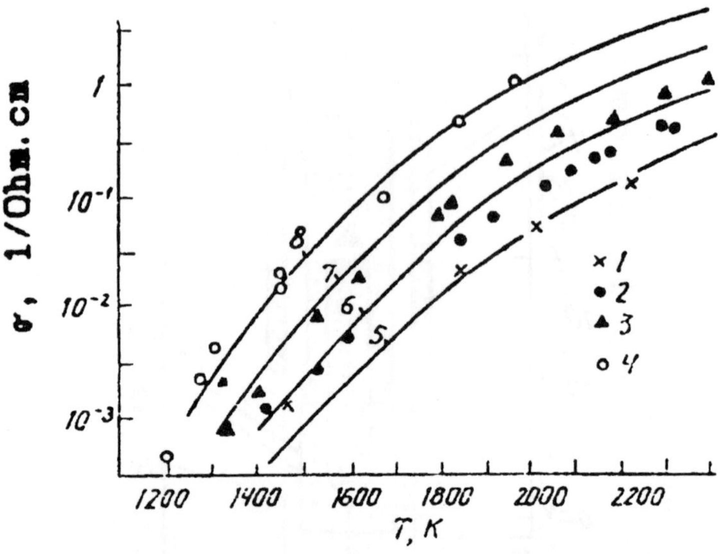

Fig. 2.11. The electrical conductivity of cesium at low pressures. Measurements: *1*, 1 bar; *2*, 10^{-1} bar; *3*, 10^{-2} bar [50]; *4*, $0.66 \cdot 10^{-3}$ bar [51]. Calculations: *5*, 1 bar; *6*, 10^{-l} bar; *7*, 10^{-2}. bar; *8*, $0.66 \cdot 10^{-3}$ bar [49].

vapors is already non-ideal. Therefore, the electrical conductivity increases as the pressure decreases, as in the case of ideal plasma.

Mercury, like cesium, has a low critical temperature. Therefore it was the first substance whose properties had been investigated in the "metal–non-metal" transition region [8, 32].

The electrical conductivity of dense mercury vapors was measured by Hensel and Frank [32]. Isochores of electrical conductivity are given in Fig. 2.12. For this purpose, the density was determined by extrapolating the experimental $p - \rho - T$ data obtained at higher densities. It can be seen from Fig. 2.12 that the electrical conductivity of mercury plasma at constant density grows with temperature if $\rho \leq \rho_c$. The temperature dependence of electrical conductivity in this region exhibits a semiconductor behavior, $\sigma \sim \exp(-\Delta E/T)$. As the density increases, the slope of the curves decreases, indicating that the "energy gap" ΔE has decreased.

Figure 2.13 shows the temperature coefficient of conductivity of mercury as a function of density, $\Delta E = -2d \ln \sigma / d\beta$, where $\beta = 1/T$. The energy gap ΔE is closed at $\rho \approx \rho_c$. An interesting, though still not properly understood, peculiarity in the $\Delta E(\rho)$ dependence was observed by Kikoin and Senchenkov

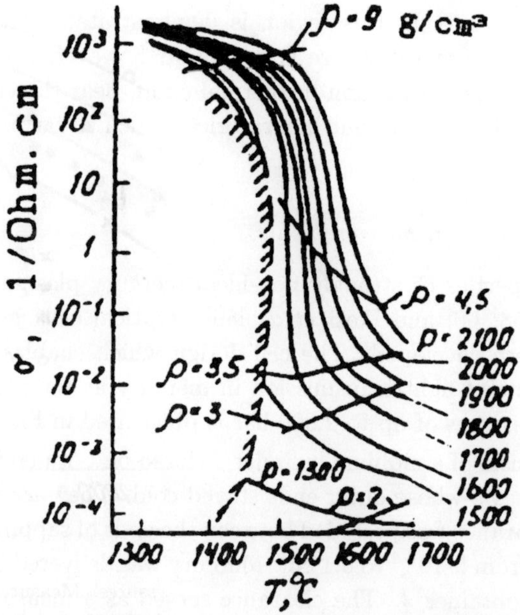

Fig. 2.12. Isochores of the electrical conductivity of mercury [32].

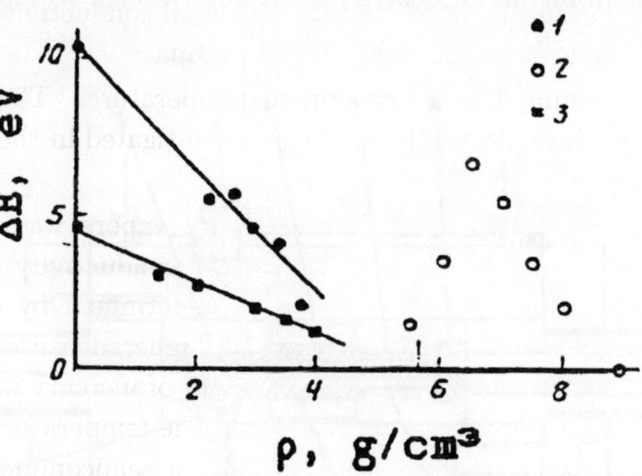

Fig. 2.13. The transport, ΔE, and optical, $\Delta E_{\rm opt}$, gaps in the energy spectrum of mercury, $T = 1,820$ K: ΔE ■ [39] and ∘ [31]; $\Delta E_{\rm opt}$ • [52].

[31] at $\rho > \rho_c$. In the liquid state region, the gap opens on further compression and finally closes again only at $\rho \approx 9$ g/cm^3. Kikoin and Senchenkov [31] concluded that mercury had metallized at only $\rho \geq 9$ g/cm^3.

A great uncertainty in this region is demonstrated by the data on the thermal e.m.f. of mercury. There is no doubt, however, that an interesting phenomenon of a sharp variation in thermal e.m.f. near the critical point [34, 53–56] has been observed in four laboratories, which so far has no theoretical interpretation.

2.1.2. *Optical Absorpton Measurements*

The optical properties of strongly non-ideal mercury plasma were measured by Hensel [57, 58, 60] and Ikezi *et al.* [59]. Optical cells in the 1.3–4.9 eV energy range were developed. The cell design which enables one to perform the measurements of light transmission in mercury at temperatures of up to 1,700°C and pressures of up to 2,000 bar is illustrated in Fig. 2.14.

The cell consists of a niobium cylinder *1* closed on both ends by windows of polished sapphires *2* whose outer ends stayed cold. They are sealed by Teflon spacers *5*. A clearance was provided between the ends of sapphires whose width could be varied from $5 \cdot 10^{-4}$ to 0.1 cm. Mercury was delivered to that clearance from a special container *4*. The clearance served as a measuring element for determining the absorption index of mercury plasma. The cell was heated by a furnace *6* inside an autoclave *7*. Argon provided the pressure-maintaining medium. Light from the autoclave passed out through windows arranged in ends *8*.

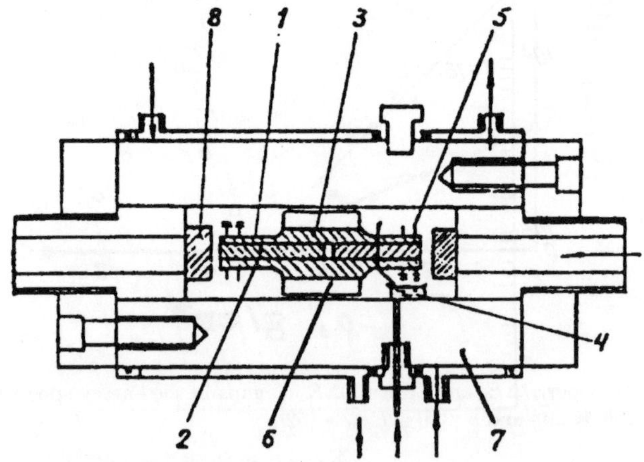

Fig. 2.14. Optical cell for determining the absorption in a dense mercury plasma [57]: *1*, niobium cylinder; *2*, optical sapphires; *3*, measuring volume; *4*, container with mercury; *5*, Teflon seals; *6*, furnace; *7*, autoclave; *8*, high-pressure windows.

The absorption index $k = (1/d)\ln(I_0/I)$ could be determined from the measured radiation transmission

$$T = (1 - R)^2 \exp(-kd)\,,$$

where d is the sample thickness, I_0 and I are the incident and emitted radiation intensities, and R is the reflection coefficient. The values of k were determined by Hensel [57] without using the values of R, but as a result of measurement of T with two values of d.

The sources of errors, such as the uncertainty in the optical path, non-uniformity of temperature distribution and inaccuracy of intensity recording, led to errors in the absorption coefficient not exceeding $\pm 20\%$. The main source of uncertainty is provided, the same as in the case of measuring the electrophysical quantities, by errors in recording the temperature and density. The stimulated radiation correction to the absorption index, made by the factor $[1 - \exp(-\hbar\omega/T)]$ in the conditions under investigation, is small.

Figure 2.15 shows the frequency dependence of the absorption index of dense mercury vapors obtained in two different laboratories. Solid curves

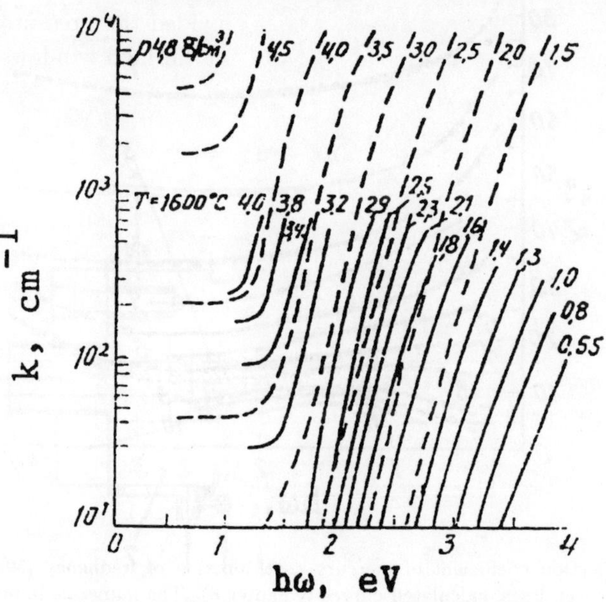

Fig. 2.15. Absorption index of mercury as a function of the incident light frequency, $k(\omega)$: solid curves [59]; broken curves [52].

indicate the results derived by Ikezi *et al.* [59] and the broken curves the data by Uchtmann and Hensel [61]. The absorption index $k(\omega)$ varies exponentially upon variation of the incident light wavelength. The first thing to strike one's eye is the abruptness of variation of $k(\omega)$. A variation of $k(\omega)$ by an order of magnitude corresponds to a variation of $\hbar\omega$ by only several tenths of an electron-volt. The data of the two experiments are in good agreement with each other.

The mercury reflection coefficient R was measured by Ikezi *et al.* [59] and Hefner *et al.* [61]. Figure 2.16 shows the results of measurements of the reflection coefficient of radiation incident, which is almost normal to the surface of the plasma layer [59]. These results cover the region in which mercury loses its metallic properties at temperatures somewhat exceeding the critical point and changes to the state of strongly non-ideal plasma. As in the case of the results of electrical conductivity measurements, the density of 9 g/cm^3 separates the metal and dielectric regions. The measured absorption and reflection coefficients were used [61–63] to obtain the frequency dependencies of dielectric permeability and electrical conductivity (cf., Chapter 8).

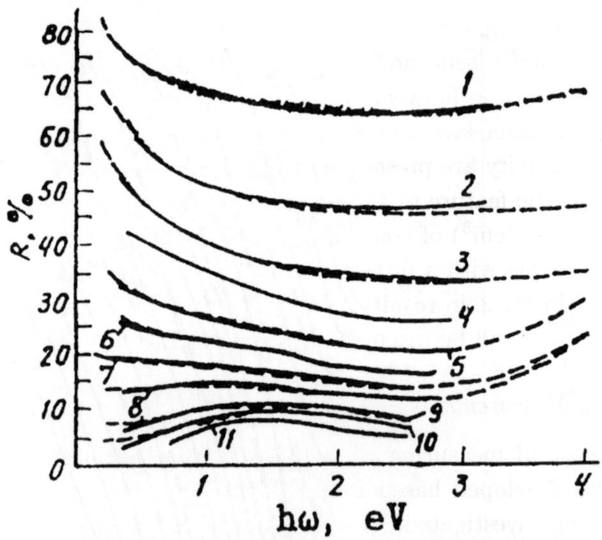

Fig. 2.16. Reflection coefficient of mercury as a function of frequency [59]: solid curves, experiment; broken lines, calculated curves (Chapter 8). The numerals indicate. ρ(g/cm^3), p (bar), T (°C.): *1*, 13.6 g/cm^3, 1,200 bar, 20°C.; *2*, 12. 1,865, 650; *3*, 11. 1,755, 1,090; *4*, 10.6, 1,690, 1,220; *5*, 10. 1,740, 1,325; *6*, 9.6, 1,770, 1,390; *7*, 9. 1,790, 1,440; *8*, 8.3, 1,810, 1,475; *9*, 6.5, 1,850, 1,510; *10*, 5. 1,875, 1,575; *11*, 4. 1,750, 1,550.

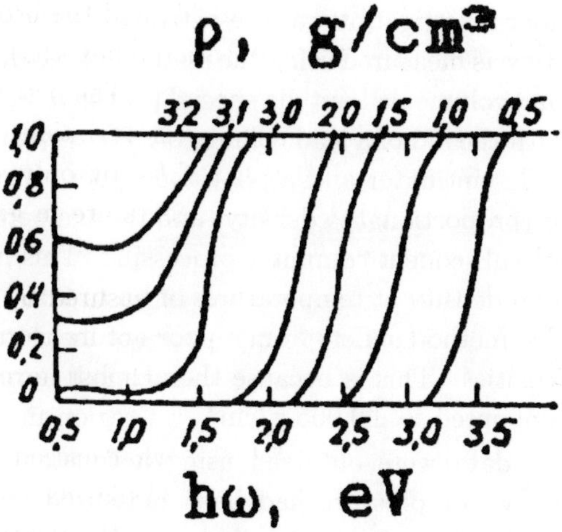

Fig. 2.17. The emissivity of mercury plasma at 1,600°C and various densities ρ [64].

The emissivity of mercury was measured by Uchtmann *et al.* [64] in an apparatus analogous to that shown in Fig. 2.14. The radiation intensity of a plane layer of plasma was determined as the difference between its values recorded in an empty cell and in a mercury-filled cell. The emissivity ε was determined as the ratio between the intensity being measured (normal to the surface of the plasma layer) and the intensity of the perfect radiator. The isochores of emissivity are presented in Fig. 2.17.

A characteristic feature of these curves resides in the occurrence (at densities exceeding 3 g/cm^3) of continuum in the infrared region, the intensity of which grows sharply with a further increase in density. This is in good agreement with the increase in results from measuring the absorption index at the same densities and will be discussed in Chapter 8.

2.1.3. *Density Measurements*

For the purpose of measuring the density of non-ideal plasma of metals, a technique was developed based on measuring the intensity of γ-radiation of the metal being investigated. The apparatus is described by Alekseev [30] and Korshunov *et al.* [65]. A thick-walled tungsten tube is defined by a thin tungsten bottom on one end and by an elastic bellows on the other. This element is filled with cesium under vacuum and sealed off. Located beneath the cell is a point measuring volume. In order to avoid convection, the vacant space is filled with a tungsten insert.

The metal under investigation was pre-activated in the reactor by slow neutrons. The density was measured using the method of recording the γ-radiation from the measuring volume, led out via special port in a high-pressure chamber and collimated with the aid of a lead collimator. For γ-quantum recording, use was made of a CsI scintillator and a photomultiplier. The number of pulses from the detector (proportional to density) and the temperature were recorded in digital form for subsequent computer processing. This method was used to measure the cesium density at temperatures of up to 2,500°C and pressures of up to 600 bar. The method suffers from a poor accuracy (ca. 10%) in measuring low vapor densities. This is because the absolute error in measuring the cesium density amounted to ± 0.005 g/cm^3.

More accurate data were obtained using a constant-volume piezometer where the cesium vapor pressure had been measured by the compensation method. Stone and Ewing [66] and Novikov and Roschupkin [67] used an elastic membrane introduced in the piezometer as a pressure cell. This reduced the range of temperatures under investigation because of the adverse effect of high temperatures on the elastic properties of the membrane. Volyak and Chelebayev [68] removed the membrane from the heated volume. This permitted $p-\rho-T$ measurements of cesium vapors at 52 bar and 1,940 K.

A schematic of the apparatus is shown in Fig. 2.18. Essentially, it comprises a piezometer consisting of a cylindrical chamber *1* with a volume of 50–100 cm^3

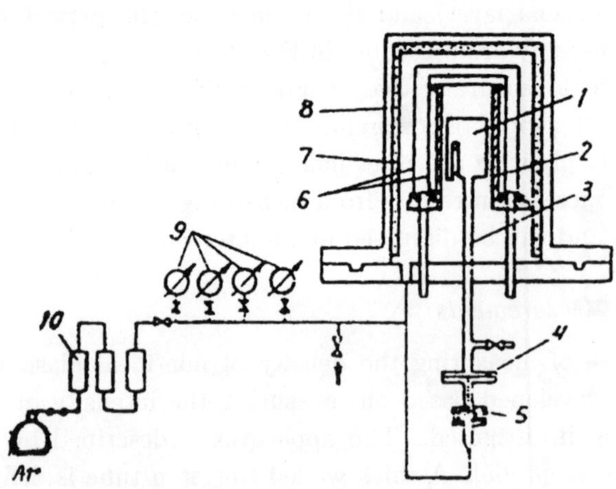

Fig. 2.18. Element for measuring the cesium vapor density [68]: *1*, piezometer; *2*, graphite heater; *3*, capillary with liquid cesium; *4*, membrane; *5*, "zero" manometer pressure cell; *6*, heat shields; *7*, heat insulation; *8*, apparatus body; *9*, pressure gauges; *10*, argon-cleaning system.

placed in a heater *2*. On heating the piezometer, the pressure of cesium vapors in the chamber is transmitted through a column of liquid cesium in capillary *3* to a membrane *4* of induction null pressure cell *5*. Consequently, the membrane deflects. By applying an argon pressure to its other side, the membrane is brought into "zero" position.

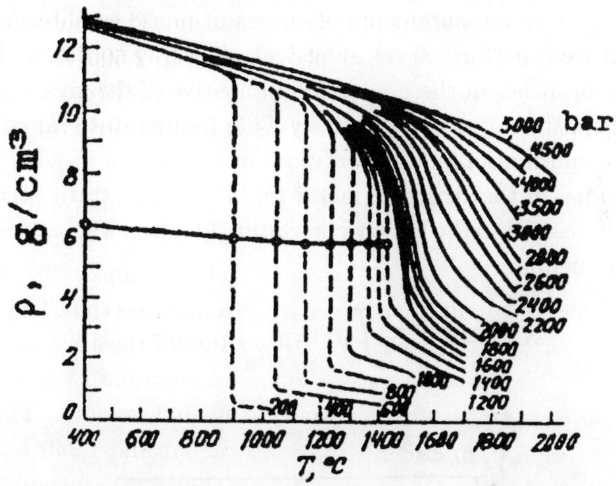

Fig. 2.19. Thermodynamic equation of state for mercury [69].

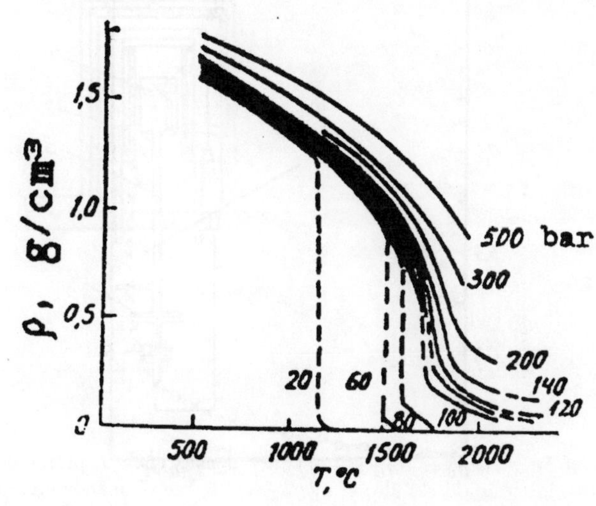

Fig. 2.20. Thermodynamic equation of state for cesium [33].

The specific volume of cesium vapor was determined as the ratio of the piezometer chamber volume (with due regard for its thermal expansion) to the mass of cesium vapor in the chamber. The accuracy of the experiment ($\pm 0.65\%$) was estimated from the error of compressibility coefficient.

The results of density, temperature and pressure measurements enable one to construct the $p - \rho - T$ equation of state [33, 69–71].

The most precise measurements of curves of phase equilibrium for cesium and rubidium were performed by Jungst *et al.* [72] (Fig. 2.21). The revealed asymmetry of branches of these curves is indicative of the invalidity of the law of the rectilinear diameter in a relatively wide temperature range.

Important information on the structure of material is provided by neutron diffraction studies of the structure factor (cf., [73]). The static structure factor $S(q)$ is directly related to the binary correlation function where N is the particle density of the medium:

$$S(q) = 1 + N \int d\mathbf{r} \exp(-i\mathbf{qr})[g(\mathbf{r}) - 1]\,. \tag{2.2}$$

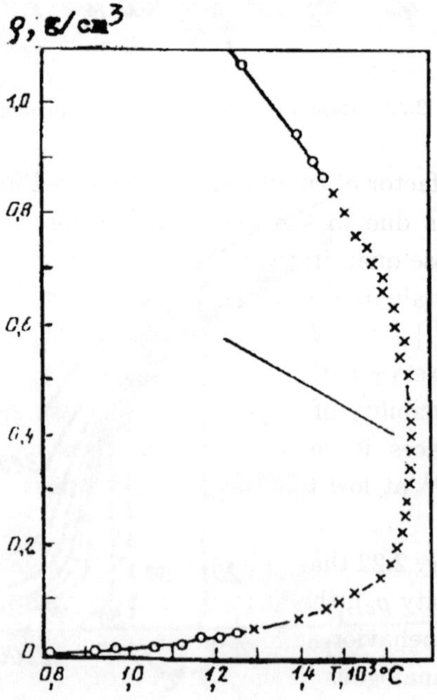

Fig. 2.21. Curve of phase equilibrium for cesium [72].

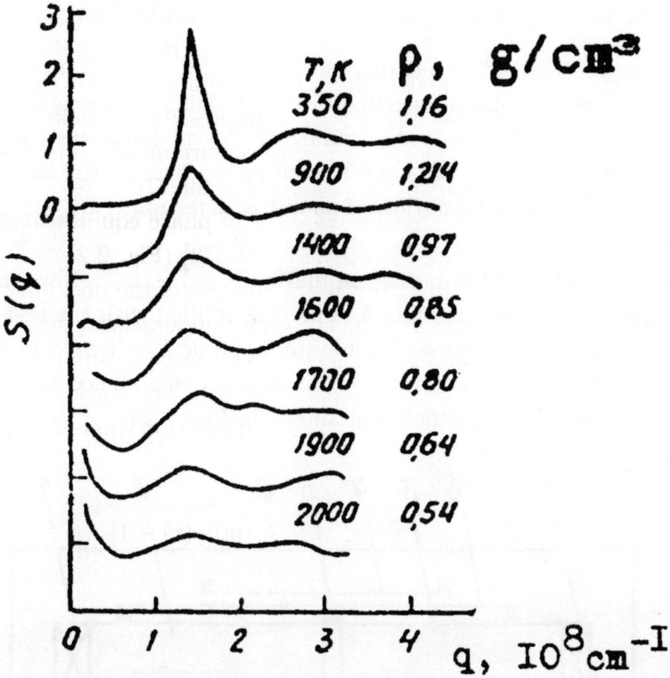

Fig. 2.22. Static structure factor of rubidium [74].

The static structure factor of rubidium was measured in [74, 75] (Fig. 2.22).

The basic error is due to the inaccuracy of discrimination of the background intensity whose quantitative evaluation is very difficult to accomplish. However, it can be evaluated from the requirements of satisfying the rules of sum $\int_0^\infty q^2 dq[S(q) - 1] = 2\pi^2 N$, the asymptotics $S(\infty) = 1$, and value at zero $S(0) = NTk\chi_T$, where χ_T is the isothermal compressibility. Proceeding from this and the reproducibility of $S(q)$, which is measured using different measuring cells and furnaces, it was concluded that the error of $S(q)$ amounts to $\pm 5\%$ for $q > 1.3$ A^{-1} at low temperatures. On heating, the error increases to $\pm 15\%$.

It follows from Fig. 2.22 that at $\rho < 0.66$ g/cm^3 (which is two times higher than the critical density ρ_c), the behavior of $S(q)$ with high values of q is fully smoothed while the behavior at low values of q points to sharply increasing scattering, which is analogous to the behavior of $S(q)$ of argon. From here, it apparently follows that the critical fluctuations are not a factor defining the beginning of the "metal–non-metal" transition.

2.1.4. *Sound Velocity Measurements*

The acoustic method of "fixed distance" consists of the following [75, 76]: Longitudinal sound oscillation excited in the medium is investigated by means of piezoelectric transducers *1*; the sound velocity is determined by the time; taken by a sound pulse to pass the known speed base in the medium under investigation; and the attenuation of sound is determined by the decrease in its amplitude.

The test assembly is shown diagrammatically in Fig. 2.23. Two sound ducts *3* are separated by a niobium ring *5*. The ring is filled with the test substance and its width is the speed base. A niobium ampoule *2* is heated to the desired temperature. As the pressure varied, the mercury introduced into chamber *6* was forced out into an expander (not shown in Fig. 2.23).

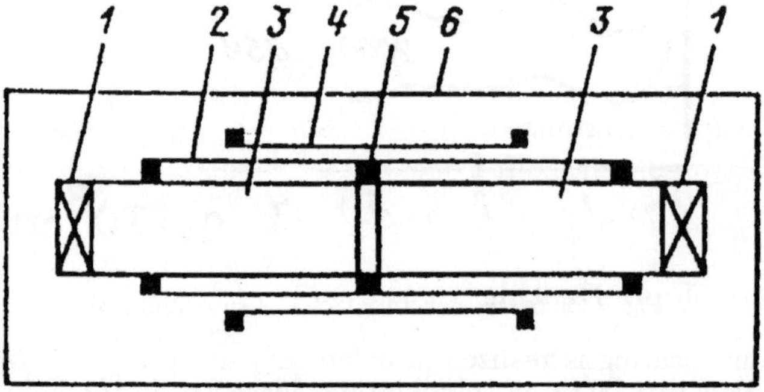

Fig. 2.23. Cell for measuring the sound velocity in mercury [77]: *1*, quartz piezoelectric transducers; *2*, niobium ampoule; *3*, sound ducts; *4*, heater; *5*, niobium ring; *6*, high-pressure chamber.

The measuring scheme enables one to register, separately, the pulse which passed through both sound ducts 3, and the pulse which was singly reflected from the internal interface between the medium and receiving duct line before returning to the radiating piezoelectric transducer. The difference between the in-phase time co-ordinates of both pulses gives rise to the time taken by the sound wave to pass the speed base. This enables one to calculate the sound velocity. This procedure is described in more detail by Kulik *et al.* [39].

The measurement results are given in Fig. 2.24. One can see the rise in the slope of isobars, especially in the liquid phase, as the critical point is approached. The behavior of supercritical isobars becomes smooth. Apparently, the smoother it is, the higher the value of p/p_c. The $v_S(\rho)$ relationship,

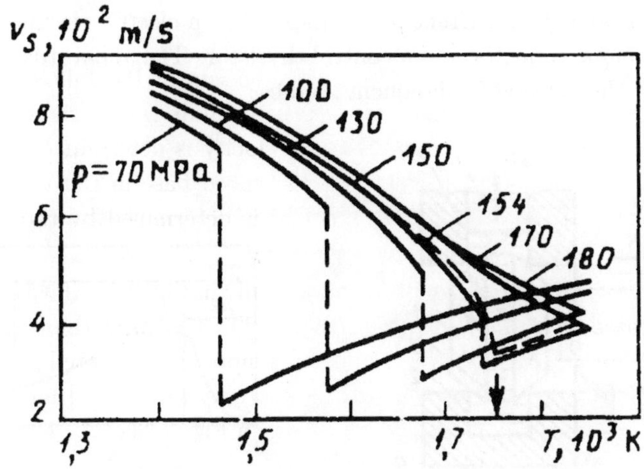

Fig. 2.24. Isobars of the sound velocity in mercury [77].

derived during the rearrangement of these data, clearly reveals a kink in the case of density of 9 g·cm^{-3}. It is known that at this density, mercury becomes metallized.

2.2. Isobaric Joule Heating

Isobaric Joule heating is realized in a capillary or in an atmosphere of inert gas. These methods help to attain a high concentration of charged particles (up to $10^{20} 1/\text{cm}^3$) and high temperatures (up to 10^5 K) at pressures of up to 1,000 atm. However, the plasma generated proves to be substantially inhomogeneous. Moreover, in this case, one fails to realize direct reliable measurements of temperature, density and pressure.

2.2.1. *Isobaric Heating in a Capillary*

This technique was developed to investigate the properties of non-ideal plasma of cesium and, later, of sodium, potassium and lithium [39]. A schematic of the apparatus is shown in Fig. 2.25. A transparent quartz (or glass) capillary 0.7 mm in diameter and 20 mm long is filled with liquid cesium (or it can accommodate a wire of solid potassium, sodium or lithium). Cesium is subjected to constant argon pressure maintained throughout the experiment. Filled with argon simultaneously, the volume serves as an expander for cesium, some part of which is forced out of the capillary upon heating. The cesium in the capillary is heated by a current pulse shaped by an electric circuit comprising a capacitor bank, inductor and a control thyristor. The apparatus was greatly improved in the course of subsequent studies.

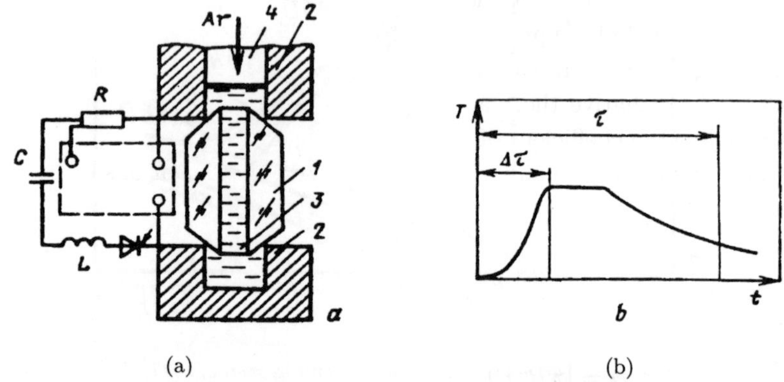

Fig. 2.25. Schematic of an apparatus for isobaric heating in a transparent capillary [44]. (a) *1*, capillary; *2*, electrodes; *3*, liquid cesium; *4*, argon purge. (b) The behavior of time dependence of plasma temperature.

In the experiment, the current-voltage characteristic is measured (from the oscillogram). In the oscillogram, the region corresponding to steady-state conditions is recorded. It is obvious that in doing so, the capillary is completely filled with plasma. By processing the data of the oscillogram, one can obtain the $F(i)$ dependence, where F is the electric field intensity and i the discharge current. This dependence provides the initial experimental data used to obtain, by calculations, the isobars of electrical conductivity $\sigma(T)$ and thermal conductivity $\chi(T)$.

The basic equations include the heat balance equation, which is known in the physics of gas discharge as the Elenbaas-Heller equation and Ohm's law. For a sufficiently long capillary, these equations take the form of

$$\frac{1}{r}\frac{d}{dr}\left(r\chi\frac{dT}{dr}\right)+\sigma F^2=0 \tag{2.3}$$

$$i=2\pi F\int_0^R \sigma r dr\,, \tag{2.4}$$

where r is the distance from the capillary axis and R is its radius. The boundary conditions are

$$dT/dr = 0\,, \qquad r = 0\,; \qquad T = T_{\rm os}\,, \qquad r = R\,,$$

where $T_{\rm os}$ is the pyrometrically measured temperature of the outside surface of the plasma.

In the plasma under investigation, the main mechanism for the extraction of heat and the shaping of temperature distribution $T(r)$ is provided by radiation transfer. Since the mean radiation path length, that is, the Rosseland length, is $l_{\rm R} \ll R$, the radiative thermal conductivity approximation is valid while $\chi(T)$ is equal to the coefficient of radiative thermal conductivity $\chi_R(T)$. The sought-for functions $\chi_R(T)$ and $\sigma(T)$ were obtained as a result of substitution in the equations

$$T(r) = T_{\rm os} + F^2 \int_r^R dr(r\chi_R)^{-1} \int_0^r \sigma r' dr'\,,$$

$$i = (\pi R^2 F) \frac{2}{R^2} \int_0^R \sigma r dr$$

of the trial functions $\sigma(T)$ and $\chi_R(T)$ followed by the iteration procedure. During the experiments, the maximum pressure of 1, 110 bar was attained and the temperature range of (4–18)$\cdot 10^3$ K was covered.

It is noted by Dikhter and Zeigarnik [78] that the main drawback of the technique resides in the impossibility of measuring either the plasma density during its expansion or the plasma temperature. The temperature gradient near the discharge column boundary is very great. Therefore, the calculation of temperature based on its boundary value undoubtedly leads, under such conditions, to great uncertainties. The situation is aggravated by the presence of a quartz wall having a temperature gradient; there is a further possibility of a reversible loss of transparency by quartz. Moreover, one cannot rule out the possibility of fusion and evaporation of quartz at the boundary with cesium plasma, as well as partial mixing of cesium plasma with oxysilicon quartz plasma, that is, a violation of the purity of the material under investigation.

Vorobiov *et al.* [79, 80] obtained the $\chi_R(T)$ and $\sigma(T)$ values for lithium, sodium, potassium and cesium at pressures of (100–1,000) atm and temperatures of (4–14)$\cdot 10^3$ K. Figure 2.26 shows the isobars of the coefficient of radiative thermal conductivity $\chi_R(T)$ for cesium. At low temperatures, that is, at maximum non-idealities, the χ_R values obtained exceed considerably the values calculated in a weak non-ideality approximation [81]. This means that a non-ideal alkali plasma proves more transparent than one might assume from ideal-gas concepts. This is especially clear from Fig. 2.27, which shows the ratios between the Rosseland lengths measured by Vorobiov *et al.* [79] and those calculated by Kuznetsova and Lappo [81].

Figure 2.28 shows the isobars and one of the isentropes of the electrical conductivity of cesium plasma.

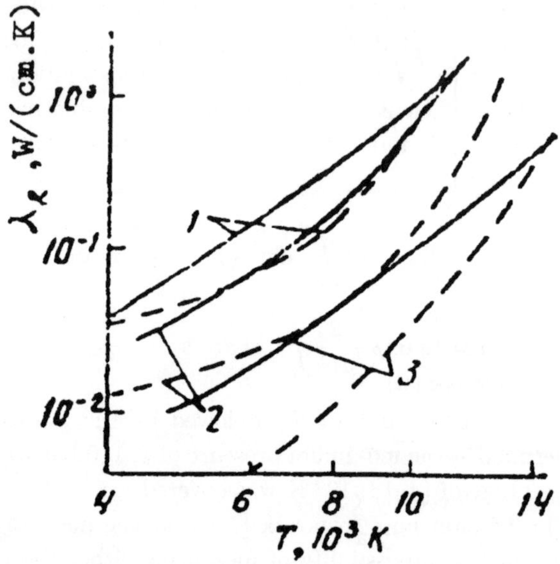

Fig. 2.26. Isobars of the coefficient of radiative thermal conductivity of cesium plasma (W/cm.deg) according to the data of [79]: *1*, 100 atm; *2*, 300 atm; *3*, 1,000 atm; broken line, calculation [81].

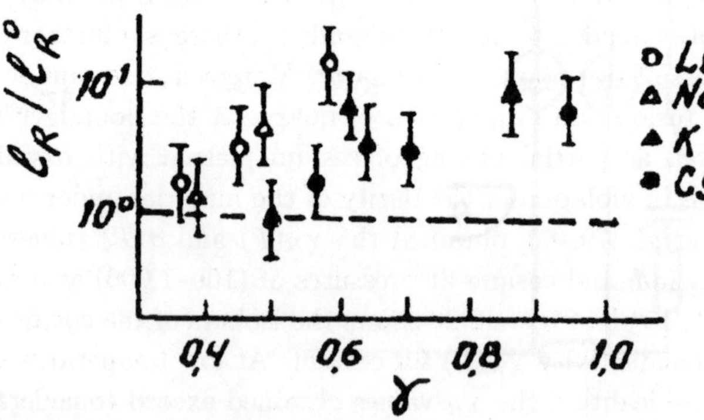

Fig. 2.27. Ratio between the measured [79] and calculated [81] values of Rosseland lengths, depending on the non-ideality parameter γ, $T = 6,000$ K.

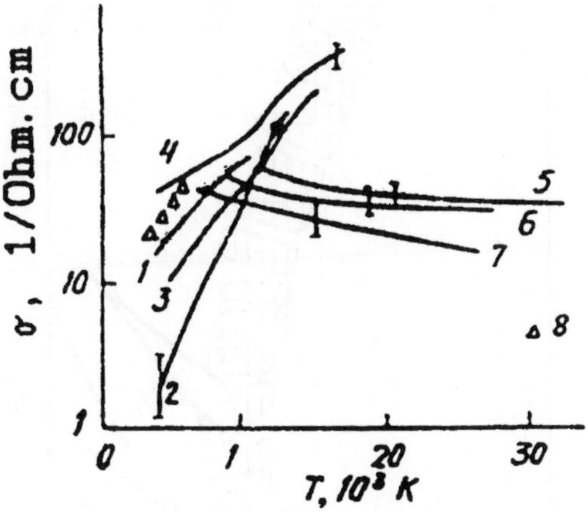

Fig. 2.28. Coefficient of electrical conductivity of cesium plasma: isobars *1*, p = 13 MPa; *2*, p = 27.5 MPa; *3*, p = 60 MPa; *4*, p = 110 MPa [39]; *5*, p = 50 MPa; *6*, p = 26 MPa; *7*, p = 126 MPa [78]; Δ, isentrope [42].

2.2.2. *Heating in an Atmosphere of Inert Gas*

This method was employed by Gathers, Shaner and Young [82, 83] to measure the electrical conductivity and the equations of state for liquid metals in the high-temperature region. Dikhter and Zeigarnik [78, 84] investigated the equation of state and electrical conductivity of non-ideal cesium and lithium plasma. The apparatus is shown schematically in Fig. 2.29 and described in

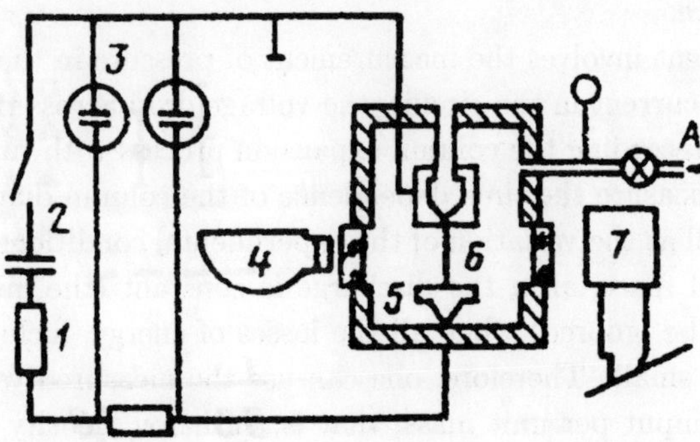

Fig. 2.29. Schematic of the apparatus for isobaric heating in an argon atmosphere [78]: *1*, ballast resistor; *2*, capacitor bank; *3*, oscillograph; *4*, high-speed photorecorder; *5*, high-pressure chamber; *6*, heated wire; *7*, spectrograph.

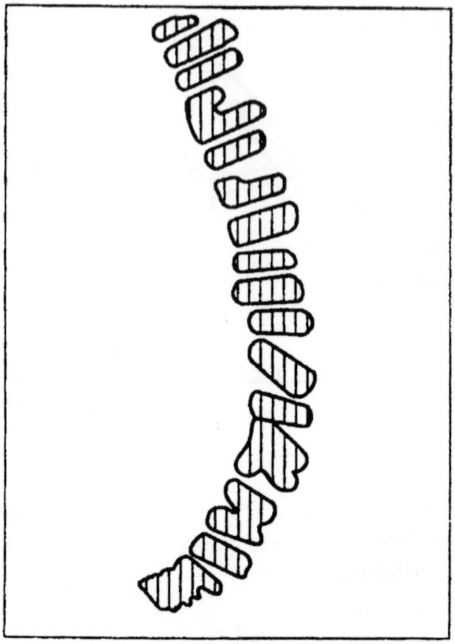

Fig. 2.30. Structure of the plasma column [78].

[78]. A cesium (or lithium) wire is placed in a high-pressure chamber *5* and a current with a density of (1–5)$\cdot 10^6$ A/cm^2 is passed through the wire. The plasma formation contained by a high-pressure inert gas expands upon heating. Whilst doing so, the pressure in the column remains constant and equal to the inert gas pressure.

The experiment involves the measurement of pressure in the chamber, the oscillography of current in the circuit, the voltage drop across the plasma column, and photorecording the column expansion process with time. The latter enables one to measure the time dependence of the column diameter. Special estimates, as well as the variation of the experimental conditions, demonstrate that the mass of material in the discharge is constant (the mixing of argon and cesium can be ignored) while all the losses of energy, including the radiation losses, are small. Therefore, one can use the measured values to calculate the energy input per unit mass, that is, enthalpy, density and electrical conductivity.

According to the estimates by Dikhter and Zeigarnik [78], the error in measuring the current and voltage does not exceed ±7% and for the diameter, ±3%. This leads to the following estimates of errors in measuring the enthalpy, density and electrical conductivity: $\Delta H = 15\%$, $\Delta \rho = 10\%$ and $\Delta \sigma = 20\%$. The scatter of experimental thermodynamic data does not exceed 22%, that is, it is in good agreement with error estimates. However, the scatter of electrical conductivity data reaches 40%.

In Fig. 2.28, the results of electrical conductivity measurements are compared with the data obtained by other authors. While leaving the discussion of the peculiarities in the behavior of electrical conductivity of non-ideal plasma for Chapter 7, we shall only draw the reader's attention to one important detail. The electrical conductivity values on isobars, obtained by the method under discussion, do not increase with heating but, on the contrary, decrease. It can be attributed (in view of the reasons given in Chapter 10) to the principal shortcoming of the method: the inhomogeneity of the plasma column caused by instability, which leads to superheating.

An interesting phenomenon was observed during the experiments: the plasma column separated into dark and light strata extending across the current. The strata are observed in the opening frames and are clearly visible during the first several hundred microseconds. There is no doubt that, given such a fairly slow heating of material, there is enough time for local thermodynamic equilibrium to set in. Therefore, these strata can be referred to as thermal, which is distinct from the ionization strata well-known in the physics of gas discharge. It was shown [85] that the observed effect could be a result of the development of thermal instability of non-ideal plasma with current. This problem is discussed in Chapter 10. At this point, we shall only emphasize that the strata revealed by Dikhter and Zeigarnik [43] still remain the most clearly defined manifestation of the instability of non-ideal plasma in external fields.

In a series of precision experiments [82, 83, 86–88], the electric discharge under an inert gas pressure was used to study the thermophysical properties of metals in the liquid phase with subsequent evaluation, on this basis, of critical point parameters. As distinct from the techniques used by Dikhter and Zeigarnik, use was made of a slower introduction of energy into the sample under investigation, which had a length of about 25 mm and a diameter of about 1 mm, thereby improving the uniformity of parameters and allowing the avoidance of the instability of the moving metal/gas interface. At the same time, the introduction of energy was sufficiently "fast" to preclude the collapse of the plasma column in the gravitational field. During the experiments, the current-voltage characteristics, surface temperature, channel diameter and, in a number of cases, the velocity of sound were recorded. The results of those measurements provided a basis for the subsequent determination of the electrical conductivity, density, enthalpy and heat capacity of about 10 metals at pressures of up to 5,000 bar and temperatures of up to $9 \cdot 10^3$ K. Whilst doing so, Gathers, Shaner and others [82, 83, 86–88], during their investigations, deliberately confined themselves to the single-phase liquid-metal region of parameters. This was because it had been established that the crossing of the boiling curve in the process of "slow" isobaric expansion causes a sharp development of non-uniformities in heated plasma parameters. In order to describe the thermodynamics of dense metal plasma, Gathers, Shaner and others [82, 86] employed the soft-sphere model, which helped in the evaluation of the parameters of critical points for copper and lead; these are in good agreement with the data of dynamic measurements [89] and are listed in Appendix 1.

In measuring the sound velocity, a column of matter was subjected to the effect of a focused laser pulse (0.1–0.5 J, 25 ns, 100 μm) [89]. An acoustic wave, which was excited in matter, was propagated across the column and caused a compression wave in gas upon exit from the latter. This enables one to determine the time of the wave exit from the column and, given the column diameter, to calculate the sound velocity. It has been found that the empirical Birch law, which stipulates that the sound velocity depends linearly on the density of matter, is quite valid. However, the reasons for this are not quite clear.

2.3. High-Pressure Electric Discharges

Because of a high density of material and high level of temperature, the charged particle concentration in a plasma of high-pressure electric arcs and discharges is as high as $(10^{18}–10^{21})\text{cm}^{-3}$. In such a plasma, the Coulomb non-ideality parameter Γ may reach substantial values. Whilst doing so, at pressures above or approximately equal to 10 atm, the plasma of high-pressure discharge is locally equilibrium. This section will briefly describe the principal approaches to the investigation of high-pressure discharge plasma. A good review of these studies was made by Asinovskii and Zeigarnik [90] in 1974.

Facilities with steady high-pressure arcs appeared as a result of improvement in the wall-stabilized Maecker arc. In the late 1960s, facilities operating at pressures of up to 20 bar were developed in a number of laboratories in the USSR and FRG. A series of investigations were also performed on the properties of non-ideal plasma. Batenin and Minayev [91] measured the electrical conductivity and emissivity of argon plasma using the following parameters: $p = 10$ bar, $T = (12 - 15) \cdot 10^3$ K, $N_e \cong 10^{18}$ cm^{-3}, non-ideality parameter $\Gamma \leq 0.5$, and a specific power input of 4.5 kW/cm to the discharge. In steady stabilized high-pressure arcs and high-pressure discharges, one manages, as a rule, to perform measurements of the electrical conductivity and emissivity of plasma. However, the possibilities of wall-stabilized arcs are greatly restricted from the standpoint of non-ideal plasma generation. The relatively low level of specific power restricts the level of temperature and, consequently, the values of the degree of ionization.

High parameters of plasma were obtained by Peters [92] who had stabilized the arc with a liquid wall. The arc was burning in water vapors which were formed due to evaporation of the liquid adjoining the discharge column. The liquid proper formed a rotating liquid wall. This was due to the rotation of the casing as a whole. The plasma parameters were as follows: $p = 1,000$ bar, $T = 18{\cdot}10^3$ K, $N_e = 8{\cdot}10^{18}$ cm^{-3}, $\Gamma = 1.25$, and a specific power of 120 kW/cm.

Considerably more intensive studies were made on pulsed discharges, which featured a number of advantages. The problem of wall cooling is non-existent. The specific power is readily increased to the 100 kW/cm level, thereby providing a high degree of ionization of the material. Quasi-steady burning conditions are maintained in a typical arc for a period of approximately one millisecond, which is sufficient to perform measurements and to establish local thermodynamic equilibrium. Because of high pressures and temperatures, the attainable concentrations of charged particles are two (and even three in the case of discharges in liquid) orders of magnitude higher than the characteristic electron concentrations in sustained arcs and reach $(10^{19}$–$10^{23})$cm^{-3}.

However, a possibility arises with the emergence of inhomogeneities in plasma: wave phenomena, and even manifestations of non-equilibrium, are possible.

Pulsed discharges in gases are generated upon discharging a tank of capacitors through an inter-electrode gap. The discharge techniques are rather varied. Figure 2.31 illustrates one possible design for a discharge tube. A

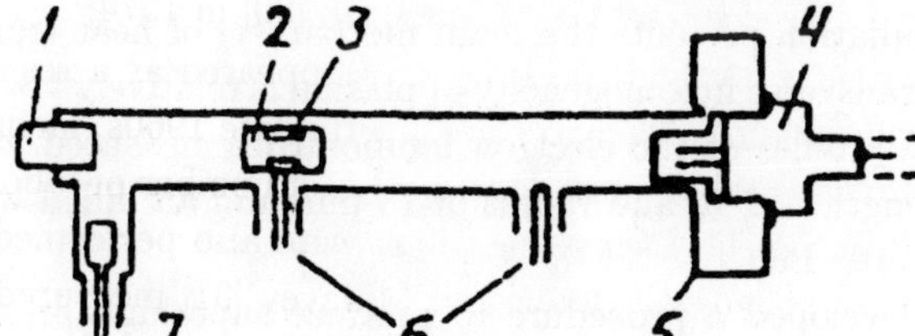

Fig. 2.31. Discharge tube [94]: *1*, quartz window; *2*, auxiliary electrode; *3*, steel ring; *4*, pressure cell; *5*, cathode; *6*, tungsten probes; *7*, anode.

detailed description of this design was given by Günter [93] and Radtke and Günter [94]. Essentially, it consists of a quartz tube with four electric leads: anode, cathode (comprising a pressure cell) and two measuring probes. Inside the tube, movable auxiliary electrode required for ignition is mounted. The tube has a length of 10 cm and a diameter of 1 cm. It is filled with an inert gas at an initial pressure of up to 1 atm.

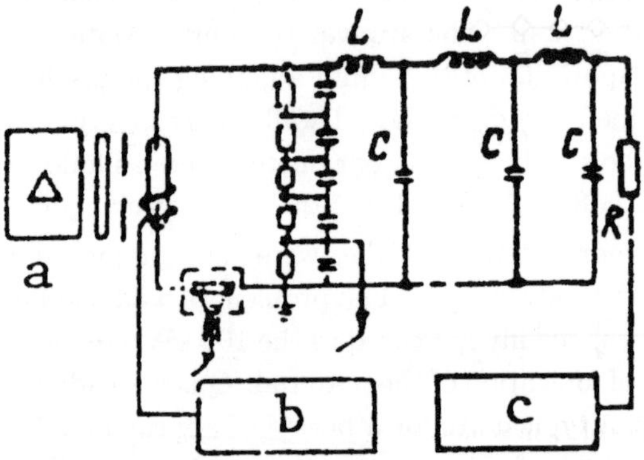

Fig. 2.32. Schematic of the experimental apparatus of Popovic *et al.* [95]: (a) Optics; (b) Start-up; (c) Power.

The basic element of the discharge circuit includes several parallel-connected *LC*-elements shaping a square pulse. Figure 2.32 is a schematic of the apparatus employed by Popovic *et al.* [95]. Measurements of the current strength i and axial field intensity F help to determine the mean electrical conductivity $\bar{\sigma}$,

$$i = \bar{\sigma}\pi R^2 F\,, \qquad \bar{\sigma} = (2/R^2)\int_0^R \sigma(r) r dr\,,$$

where R is the radius of the tube.

Since the radiation presents the main mechanism of heat transfer at high pressures, the transverse inhomogeneity of plasma is relatively low. Figure 2.33 gives the radial profiles of the electron temperature measured in flash-lamps which have a length of 1 m and radius of 13 mm and are filled with xenon at a pressure of $4 \cdot 10^3$ Pa.

Mitin [97] developed a procedure to produce superhigh pressures (about 0.1 GPa) in the plasma of a high-pressure pulsed free arc. As distinct from stationary discharges, the power of a pulsed arc may be high: up to 10 kW/cm. The discharge has a narrow central high-temperature zone (up to $7 \cdot 10^4$ K)

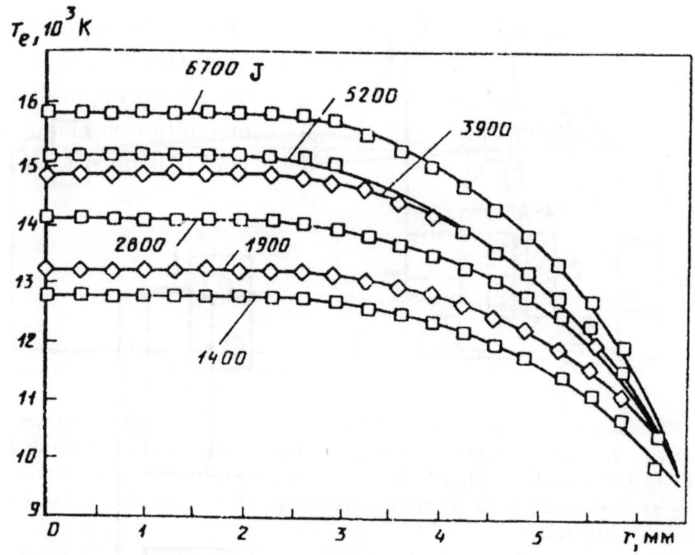

Fig. 2.33. Radial profiles of T_e in xenon plasma in a flash-lamp for different values of discharge energy (numerals by the curves) [96].

and a peripheral low-temperature zone (approximately 10^4 K). Under these conditions, conventional optical diagnostic methods provide only information about the discharge surface and the registered radiating power characterizes the integral properties of inhomogeneous plasma. A pulsed cascade arc, unlike a free one, makes it possible to produce a plasma column that is homogeneous along the axis. However, the inhomogeneity over the column cross-section persists. The coefficient of electrical conductivity of such a plasma was measured for the non-ideality parameters of $0.1 \leq \gamma \leq 0.3$.

The density of material was measured using X-ray. An X-ray beam from a standard X-ray tube passed through a top hollow electrode whose end, which turned into the chamber, had been sealed with a tungsten-tipped beryllium plug, axial region of the arc and a bottom electrode, and was recorded by a NaI photomultiplier.

In order to study the physical properties of plasma at pressures above 100 atm and temperatures of up to 20,000 K, special plasma generating sources must be provided. Andreyev and Gavrilova [98, 99] described a plasma source which was used to measure the electrical conductivity and emissivity of air. Its design is shown in Fig. 2.34. This source allows the derivation of an optically transparent plasma at pressures above 100 atm with uniform and controlled

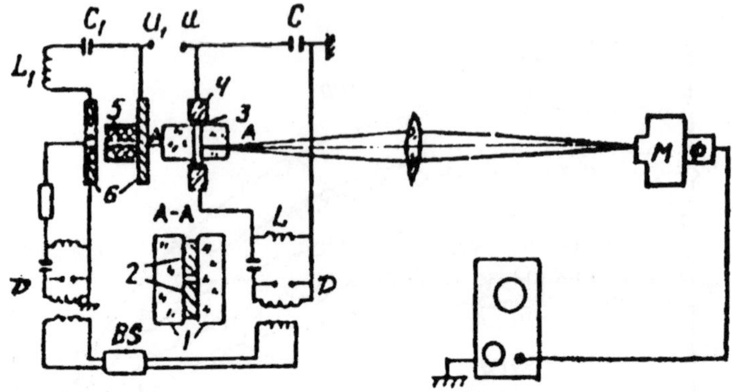

Fig. 2.34. Plasma source design and optical scheme of temperature measurements [98]. Plasma source: *1*, quartz plates; *2*, quartz spacers; *3*, platinum electrodes; *4*, tungsten cylinder ($C = 3.75\ \mu$F, $L = .51\ \mu$H, $U = 1.0 - 2.2$ kV). Transmission source: *5*, Teflon capillary; *6*, steel electrodes; M, monochromator; Φ, photomultiplier; O, oscillograph; П, ignition; BS, synchronizer.

distribution of temperature and density. The apparatus developed by Andreyev and Gavrilova [98–100] offers great possibilities for investigations in a high region of parameters.

The plasma volume of $1.0 \times 1.1 \times 10$ mm^3 is defined by quartz plates *1* and spacers *2*. Airtightness was ensured by the high quality of machining performed on the plates and spacers and by tightly pressing the electrodes against the quartz. The discharge mode was aperiodic with a current amplitude reaching 300 A and a duration of about 50 μs. The radial fluctuations of plasma density cease after ignition in less than 5 μs.

Measurements of the brightness of the transmission source and transparency of the plasma helped to determine the temperature. Spatial resolution of the optical scheme allowed the derivation of the temperature (and density) distribution in the plasma volume. The regress of wall and electrode materials into the plasma was monitored spectroscopically.

Andreyev and Gavrilova [98] used this source to measure the electrical conductivity of air at $13,400\text{ K} \le T \le 18,100$ K and $95 \le p \le 150$ atm, thus corresponding to $N_e \approx 10^{18} - 10^{19}$ cm^{-3} and $\Gamma \approx 0.7 - 1$. The plasma gap resistance was

$$R(t) = \left(U_0 - \frac{1}{C} \int_0^{t_m} i dt - L\frac{di}{dt} \right) i^{-1} .$$

At the moment of current passage through the maximum $t_{\rm m}$, it is sufficient to know U_0, $i(t_{\rm m})$, $C^{-1} \int_0^{t_{\rm m}} i dt$. The measured values of $R(t_{\rm m})$ were of the order of 1 Ohm. The error in subsequent determination of the values of electrical conductivity did not exceed 15%, with the values of σ in the central region of the plasma volume differing from $\bar{\sigma}$ by only 8–10%. The measured values proved to be approximately two times lower than those measured by Spitzer. These results are discussed in Chapter 7.

The highest values of charged particle concentrations, ca. 10^{21} cm^{-3}, were attained in pulsed discharges in liquids. In doing so, pressures of tens of kilobars are attained. The plasma is generated by discharging a bank of capacitors between two electrodes immersed in a liquid, usually in water. The current-heated plasma tends to expand the discharge column. This is impeded by the mechanical inertia of ambient water. Consequantly, high pressures develop. For this purpose, the bank of capacitors must be discharged as soon as possible while the plasma column volume is still small. The discharge is shaped over a period of time of about 1 μs. The plasma generated upon discharge in water has the following parameters: $T = (1 - 5) \cdot 10^4$ K, $p = 1 - 30$ kbar. Therefore, high pressures are combined with high temperatures, which may hardly be accomplished in other types of facilities. However, in view of the high temperature values, the parameter of non-ideality is not very high, being $\Gamma \leq 1.5$. The discharge is usually initiated by electric explosion of a thin wire interconnecting the electrodes. Therefore, the plasma consists of a multi-component mixture of the products of water vapor decomposition with a small admixture of metal from which the wire was made.

The first investigations of a plasma of pulsed discharge in water were performed by Martin [101] and Robinson [102]. At present, such discharges are employed in a wide scope of investigations related to the development of advanced technologies (cf., [27, 103]). Naturally, such investigations may yield interesting physical results as well.

The capillary discharge with evaporating wall is a special type of pulsed discharge. The characteristic feature of this discharge resides in the purging of discharge channel by vapors of the wall material evaporating due to Joule heat of the discharge. Ogurtsova *et al.* [104, 105] obtained a plasma with a charge concentration of up to 10^{20} cm^{-3} at a temperature ranging from $3 \cdot 10^4$ to 10^5 K and at a pressure of 200–500 bar. The chemical composition of plasma depends solely on the composition of wall material. The capillaries were manufactured from textolite. Therefore, the plasma had a complex elementary composition, namely, 47% hydrogen, 37% carbon and 16% oxygen. The plasma was investigated under quasi-stationary conditions. To this end:

(i) The discharge was maintained by fairly long current pulses of up to 10^{-3} ms.

(ii) Use was made of capillaries with open ends, through which the discharge current had been introduced, while the sink of the plasma generated under conditions of wall evaporation occurred in the form of plasma jets.

(iii) The relative variation of the capillary diameter due to its burning out was small. Investigations have shown that the capillary discharge with evaporating wall allows the derivation of a quasi-stationary homogeneous plasma of a desired chemical composition.

The capillary discharge with evaporating wall is characterized by a high level of temperatures at moderate pressures. Depending on the level of energy contribution, the non-ideality parameter ranges from 0.5 to 1.25. The capillary discharge with evaporating wall was used to measure the spectral absorption factor of continuous radiation and the electrical conductivity.

Three

Dynamic Methods in the Physics of Non-Ideal Plasma

The highest plasma parameters were recently obtained using dynamic techniques [106] which produced, under controlled conditions, an unprecedented level of local concentrations of energy. These techniques were based on the cumulation of energy in the material under investigation, either as a result of viscous dissipation in the front of the shock waves which propagate throughout the material to cause its compression, acceleration and irreversible heating, or as a result of adiabatic variation of pressure in the material. Without going into a detailed comparison between the electric and dynamic techniques, we shall emphasize that the high purity and homogeneity of the volume under investigation, the absence of electric and magnetic fields (which hamper the diagnostics and cause the development of various instabilities in the plasma), the high reproducibility of results, and the possibility of attaining extremely high parameters render the dynamic techniques a convenient means to attain and investigate the physical properties of highly non-ideal plasma under extreme conditions. In addition, the application of the general laws of conservation of mass, momentum and energy enables one to reduce (see Section 3.2) the recording of the thermodynamic characteristics of plasma to a registration of the kinematic parameters of motion of shock discontinuities and interfaces (that is, to the measurement of times and distances), which presents an additional important advantage of the dynamic techniques.

The use of shock waves in high-pressure physics enables one to attain the pressures of condensed matter, which are hundreds of billions of atmospheres, and, whilst doing so, to perform extensive thermodynamic, optical and electrophysical [107–111] investigations. The use of such techniques in the physics of non-ideal plasma offered the possibility of expanding, considerably, the range of plasma pressures and temperatures to be investigated, as well as laboratory investigations of states with extremely high energy concentrations. Hence physical measurements in the regions of the phase diagram of matter, which used to be inaccessible to the traditional methods of plasma experiment, are possible.

In this chapter, we consider three methods of dynamic generation of plasma, namely, shock and adiabatic compression and the method of adiabatic expansion of shock-compressed matter.

The shock compression technique appears to be the most effective from the standpoint of studies into the non-degenerate plasma of materials which, in their initial state, are gases. A combination of pre-heated pneumatic, electric discharge and explosion tubes provides information on the thermodynamic, electrophysical and optical properties of plasma at pressures of up to 300 kbar, electron concentrations of about 10^{23} cm^{-3} and densities of up to 4.5 g/cm^3 (1.5 times higher than the crystallographic density of xenon). Under these conditions, the Coulomb energy is an order of magnitude higher than the kinetic energy of particle motion so that, from the physical viewpoint, such plasma resembles a liquid whilst differing from the latter by a richer and complex spectrum of interparticle interactions. Currently, the plasma behavior at ultramegabar pressures is of special interest from the standpoint of ascertaining the role of band structure effects and determining the lower range of application of quasi-classical models of the equation of state. In order to reach this pressure range, nuclear explosions, laser radiation and rapid electric explosion of metal foils are presently used to advantage (see Section 3.5).

The shock and isentropic compression techniques make it possible to attain high pressures and temperatures in a medium of increased density and, because of thermodynamic restrictions, do not allow an investigation of the boiling curve and near-critical states of metals. The most effective method, from the standpoint of obtaining plasma with a less-than-solid-state density, is isentropic expansion of metals (Section 3.5) pre-stressed in a powerful shock front. This method helps in the investigation of an extremely wide region of the phase diagram of metals from a highly condensed metallic liquid to ideal gas, including the region of non-ideal degenerate and Boltzmann plasma and the neighborhood of the critical point.

The results of the registration of release isentropes served as a basis for the construction of modern wide-range equations of state and, in addition, helped form a more conclusive opinion on the general qualitative form of the phase diagram of matter at high pressures and temperatures. This problem is not trivial for a highly non-ideal plasma, in view of the existence of numerous predictions of exotic phase transitions due to the "metal-dielectric" transformation, as well as to a strong interaction of charged particles with each other and with neutrals. We shall discuss the relevant experimental studies in Sections 3.4 and 5.8.

3.1. The Principles of Dynamic Generation and Diagnostics of Plasma

The classification of plasma states in Chapter 1 shows (cf., Fig. 1.1) the plasma states with extremely high pressures and temperatures located at the periphery of the phase diagram of matter, where the interparticle interaction is low, to be accessible to consistent theoretical analysis.

The range of existence of non-ideal plasma appears restricted and it is located in Fig. 1.1 within the triangle defined by the $\Gamma = 1$ and $\varepsilon_F \sim e^2 N_e^{1/3}$ curves. The upper portion of this triangle relates to degenerate plasma and the lower portion to Boltzmann plasma. The maximum attainable values of the non-ideality parameter are finite and do not exceed several units. Figure 1.1 demonstrates that optimum conditions exist for the generation of highly non-ideal plasma. The conditions were further complemented by their optima dictated by the processes of ionization under conditions of the method selected for the realization of those states (cf., Sections 3.2 and 3.3). Given the isochoric heating of Boltzmann plasma in the initial stages of ionization, the electron concentration grows sharply with temperature, $N_e^2 \sim NT^{3/2} \exp\{-I/T)$, thus leading to a faster growth of the Coulomb energy as compared with kT. A further temperature increase causes a full ionization of the given electron shell ($N_e \sim$ const), and a decrease Γ in view of the increase in kinetic energy. While putting off the qualitative assessment of this effect until Sections 3.2 and 3.3, we shall note that these optima formed the basis for the design of most devices for dynamic compression of plasma.

Although it appears restricted in Fig. 1.1, the non-ideal plasma takes an exceedingly wide region in the real $p - V$ diagram of state of matter (Fig. 3.1). It adjoins directly and actually penetrates [7, 106] the region of condensed state where the physical description is extremely difficult because it depends on the specific electron spectra of atoms and molecules. Therefore, priority is given in this case to the experimental, primarily dynamic [107–111], methods which enable one to perform elegant measurements of the thermodynamic, optical and electrophysical properties of condensed matter in the megabar pressure range. Modern high-temperature experimental techniques have facilitated the advancement into the region of the boiling curve and the neighborhood of

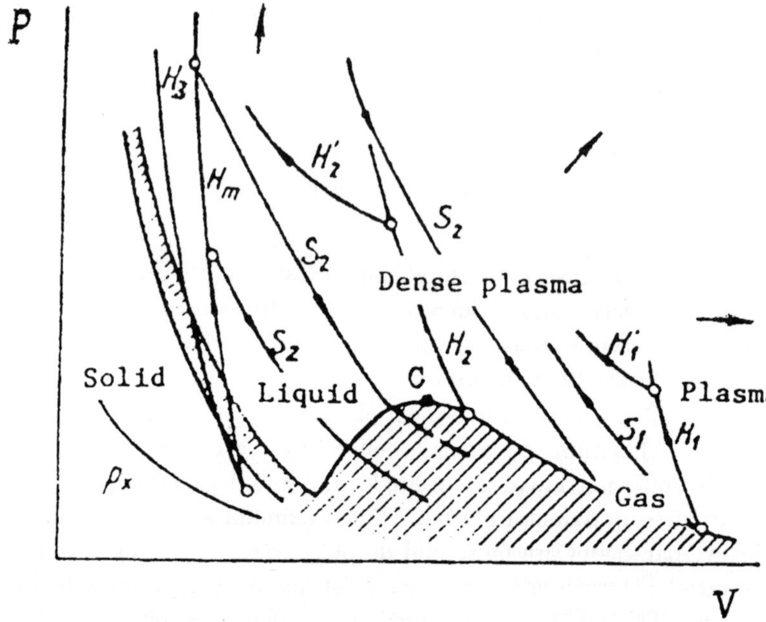

Fig. 3.1. Schematic illustrating the principles of dynamic generation of plasma. p_χ is the boundary of the maximum compressions of material, that is, "cold" ($T = 0$ K) compression curve. The shaded areas show two-phase regions of melting and evaporation. K_C is the critical point. Circles indicate the initial states of the medium. H_1, H_2 are the curves of cesium and inert gas compression by incident and reflected (prime) shock waves. H_3, H_m is shock-wave compression of solid and porous metals. S_1 is the curve of adiabatic compression of cesium while S_2 is the release adiabats for shock-compressed metals.

the critical point of low-boiling metals, such as mercury, cesium and rubidium, and the investigation of the liquid phase region of a number of other elements at $p \leq 4$ kbar, $T \leq 5 \cdot 10^3$ K [112] (Chapter 2). However, not only have the critical parameters not been found for the great majority of other metals (amounting to about 80% of all elements in the periodic system), but the qualitative form of the phase diagram has also not been clarified. This is because the greater part of the latter is taken by the region of non-ideal plasma, which is inaccessible to traditional experimental techniques.

In order to produce a plasma with strong interparticle interaction, a considerable amount of energy release is required in a higher density medium. The concomitant pressures and temperatures, as a rule, exceed appreciably the thermal-strength limits of the structural materials of the respective facilities. Therefore, such plasma can only be maintained for a short period of

Table 3.1. Sources of energy and experimental facilities employed in the physics of high energy concentrations.

Primary Energy Source	Final Form	dE/dV, MJ·cm^{-3}	T, eV	p, bar	E_{total} MJ	t, s	W, W
Chemical explosive		10^{-2}	0.5	$5 \cdot 10^5$	10^2	10^{-7}	10^{10}
	Metal plates	0.3	60	10^7	3	10^{-6}	10^{12}
	1-MOe magnetic field	$4 \cdot 10^{-3}$	0.3	$5 \cdot 10^4$	5	10^{-6}	$5 \cdot 10^{12}$
	25-MOe magnetic field	2.5	200	$2.5 \cdot 10^7$	1	10^{-7}	10^{13}
	Explosion plasma generators	10^{-2}	60	10^5	30	10^{-6}	10^{12}
Nuclear explosive		10^4	10^7	10^{10}	10^{11}	10^{-6}	10^{22}
	Neutron heating	10	50	$2 \cdot 10^7$	10^3	10^{-6}	10^{15}
	Shock waves in solid	5	50	$5 \cdot 10^7$	10^4	$3 \cdot 10^{-6}$	10^{15}
	Shock waves in gas	0.3	40	$2 \cdot 10^5$	10^7	10^{-5}	10^{18}
Compressed gas	Adiabatic compression	$2 \cdot 10^{-5}$	0.3	150	10^3	$6 \cdot 10^{-3}$	10^5
	Pneumatic shock tubes	10^{-4}	1	250	10^{-2}	10^{-4}	$3 \cdot 10^8$
	Combustion shock tubes	10^{-6}	2	10	$2 \cdot 10^{-2}$	$3 \cdot 10^{-4}$	10^8
	Electric discharge shock tubes	10^{-7}	2	1	10^{-2}	10^4	10^8
Capacitor		10^{-8}	–	–	40	10^{-5}	10^{12}
Rotor machine		10^{-3}	–	–	100	10^{-4}	10^1
Induction storage		10^{-4}	–	–	100	10^{-4}	10^{12}
Storage cell		$5 \cdot 10^{-4}$			1,000	10^{-3}	10^{12}
	Rapid explosion of wires	$5 \cdot 10^{-2}$	4	10^5	10^{-3}	10^{-6}	10^{-9}
	Slow explosion of wires	$2 \cdot 10^{-2}$	0.5	$5 \cdot 10^2$	10^{-3}	10^{-4}	10^7
	Pulsed discharges	10^{-3}	10	10^4	10^{-4}	10^{-3}	10^9
	Plasma focus	10^{-2}	1,0002	10	10^{-4}	10^{-5}	10^{10}
	High-pressure arcs	10^{-5}	0.3	10^4	10^{-4}	Steady	10^4
	Furnace	10^{-3}		$5 \cdot 10^3$	10^{-3}	Steady	10^3
Laser		10^{-6}	–	–	$0.5 \cdot 10^{-3}$	10^{-10}	10^{13}
	Target	10^4	$5 \cdot 10^3$	10^8	0.5	10^{-10}	10^{13}
Electron beam		10^{-6}	–	–	1	10^{-8}	10^{13}
	Target	$5 \cdot 10$	$5 \cdot 10^3$	10^7	0.1	10^{-8}	10^{13}

Note: The following nomenclature is used in the table: dE/dV, energy density; E_{total}, total energy; W, power.

time, which is defined by its inertial expansion. This compels one to conduct the experiment under heavy-duty pulsed conditions at a high power level. The appropriate experimental facilities must provide a fast energy supply to the material under investigation, and should feature geometric dimensions that are sufficient for reliable diagnostics. The existing possibilities for plasma generation are listed in Table 3.1, which recites the characteristic (not necessarily maximum) parameters of energy sources used in the compression and heating of the material. Whilst doing so, because of the above restrictions due to plasma degeneracy and superheating, the extreme parameters listed in Table 3.1 do not correspond at all to the maximum Coulomb non-ideality effects.

The dynamic methods of plasma generation are based on hydrodynamic methods of heating a material as a result of viscous dissipation of energy in a shock front or adiabatic compression of the medium [113]. By now, three techniques have come to be most widely used in plasma investigations, namely, isentropic and shock compression, as well as isentropic expansion of pre-shock-compressed material. The use of various energy sources to excite shock and compression waves, such as compressed and electric discharge-heated gas, condensed and nuclear explosives, and powerful laser and neutron fluxes enables one to produce a non-ideal plasma of different elements in a wide and continuously varying range of densities: from solid-state to gas at pressures of tens of millions atmospheres and temperatures of tens of kiloelectron-volts.

A considerable amount of data were obtained by propagating shock waves over the material under investigation to cause irreversible compression, heating and acceleration. In doing so, the initial states of material may be either in the solid or gas phase. The latter case is especially convenient to obtain a highly non-ideal Boltzmann plasma (Sections 3.2 and 3.3). It is generated as a result of shock compression of high-pressure gases whose initial states are (cf., Fig. 3.1) in the neighborhood of the saturation curve (cesium, argon) or under supercritical conditions (xenon). By recording the state of single or double compressions, it is possible to obtain a plasma with supercritical parameters in a wide range of pressures (up to 110 kbar) and temperatures (up to 10^5 K), and to penetrate, from the "gas" phase side, the condensed state region. The maximum density of xenon plasma amounted to 4.5 g/cm^3. This is about 1.5 times greater than the density of solid xenon and the order of the density of aluminum. The compression of liquid xenon using light-gas "guns" in the pressure range of up to 1.4 Mbar yields valuable information on the electron spectrum of superdense plasma in the megabar range. Adiabatic compression of saturated cesium vapors, S_1 (Fig. 3.2), makes it possible (Section 3.2) to attain lower degrees of plasma heating where the charge-neutral interaction

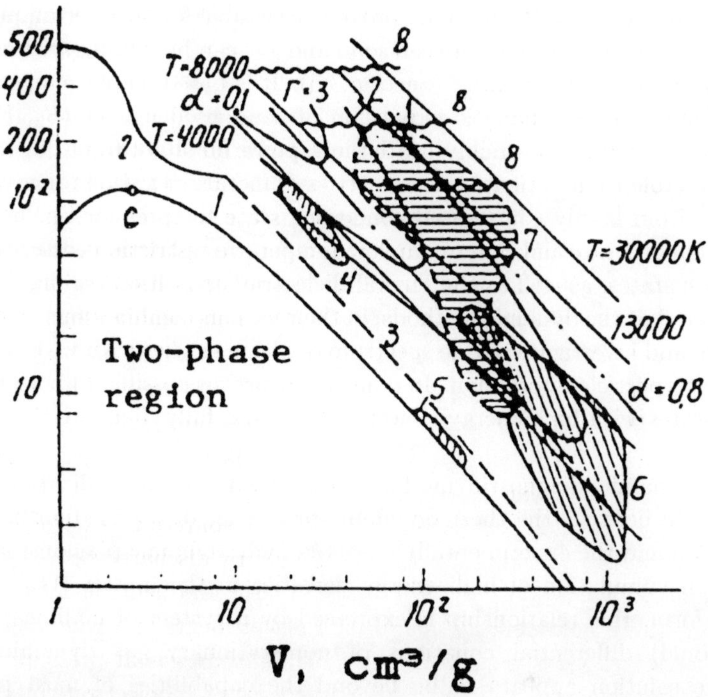

Fig. 3.2. Phase diagram for cesium: saturation curve, *1* (Chapter 2); static experiments; *2*, *3*; region of isentropic compression *4* [40, 123] from the initial states *5*; compression by incident (*6*) and reflected (*7*) shock waves; electric explosion, *8*; α, degree of ionization; K_C, critical point; γ, non-ideality parameter.

prevails. The compression of metals by powerful shock waves [107–111], a result of the detonation of condensed explosive, transforms the metal into a state with a pressure of up to 5–10 Mbar and temperature of tens of thousands degrees when the metal is molten. As a matter of fact, a disordered electron-ion plasma, in which the electron component is degenerate or partly degenerate, is realized.

The use of powerful underground explosions [114–119], neutron radiation resulting from the blasting of nuclear charges [11], and concentrated laser radiation increases the pressure by up to tens and hundreds of megabars. It also serves as a basis for extrapolation checks on quasi-classical theories [121].

One characteristic feature of the shock wave technique is that it helps in the attainment of high pressures and temperatures in compressed media while the region of lower plasma densities (including the boiling curve and the neighborhood of the critical point) prove inaccessible for such techniques. The intermediate plasma states between solid and gas can be investigated using the isentropic expansion technique (Section 3.5). It is based on plasma generation under conditions of adiabatic expansion of condensed matter (S_2, Fig. 3.2) precompressed and irreversibly heated in a powerful shock front. In this way, it was possible to investigate the properties of metals in a wide range of phase diagram, from highly compressed condensed state to ideal gas, including the region of degenerate and Boltzmann low-temperature plasma, near-critical and two-phase states, as well as the "metal-dielectric" transition region.

We see that the dynamic methods, in their various combinations, enable one to realize and investigate a wide spectrum of plasma states with variegated and strong interparticle interaction. In doing so, it proves possible to both actually realize states with high energy concentrations and fully diagnose those exotic states.

The dynamic diagnostic methods are based on the utilization of the relationship between the thermodynamic properties of the medium under investigation and the experimentally observed hydrodynamic phenomena occurring upon cumulation of high energy densities in the matter [113]. In the general form, this relationship is expressed by a system of nonlinear (three-dimensional) differential equations of non-stationary gas dynamics. Its complete solution appears to be beyond the capabilities of most powerful modern computers. For this reason, one tends to use in dynamic investigations [107, 113] self-similar solutions in the form of stationary shock wave or centered Riemann expansion wave, which express the conservation laws in a simple algebraic or integral form. Whilst doing so, the conditions of self-similarity of flow must be observed in the experiment in order to use these simplified relationships.

Upon propagation of a stationary shock wave discontinuity through the material, the laws of conservation of mass, momentum and energy are satisfied in the front of the latter [113]:

$$\left.\begin{aligned} V/V_0 = (D-u)/D\,, \qquad p = p_0 + Du/V_0\,, \\ E = E_0 = (1/2)(p+p_0)(V_0 - V)\,, \end{aligned}\right\}, \tag{3.1}$$

which allows the derivation of the hydrodynamic and thermodynamic characteristics of the material from the recording of any two out of five parameters characterizing the shock wave discontinuity (E, p, V, D, u). The shock velocity D is measured most readily and accurately using baseline techniques. The choice of the second parameter to be measured depends on the actual experimental conditions.

An analysis of errors in the relationships included in Eq. (3.1) indicates [122, 123] that in the case of readily compressible ("gas") media, it is practical to perform the recording of density $\rho = V^{-1}$ of shock-compressed material. A procedure has been developed for such measurements based on the registration of absorption by cesium (Section 3.2), argon (Section 3.5) and the air (Section 3.6) plasma of "soft" X-ray radiation. In the case of lower compressibility of the system (condensed media), tolerable accuracies are ensured [107] by recording the mass velocity of travel u.

In the experiments which involve the recording of the isentropic expansion curves for shock-compressed matter (3.5), the states in a centered release wave are described by the Riemann integrals [113]:

$$V = V_{\mathrm{H}} + \int_{p}^{p_{\mathrm{H}}} \left(\frac{du}{dp}\right)^{2} dp\,; \qquad E = E_{\mathrm{H}} - \int_{p}^{p_{\mathrm{H}}} p\,(fracdudp)^{2}\,dp\,, \tag{3.2}$$

which are calculated along the measured isentrope $p = p(u)$. By recording under different initial conditions and shock wave intensities, one can determine the caloric equation of state $E = E(p, V)$ in the region of the $p - V$ diagram, which is overlapped by Hugoniot and/or Poisson adiabats. In the experiments involving dynamic stimulation of plasma performed today, the variation of the shock wave intensity was effected by varying the power of excitation sources, such as the pressure of propelling gas (Section 3.2), types of explosives, projectile launchers and targets (Section 3.5). In addition, use was made of the various ways to vary the parameters of the initial states, such as the initial temperature and pressure (a plasma of inert gas (Section 3.5), cesium (Section 3.4), liquid [124]), as well as the use of finely divided targets with a view to increasing the effects of irreversibility [113].

Therefore, the dynamic diagnostic techniques, which are based on the general conservation laws, enable one to reduce the problem of determining the caloric equation of state $E = E(p, V)$ to a measurement of the kinematic parameters of motion of shock waves and contact surfaces, that is, to the registration of distances and times which can be done with a high level of accuracy. The internal energy, however, is not the thermodynamic potential in relation to the p, V variables and, in order to construct closed thermodynamics of the system, one needs additional temperature dependence $T = T(p, V)$. In optically transparent and isotropic media (gases, Section 3.5), the temperature is measured jointly with other shock-compression parameters. Condensed media and, primarily, metals are, as a rule, non-transparent. Therefore, the light radiation of shock-compressed medium is inaccessible to registration.

Based on the first law of thermodynamics and the experimentally known relationship $E = E(p, V)$, a thermodynamically complete equation of state can be constructed directly from the results of dynamic measurements without introducing *a priori* considerations of the properties and nature of the material under investigation [113, 125, 126]. This leads to the linear inhomogeneous differential equation for $T(p, V)$:

$$\left[p + \left(\frac{\partial E}{\partial V}\right)_p\right]\frac{\partial T}{\partial p} - \left(\frac{\partial E}{\partial p}\right)_V \frac{\partial T}{\partial V} = T\,. \tag{3.3}$$

The solution is constructed by the method of characteristics:

$$\frac{\partial p}{\partial V} = \frac{p + (\partial E/\partial V)_p}{(\partial E/\partial p)_V}\,; \qquad \frac{\partial T}{\partial V} = -\frac{T}{(\partial E/\partial p)_V} \tag{3.4}$$

$$E = E_0 \exp\left\{-\int_{V_0}^{V} \gamma(V, E)\, d\ln V\right\}\,;$$

$$T = T_0 \frac{pV}{p_0 V_0} \exp\left\{-\int_{V_0}^{V} \left(\frac{\partial \ln \gamma(E, V)}{\partial \ln V}\right) d\ln V\right\}\,. \tag{3.5}$$

Equations (3.3) to (3.5) are complemented by boundary conditions, namely, the temperature is assigned in the low-density region where its reliable theoretical calculation is possible (cesium plasma (Section 3.2)) or the temperature is known from experiment [125].

The $E(p, V)$ or $\gamma(p, V)$ relationships required in the calculation of the right-hand sides of Eqs. (3.4) and (3.5) were determined from the experimental data in the form of power polynomials and rational functions. The accuracy of the obtained solution, as a function of the experimental errors and inaccuracy of initial data, was determined using the Monte Carlo method by computer modeling of the probability structure of the measuring process [126].

This method is free from restrictive assumptions on the properties, nature and phase composition of the medium under investigation. This is because it makes use of the first principles of mechanics and fundamental thermodynamic identity (3.3). This thermodynamic universality enables one to construct, using a unified procedure, the equation of state for a wide spectrum of condensed media and to use it to describe phase transformations [127]. The method proved to be especially effective in the study of the thermodynamics of non-ideal cesium plasma on the basis of experiments in shock [122, 128] and adiabatic [123] compression of saturated vapors.

3.2. Dynamic Compression of the Cesium Plasma

Cesium has the lowest ionization potential, ca. 3.89 eV, of the practically accessible elements, which helps attain a high charge concentration N_e at moderate temperatures, thereby ensuring a substantial value of the non-ideality parameter at relatively low energy contributions. Therefore, this element is the most popular choice for use in non-ideal plasma experiments. Its phase diagram indicating regions of parameters accessible for various techniques is shown in Fig. 3.2. The dynamic generation of cesium plasma was accomplished by way of adiabatic (state 4) and shock compression of saturated vapors in the incident (6) and reflected (7) shock fronts.

Experiments in dynamic compression of cesium vapors were performed [122, 128] in a pneumatic, diaphragm shock tube, as illustrated schematically in Fig. 3.3. In order to attain initial high pressures of saturated vapors, the experimental apparatus having a length of 4 m and an inside diameter of 4.5 cm was heated to 700°C. Along with the marked aggressive properties and chemical activity of cesium, these high temperatures presented the main reason for the difficulties involved in such an experiment. An ionizing shock wave emerged

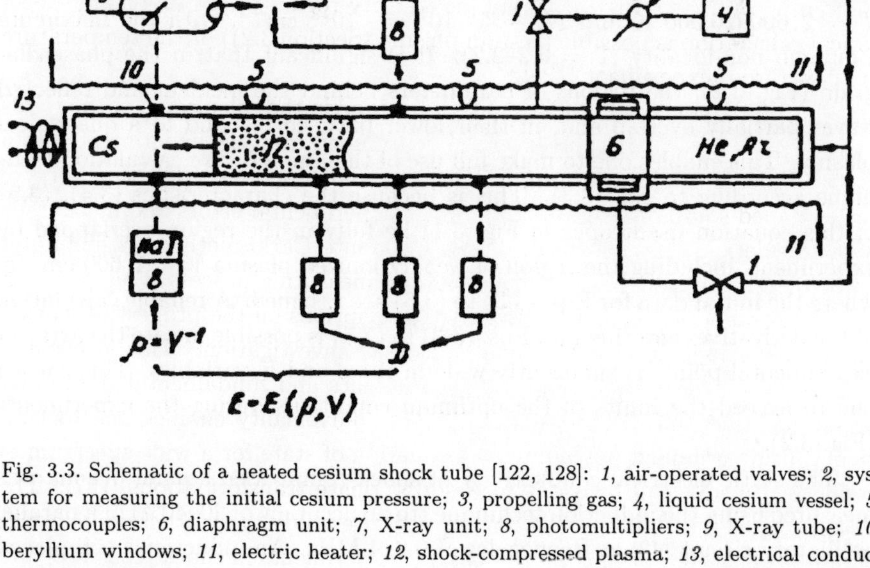

Fig. 3.3. Schematic of a heated cesium shock tube [122, 128]: *1*, air-operated valves; *2*, system for measuring the initial cesium pressure; *3*, propelling gas; *4*, liquid cesium vessel; *5*, thermocouples; *6*, diaphragm unit; *7*, X-ray unit; *8*, photomultipliers; *9*, X-ray tube; *10*, beryllium windows; *11*, electric heater; *12*, shock-compressed plasma; *13*, electrical conductivity measuring coil.

upon expansion to saturated cesium vapor of helium, argon or their mixtures pre-compressed to about 10^3 bar. An additional increase in shock-compressed plasma parameters was attained with the aid of shock waves reflected from the tube end.

In accordance with the dynamic approach, two mechanical parameters characterizing shock compression were independently registered in each experiment. The optical and X-ray [122] baseline techniques were employed to determine the shock front velocity (to an accuracy of about 1%) and to monitor the steadiness of flow of shock-compressed plasma. The plasma density was recorded using the flash radiography technique based on measuring the plasma attenuation of "soft" X-ray radiation. The wavelength of the latter radiation was selected for reasons of maximum sensitivity and minimum statistical error of the measuring scheme. It amounted to 0.2–0.5 Å. The X-ray photographs obtained point to the absence of marked condensation of shock-compressed cesium and permit the determination of the plasma density with an error of 5–10%. This is supported by the results of radiography of reliably calculated shock waves in xenon [128].

Experiments in shock compression of cesium were conducted in the optimum (as regards non-ideality effects) region of parameters at initial pressures of 0.04–0.54 bar and shock velocities of $(0.8\text{–}3) \cdot 10^5$ cm/s. These correspond to a sizable region of the phase diagram (Fig. 3.2), namely, $p \sim 1.4\text{–}200$ bar, $T \sim 2,600\text{–}20{,}000$ K and $N_e \sim 5 \cdot 10^{15}\text{–}5 \cdot 10^{19}$ cm^{-3}, with the maximum Coulomb non-ideality ($\Gamma = 0.2\text{–}2.2$). It is significant that in the phase diagram (Fig. 3.2), the regions of parameters behind the incident and reflected waves partially overlap and, in their lower part, correspond to a quasi-ideal plasma. This enables one to make full use of the temperature calculation technique according to Eq. (3.3). This is because the characteristics (3.4), (3.5) of this equation (isentropes in Fig. 3.1) lie fully in the region overlapped by experiment, including the region of weakly ionized plasma $V = 1,600$ cm^3/g, where the initial data for Eqs. (3.3) to (3.5) are assigned. A reliable calculation of the derivatives entering into Eqs. (3.3) to (3.5) is possible in case the array of experimental points is sufficiently wide in the p and V variables. This caused one to exceed the limits of the optimum conditions during the experiments (Fig. 3.2).

The static electrical conductivity of shock-compressed cesium plasma was measured using the induction technique (to an accuracy of 20–40%) in a parallel oscillatory circuit [42] at frequencies of 0.2–4 MHz. An inductance coil placed as a flat coil in a shock tube end (Fig. 3.3) changed its inductance under the effect of plasma formed behind a reflected shock front. The experiments were performed at $10 \le p \le 150$ bar and $4,000 \le T \le 25,000$ K, with the prevailing Coulomb interaction ($0.3 < \Gamma < 2$). Under these conditions, the effects of unpairing and correlation of atoms are small, as well as cluster effects, whereby the Coulomb component of electrical resistance can be isolated and compared with plasma theories (Fig. 7.5).

In the case when a weakly ionized plasma with strong charge-neutral interaction is of great interest, the adiabatic compression technique proves effective. This technique is free from irreversibility effects. Therefore, it is possible to attain lower (as compared to the shock wave technique) temperatures and appreciable densities of material [40, 48, 123, 129]. In doing so, compression must be rather "smooth" so as to preclude shock wave generation and, at the same time, sufficiently "rapid" to eliminate heat transfer effects. Suitable

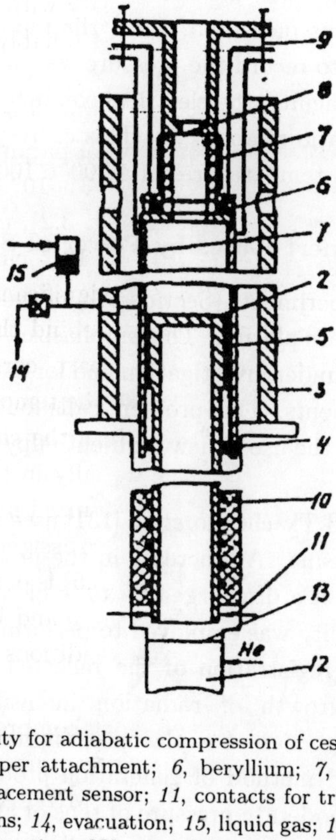

Fig. 3.4. Schematic of a facility for adiabatic compression of cesium [40]: *1*, barrel; *2*, piston; *3*, heater; *4*, Teflon; *5*, copper attachment; *6*, beryllium; *7*, conductivity sensor; *8*, foil; *9*, seal adjustment; *10*, displacement sensor; *11*, contacts for triggering the instruments; *12*, propelling gas; *13*, diaphragms; *14*, evacuation; *15*, liquid gas; *16*, X-ray tube; *17*, scintillator and photomultiplier for registration of X-ray radiation (broken line).

apparatuses were developed in the 1940's [129] and permitted unique measurements of the characteristics of overdense gases at temperatures of up to $9 \cdot 10^3$ K and pressures of up to 10 kbar to be carried out. The use of such techniques in the physics of non-ideal plasma called for substantial changes in the basic arrangement of adiabatic compression facilities.

In a two-stage adiabatic tube, compression of cesium is effected [123] by means of an unloaded lightweight piston *7* moving in a chamber *5* pre-heated to 1,000°C. The fact is that in this apparatus, the leakage of cesium through the lightweight piston renders the realization of the principal advantage of the adiabatic technique impossible, that is, the high compression ratios which in this case did not exceed the value of six, this being close to the values characteristic of shock waves. This shortcoming was eliminated in a single-stage tube [40] using a heavy piston *2* with an extended seal surface (Fig. 3.4). The compressed plasma density was measured using the "soft" X-ray technique. Electrical conductivity was measured, using the induction method. The use of the crusher technique to record the pressure was also proposed. During the first experiments in this facility, the electrical conductivity of plasma was measured at a compression ratio of about 25. This corresponds to rated pressures of up to 330–250 bar and temperatures of 4,900–6,100 K.

3.3. Compression of Inert Gases by Powerful Shock Waves

An analysis of cesium experiments (Section 3.2) revealed a sizable contribution of bound states in the dense plasma. This emphasized the need for extension of the region of parameters under investigation and for a transition to experiments with other chemical elements. This problem calls for a substantial increase in shock wave intensity and the use of new high-enthalpy working media for shock wave generation.

Volkov *et al.* [130] and Tkachenko *et al.* [131] used a shock tube to generate argon, xenon and air plasma. An increase in the propelling gas enthalpy had been attained by means of a discharge of a tank of capacitors with an energy of about 60 kJ. This facility was employed to perform precision measurements of the thermodynamic (registration of the reflected shock velocity), optical (measurement of the growth of radiation intensity) and electrophysical (induction method) properties of a plasma compressed in a reflected shock wave. Since the selected method of generation produces a weakly non-ideal plasma of $\Gamma \leq 0.2$ at a moderate pressure of tens of atmospheres, these experiments register only initial manifestations of plasma non-ideality and a slight shift of the ionization potential and photoionization continuum, as well as minor deviations in the electrical conductivity values from those derived by Spitzer.

Substantially higher parameters can be attained by using powerful condensed explosives in view of their high specific energy content and rate (ca. 10^{-7} s) of detonation transformation, which enabled the development of high-power facilities (10^{10}–10^{12} W). Cristian and Yarder [132] were the first to employ directly the explosion technique for the registration of shock adiabat of gaseous argon. A similar technique was thereafter used by Deal [133] to register shock adiabats of air at atmospheric pressure with subsequent determination, on that basis, of the nitrogen dissociation energy. In subsequent studies, explosion shock waves in inert gases and air were used as a source of intense optical radiation for high-speed photography, laser pumping, excitation of detonation, the study of the effect of radiation on matter, spectroscopic investigations and so forth [134]. Since the initial gas pressures in the respective facilities did not exceed one bar, there was not enough time for the non-ideality effects in such plasma to show up vividly. Hence the non-ideality effects did not serve the subject of investigation.

Special experiments aimed at studying the effect of non-ideality upon the physical properties of explosion plasma were staged in the early 1970s. Inert gases were chosen as the subject of investigation in view of the possibility of their efficient heating in the shock front. This is due to the elimination of dissociation losses of energy and high molecular weight, as well as the much more simple interpretation of the measurement results. The latter is attributed to the absence of complex molecular and ion-molecular clusters and to prior detailed studies of elementary processes.

In order to assess the optimum experimental conditions and the choice of plasma generator designs, machine computations were performed [135] on the thermophysical and gas-dynamic properties of shock waves in dense inert gases. It has been revealed that the optimum values of the non-ideality parameter are attained at shock velocities D of about $9 \cdot 10^5$ cm/s in argon and $5 \cdot 10^5$ cm/s in xenon. A further increase in D leads to overheating and an increase in the ionization multiplicity of plasma, while a rise of initial pressure leads to plasma degeneracy. It is important that the optimum conditions can be realized by using a simple linear scheme of shock wave excitation, while to produce a highly heated, multiply ionized plasma one must employ reflected shock waves or geometric cumulation effects.

In a linear explosion shock tube (Fig. 3.5) [136–139], an ionizing shock wave is generated upon expansion into the gas under investigation of the

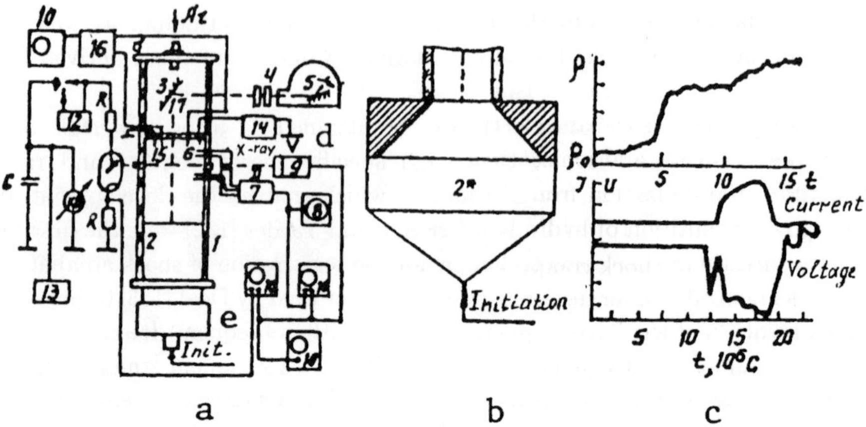

Fig. 3.5. Scheme of measurements in an explosion generator of non-ideal plasma: *1*, generator channel; *2*, explosive charge; *3*, mirror; *4*, filter and attenuator; *5*, high-speed camera; *6*, probes for measuring the shock front velocity; *7*, supply to electric contacts; *8*, oscillograph for velocity recording; *9*, photomultiplier with NaI scintillator; *10*, oscillographs; *11*, X-ray tube; *12*, X-ray control unit; *13*, supply to X-ray tube; *14*, d.c. source; *15*, potential and current probes for registration of electrical conductivity; *16*, differential amplifier. In experiments with reflected shock wave, a Plexiglas obstacle *17* was installed. In experiments with cumulative devices, the structure of the active charge *2* was changed: scheme (*a*); X-ray photograph of density (*b*); current and voltage (*c*); X-ray; current; voltage.

hexogen detonation products (hexogen is a condensed explosive with a specific energy content of approximately 10^4 J/cm^3), which feature, after the completion of the detonation transformation reaction, high dynamic characteristics of $p \sim 370$ kbar, $T \sim 5 \cdot 10^3$ K and $\rho \sim 2.3$ g/cm^3. An additional (as compared to conventional shock tubes) increase in the propelling gas enthalpy is caused by the fact that the disintegration of pressure discontinuity occurs in a co-ordinate system moving at a rate of $2.1 \cdot 10^5$ cm/s. The use of a specially shaped detonation lens and proper choice of the size of active explosive charge accounted for the one-dimensional nature and stationary position of the detonation front upon its exit from the explosive and into the test gas. The total energy release in each experiment amounted to approximately $3 \cdot 10^6$ J with a power of about 10^{11} W. Naturally, this leads to the destruction of the entire apparatus and the need to use premises with special protection and to take the appropriate safety measures. The results of photographic, electrophysical and X-ray measurements revealed the one-dimensional nature and quasi-steadiness of plasma flow, which are accomplished through the inertial confinement of shock-compressed plasma by the massive walls of the shock tube channel. The explosion shock tube (Fig. 3.5) was fully equipped for independent thermodynamic and electrophysical measurements. The shock front velocity was measured [137] by the optical and electrocontact baseline techniques with the aid of high-speed cameras and ionization sensors at an accuracy of 1–1.5%. Within this error, the ionization front coincided with the shock wave luminescence front and the position of hydrodynamic shock.

The density of shock-compressed argon plasma was registered at an accuracy of about 8% by flash radiography (Section 3.2), which is characterized by high time (ca. 10^{-7} s) and space (ca. 2 mm) resolution and the absence of perturbation in the plasma flow.

In view of the transparency of plasma before the shock front and small dimensions of viscous shock, thermal radiation is capable of leaving the plasma volume without any obstacles and yielding of experimental data on equilibrium temperature and absorption factors of shock-compressed plasma [134, 140]. The intensity of this radiation, in the case of the brightness method (at an accuracy of 5–10%), for temperature recording was determined [137, 141] by photometric comparison of time scans of the luminescence of shock-compressed plasma and reference light sources, such as a flash lamp with a brightness temperature of $8,600 \pm 200$ K, pulsed capillary light source with $T = 39,700 \pm 700$ K, and shock wave in air with $T = 11,800 \pm 600$ K.

As demonstrated in a special series of test experiments, as well as by estimates, the self-shielding effect is absent in view of the heating of gas by ultraviolet plasma radiation. According to the data by Tsikulin and Popov [134], this effect becomes conspicuous for argon ($p_0 = 1$ bar) beginning at $D \geq 15$ km/s and from $D > 8$ km/s for xenon, and decreases with a rise in the initial gas pressure.

The registration of the shock front velocity and the density yields, in accordance with the conservation laws of Eq. (3.1), the caloric equation of state for non-ideal plasma, $E = E(p, V)$, which, together with the temperature measured, $T = T(p, V)$, provides one with thermodynamically complete information on argon plasma at $p \sim (1.6\text{–}5.8) \cdot 10^3$ bar, $T \sim (15.5\text{–}23) \cdot 10^3$ K, $\Gamma \sim 1.3\text{–}2.2$ and $N_e \sim (1.5\text{–}3.5) \cdot 10^{20}$ cm^{-3}.

Electrical conductivity is an important characteristic of plasma which offers valuable information on its structure and elementary processes. In view of the high level of electrical conductivity, its measurements [138, 139, 142, 143] were performed using the probe technique (Fig. 3.5). It is characterized by high space resolution and a relative simplicity of realization under conditions of single dynamic experiment. A characteristic oscillogram of transport current and voltage is given in Fig. 3.5 where one can clearly see a "plug" of shock-compressed plasma.

Extremely high densities of plasma were attained by two-stage compression of near-critical states of xenon in the incident and reflected shock waves [139]. To this end, a Plexiglas obstacle was placed at a distance of 7 cm from the charge end face (Fig. 3.5). The interaction with the obstacle causes a reflected shock wave in the plasma cluster. This leads to additional compression and heating of the xenon plasma. Whilst doing so, a shock wave emerged in the Plexiglas obstacle. Its velocity enabled one to determine the pressure, density and enthalpy of xenon following its double compression (cf., point 10 in Fig. 3.6). A consistent description of these thermodynamic data was

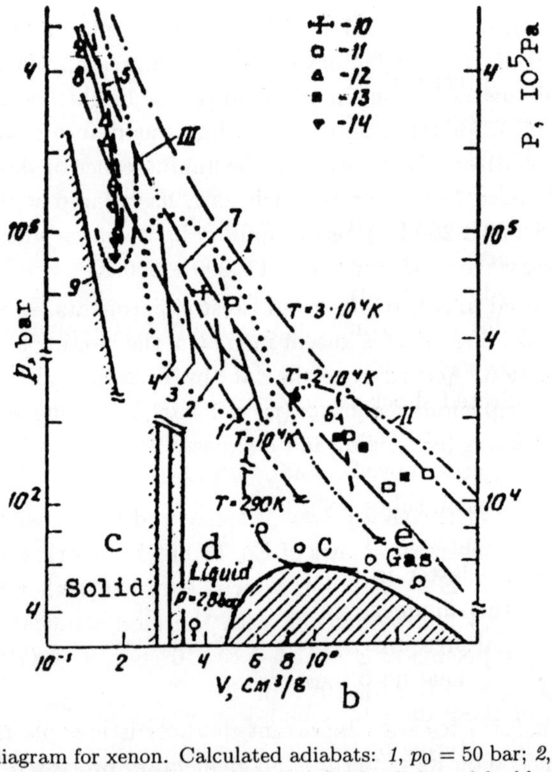

Fig. 3.6. Phase diagram for xenon. Calculated adiabats: *1*, $p_0 = 50$ bar; *2*, 60 bar; *3*, 70 bar; *4*, 80 bar; *5*, 2.8 bar, $T_0 = 165$ K; *6*, $p_0 = 2$ bar; *7*, shock adiabat of double compression [139]; *8*, calculation under the band model [147]; *9*, $T = 0$ K isotherm. Thermodynamic measurement data: *10*, "reflection" method [139]; *11*, [145]; *12*, [147, 148]. Electrical conductivity measurement data: *13*, [138]; *14*, [148]. K_C, critical point.

accomplished using a "chemical" (Chapter 5) model of plasma. In this model, the Coulomb interaction is described by the modified Debye approximation and the repulsion at small distances, by virial corrections. This thermodynamic model was then used to interpret (Section 3.5) the results of measurements of the electrical conductivity of plasma under supercritical conditions.

These experiments have yielded data on the electrical conductivity of xenon [139] under the highly supercritical conditions of $\rho \sim$ 1–4 g/cm^3 ($\rho_c \sim$ 1.1 g/cm^3) at high pressures and temperatures ($p \sim$ 20–110 kbar and $T \sim$ (1–2) $\cdot\, 10^4$ K), where a wide spectrum of strong interparticle interactions involving the participation of neutral and charged particles is realized. Note that in this manner, one manages to generate plasma under unusual conditions, that is, the plasma density is 1.5 times higher than that of solid xenon and is comparable to the density of solid metals. The investigated parameters region I in Fig. 3.6 extends from the lower density states II, where the thermodynamic and electrophysical properties of material are described by plasma models, to adjoin directly the region III of solid-state densities described by the band theory of solids. The latter is formed as a result of the dynamic compression of liquid xenon.

The urge to obtain a plasma with high non-ideality parameters compels one to perform experiments at relatively low temperatures of the order of 20,000 K, at which the plasma is as a rule, not fully ionized [123, 128, 137, 140]. At the same time, the study of a multiply ionized non-ideal plasma is of considerable interest and enables one to investigate purely Coulomb effects in a highly heated medium.

The use of reflected shock waves makes it possible to advance, at reduced initial gas pressures, into the region of multiple ionization and to produce plasma at temperatures above 20,000 K [142, 143]. The same purpose is served by cumulative explosion tubes [142, 143], which depend for their action on the increase in the parameters of propelling gas under conditions of centripetal motion [144] of the latter in a conical chamber (Fig. 3.5) with apex angles of 60–120°. In this manner, shock velocities in xenon of 8–15 km/s were attained with initial pressures of up to 10 atm. This corresponds to extremely high (3–10) $\cdot\, 10^4$ K, heating of non-ideal ($\Gamma \sim 2$) plasma. While leaving the interpretation of these data to Chapter 5, we shall point out that it is the thermal electron scattering by the inner shells of ions which is important under the given conditions.

In order to measure the thermodynamic and optical characteristics of plasma at higher (as compared to shock tubes) pressures, use was made of linear explosion generators of step shock waves (Fig. 3.7) of varying intensity [145, 146] and duration, as well as of light-gas propulsion devices [147, 148]. In such apparatuses, the ionizing shock wave emerged upon expansion of metal or polymer targets, pre-compressed to about 10^6 bar, into the gases under investigation (argon, xenon). Powerful shock waves in the targets were excited by linear explosion propulsion devices [106, 107] operated by acceleration of flat metal projectiles to 2–6 km/s by the detonation products. The energy release in these experiments amounted characteristically to $(2\text{–}30) \cdot 10^6$ J at a power of the order of 10^{11} W.

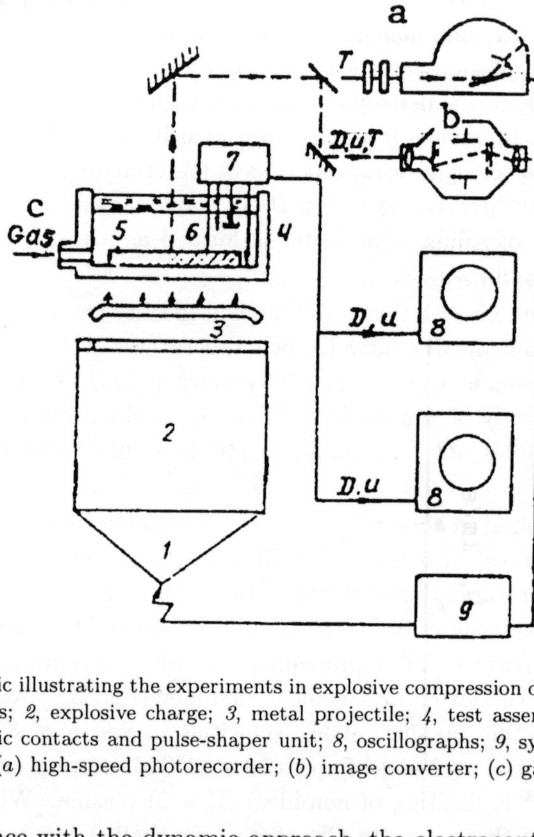

Fig. 3.7. Schematic illustrating the experiments in explosive compression of inert gas plasma: *1*, detonation lens; *2*, explosive charge; *3*, metal projectile; *4*, test assembly; *5*, target; *6*, test gas; *7*, electric contacts and pulse-shaper unit; *8*, oscillographs; *9*, synchronization and detonation unit; (*a*) high-speed photorecorder; (*b*) image converter; (*c*) gas.

In accordance with the dynamic approach, the electrocontact and optical baseline techniques were employed [145, 146] in the experiments in these generators for simultaneous independent registration of two kinematic parameters: the front velocity D and the plasma mass velocity u. Open electrocontact pins recorded the front velocity D at an accuracy of the order of 1% with the aid of fast oscillographs. The mass velocity u was registered (at an accuracy of 1.5–2%) by enclosed pins of special design which are unresponsive to the shock wave in plasma. They also operate upon the arrival of the plasma/target heavy contact interface. The system of sensor pins was arranged to allow for the possible shock wave tilt and to monitor the parameters of the driving pressure pulse from the projectile accelerated by the detonation products. In the case of the optical registration technique [146], a Plexiglas obstacle (Fig. 3.7) was placed at a pre-set distance from the target; the emission of shock-compressed plasma was registered through the obstacle with the aid of high-speed cameras or image converters. The escape of the shock wave from the target into the test gas leads to the appearance of luminescence which intensifies the moment the shock wave is reflected from the obstacle (determination of D), and disappears when the heavy contact interface arrives at the target (registration of u). In addition to shock compression parameters, the temperature of shock-compressed plasma was also recorded in a number of experiments using the brightness method.

The results of recording the shock adiabats of argon and xenon in kinematic variables are illustrated in Fig. 3.8, where one can clearly see a good

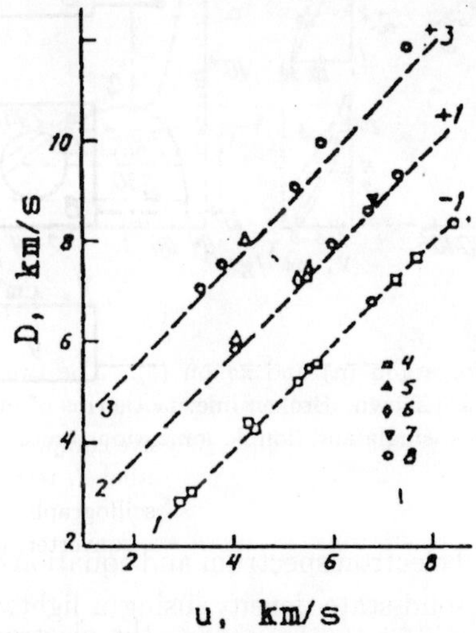

Fig. 3.8. Shock adiabats of argon (*1*, p_0 = 0.78 bar; *2*, 5 bar) and xenon (*3*, p_0 = 10 bar) registered using different techniques (*4*, optical [132]; *5*, electrocontact [145]; *6*, X-ray [147]; *7*, optical [142]; *8*, electrocontact and optical [146]; broken line, calculation [135]). The numerals on the right indicate the values of shifts of each plot with respect to the front velocity D (km/s); u, mass velocity.

agreement between the data obtained by the optical, electrocontact and X-ray techniques in explosion shock tubes and step-wave generators using metals, polymers and condensed explosives as targets. The results obtained in determining the equation of state for non-ideal argon plasma cover a wide range of parameters (Fig. 3.9), namely, $p \sim 0.3$–40 kbar, $T \sim (5.2$–$60) \cdot 10^3$ K and $N_e \sim 10^{14}$–$3 \cdot 10^{21}$ cm^{-3}, which is characterized by developed ionization of $\alpha \sim 3$ and strong Coulomb interaction of $\Gamma \sim 10^{-2}$–5.2. It is interesting to note that in the experiments described above, the equation of state for a plasma with $\rho \sim 1.3$ g/cm^3, was studied. It exceeded the critical density of xenon. Note that under these conditions, the interaction proves so strong that it causes a perceptible shift and deformation of the energy levels in the plasma.

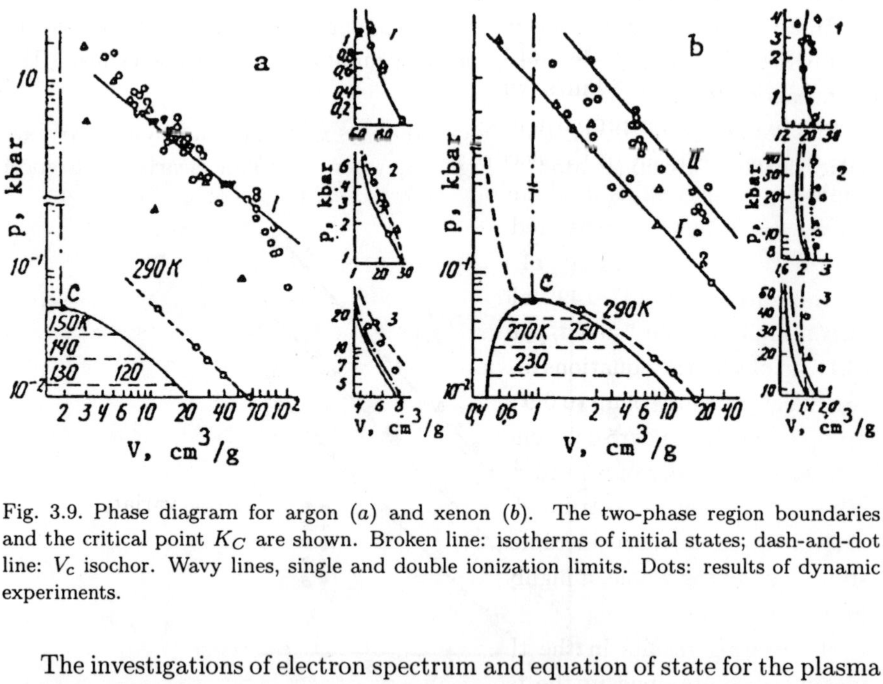

Fig. 3.9. Phase diagram for argon (a) and xenon (b). The two-phase region boundaries and the critical point K_C are shown. Broken line: isotherms of initial states; dash-and-dot line: V_c isochor. Wavy lines, single and double ionization limits. Dots: results of dynamic experiments.

The investigations of electron spectrum and equation of state for the plasma of argon and xenon of solid-state density, using a lightweight ballistic facility [150] for shock wave compression, are found in [147–149]. In a two-stage light-gas "gun" (Fig. 3.10), powder gases *1* resulting from the charge combustion accelerate a heavy piston *2* which moves in a chamber *3*, having a length of 10 m and a diameter of 90 mm on its inside, to compress hydrogen *4*. Following the rupture of the diaphragm, the expanding hydrogen acts to accelerate

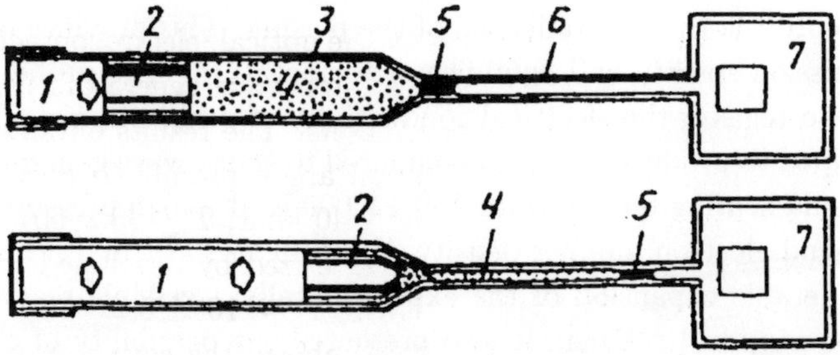

Fig. 3.10. Schematic of a two-stage ballistic facility [150] (top, prior to experiment; bottom, during experiment): *1*, powder gases; *2*, heavy piston; *3*, high-pressure chamber; *4*, hydrogen; *5*, impactor; *6*, acceleration channel; *7*, sample.

a light impactor *6*, which measures 28 mm in diameter, in a 9 m long evacuated acceleration channel *5*. In this manner, a highly symmetric "smooth" and heating-free acceleration of the impactor is effected with a mass of about 20 g to velocities of 7–8 km/s, which are somewhat higher than those attained in the case of plane propulsion in explosion devices such as those found in Fig. 3.7. The impact of the accelerated impactors leads to the generation in liquid argon, as well as in aluminum shields, of plane stationary shock waves whose velocities were registered by a system of electrocontact sensors with a time resolution of hundreds of picoseconds. These measurements were used by the authors of [147–149], along with the regularities of disintegration of discontinuity on the "shield/plasma" interface and with the conservation laws (3.1), to obtain the equations of state for a highly compressed argon and xenon plasma at pressures of up to 0.9 and 1.3 Mbar and at maximum temperatures of up to 30,000 K. Under such conditions of high compression (maximum nuclear density of the order of $5.7 \cdot 10^{22}$ cm^{-3}), one observes a marked deformation of the electron energy spectrum of plasma which causes a triple variation of pressure. In doing so, a considerable number of electrons are in a thermally excited state and, on the whole, a highly compressed plasma possesses "semiconductor properties" [148].

Interesting results in the thermal physics of non-ideal argon and xenon plasma were obtained in the facility for adiabatic compression. These were described by Hess [153]. At an initial pressure of about 2,000 Pa, the test gas was compressed by a heavy (4.9 kg in mass) piston in a steel channel measuring 7.5 m long and 150 mm in diameter. The pressure of the adiabatic compression was registered using a piezoelectric transducer and the compression ratio was defined by the position of the piston. In addition, an optical method was used to record the thermal radiation of the plasma. This provides information on the emission spectra and equilibrium temperature. The probe techniques were used to register the electrical conductivity of dense plasma. This unique facility helped to produce lower (as compared to shock wave generators) values of the plasma heating temperature $T \sim 14,000$ K at a high pressure p of up to 14.5 GPa and electron number density N_e of up to 10^{20} cm^{-3}. This resulted in a considerable expansion of the experimentally accessible region of phase diagram of non-ideal plasma. It also presented the possibility of checking on the validity of asymptotic theories.

The optical properties are of considerable interest from the standpoint of plasma physics. This is because they enable one to follow the effect of non-ideality on the dynamics and energy spectrum of electrons in a dense disordered medium. The plasma absorption coefficient χ_ν was measured by registering the rise in time of the intensity of optical radiation, which emanates from a plane plasma layer enclosed between the shock front and contact surface [140, 151]. In view of the small time of photon relaxation as compared to the characteristic gas-dynamic time, the spectral radiation intensity of such a layer, $J_\nu(t)$, in a quasi-stationary approximation [9] takes the form of

$$J_\nu(t) = J^0_\nu(1 - \exp\{-\chi'_\nu(D - u)t\}), \tag{3.6}$$

where $J^0_\nu = J_\nu(\infty)$ is the Planck radiation intensity, $\chi'_\nu = \chi_\nu(1 - e^{-h\nu/\mathrm{kT}}))$ is the absorption coefficient corrected for induced radiation, and D and u are the rates of travel of the front and rear boundaries of the radiating layer, respectively.

An important requirement of this technique is the provision of the conditions of one-dimensional and steady flow of plasma in the zone of recording. This enables one, when interpreting the measurement results, to use the simple relationship (3.6) instead of solving the integral-differential equation of radiation transfer. During experiments [151], these conditions were monitored by electrocontact and optical measurements, as well as by gas-dynamic calculation.

A facility utilizing explosion propulsion devices is shown schematically in Fig. 3.11. Light radiation emanating from a plasma is diverted with the aid of mirrors and optical system to polarization filters and photomultipliers. The respective signals are recorded by high-speed oscillographs. The kinetic parameters of the plasma, which are essential in determining the absorption coefficients, were determined by electrocontact and optical methods. The exit of the shock wave from the target and into the argon led to a smooth rise

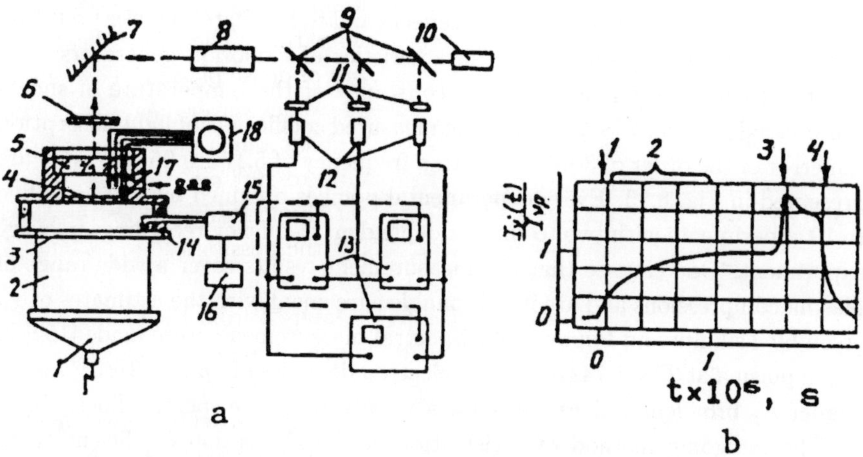

Fig. 3.11. Schematic illustrating the experiment in registering the absorption coefficient of shock-compressed plasma [151], *a*: *1*, detonation lens; *2*, explosive charge; *3*, impactor; *4*, target; *5*, transparent obstacle; *6*, diaphragm; *7*, mirror; *8*, optical system; *9*, semi-transparent mirrors; *10*, adjustment laser; *11*, interference light filters; *12*, photomultipliers; *13*, oscillographs; *14*, electrocontacts for measuring the shock velocity; *15*, pulse-shaping circuit; *16*, delay line; *17*, electrocontacts for measuring the shock velocity; *18*, oscillograph for measuring the shock velocity. Oscillogram of the experiment, *b*: *1*, shock wave entry into the gas; *2*, zone of radiation rise; *3*, shock wave collision with the obstacle; *4*, beginning of the destruction of obstacle by the target material.

(Fig. 3.11b) in light radiation until its saturation in accordance with Eq. (3.6). A flash-up of brightness marks the time (and, consequently, the velocity D) of the front arrival to a transparent obstacle. The subsequent radiation cutoff is caused by the destruction of the obstacle by a massive target moving at a velocity u.

Note that such an organization of experiments enables one to obtain diversified physical information in a single experiment. The rise of intensity $J_\nu(t)$ in the initial portion of flow determines, from Eq. (3.6), the light absorption coefficient and the level of radiation at the stage of saturation, $J_\nu(\infty)$, determines the brightness temperature. The registration of shock wave reflection from the obstacle helps in the measurement of D and u which, with due regard for the conservation laws (3.1), is equivalent to the direct measurements of pressure, density and internal energy. Thus, the data obtained are consistent and complement the results of special thermodynamic measurements [137, 145, 146].

In order to reveal the compression effects and to facilitate the interpretation of the results, the gas-dynamic characteristics of explosion generators were selected to provide for approximate constancy of the temperature of shock-compressed plasma, $T \sim 2 \cdot 10^4$ K. The measured coefficients of light absorption (an error of the order of 30–40%), with a frequency of $5.17 \pm 0.05 \cdot 10^{14}$ s^{-1}, are presented in Fig. 8.13. Every experimental point is obtained by averaging five to 10 experiments with two to three independent frequency recordings in every experiment. One can see that the measurement results cover a wide range of plasma compressions and are in reasonable agreement with the estimates of χ_ν given by Tsikulin and Popov [134]. Spectral measurements were made [152] in argon plasma at $T \sim 1.74 \cdot 10^4$ K and electron densities of up to $1.2 \cdot 10^{19}$ cm^{-3}. Higher N_e broadening does not allow a separation of the spectral lines.

The foregoing method of registration of the absorption coefficient turns out to be restricted on the side of low plasma densities by the finite time of existence of the steady flow of plasma. Therefore, when registering low absorption coefficients, Bespalov *et al.* [141] performed a somewhat different version of such experiments by involving the use of explosion shock tubes as the source of air plasma. A shock wave decay in such systems led to the emergence of a maximum for the relationship $J_\nu(t)$ being registered. The position of the latter had helped to derive the value of χ'_n in every experiment and to evaluate the other parameters of shock compression on the basis of gas-dynamic data. The results obtained provided a basis for evaluating the photodisintegration cross-section of the negative ion of nitrogen.

The reflection of laser radiation from non-ideal plasma yields interesting information on the physical properties of a shock-compressed medium, whereby one can study the high-frequency conductivity of this medium in the field of electromagnetic wave. This method is fairly widely used in the physics of metals and semiconductors to determine the energy spectrum, concentration and effective mass of charge carriers. Of special interest is the region of resonance interaction, in which the frequency ω_λ incident electromagnetic wave is comparable to the plasma frequency $\omega_p = \sqrt{4\pi N_e e^2/m}$. In this case, the resonance condition $\omega = \omega_p$ corresponds to a strong reflection of incident radiation and defines the electron number density, and the dispersion of the reflection curve characterizes the losses of electron energy, thereby enabling one to find the electron collision frequency.

The experiments in measuring the reflection of laser radiation from non-ideal plasma [144] were performed using explosion generators of square shock waves in the scheme of Fig. 3.11. The electron number density of xenon plasma was varied by varying the initial parameters of xenon, namely, $p_o \sim 1 \div 5.7$ MPa, $\rho_o = 0.06 \div 0.8$ g/cm^3 and $T_o = 270$ K. In view of considerable proper thermal radiation of shock-compressed plasma with $T \sim 3 \cdot 10^4$ K, the probing pulse was developed by a pulsed (with $\tau \sim 10^{-8}$ s) aluminum-garnet laser producing a high ($T \sim 6 \cdot 10^7$ K) spectral temperature and a small angular divergence of the light flux. The laser system was equipped with an electronically controlled Pockels shutter which enabled one to synchronize the explosion generation and laser probing with the difference in time of not worse than 10^{-8} s. The sounding radiation was introduced into the explosion chamber by means of a special optical system and, after being reflected from shock-compressed matter, it was applied to fast photomultipliers equipped with interference filters with $\Delta\lambda \sim 0.02$ μm. For the focal spot diameter of about 0.6 mm, the specific intensity of laser radiation in the shock wave front amounted to (1–7) 10^5 W/cm^2, which is insufficient for substantial heating of matter.

The optical reflectivity of plasma R was determined by comparing the intensities of incident radiation with the radiation reflected to the aperture of the receiving ring-shaped lens. In order to perform an overall check on the procedure in the dynamic mode, experiments were carried out to register the coefficient R for shock wave compressed carbon tetrachloride and silicon. The values obtained for R are in good agreement with the results of previous measurements in the region of metal-dielectric transition. In each experiment, in order to determine the thermodynamic parameters of shock-compressed plasma, the electrocontact baseline method was used to measure the velocity of shock wave propagation in plasma and the impactor approach velocity.

The measurements cover a wide range of values of plasma pressure, $p \sim 1.6 \div 17$ GPa, and density $\rho = 0.5 \div 4$ g/cm^3 that exceed considerably the values of the density of xenon at the critical point. Under these conditions, a highly heated ($T \sim \cdot 10^4$ K) disordered plasma with considerable Coulomb interaction ($\Gamma \sim 2 \div 7$) is realized. In doing so, the electron number density, calculated within the ring Debye approximation, exceeds by an order of magnitude the critical (for $\lambda \sim 1.06$ μm) value of $N_c \sim 10^{21}$ cm^{-1}. The reflectivity also reaches the value of 50%, which is characteristic of metals. The estimates of the space structure of the shock front, caused by the ionization kinetics and radiant heat transfer, have shown that it is the resonant properties of plasma electrons that govern the reflection of radiation. These measurements can be used to analyze the physical properties of strongly non-ideal plasma.

3.4. Isentropic Expansion of Shock-Compressed Metals

The shock wave technique described in Sections 3.2 and 3.3 enables one to attain high pressures and temperatures in compressed media whereas the region of reduced pressures (including the saturation curve and the neighborhood of the critical point for metals) proves inaccessible to such investigation techniques. Steady-state experiments performed under conditions of normal pressure and temperatures below 2,500 K allow the determination of heat capacity and isothermal and adiabatic compressibility, as well as the entropy and density jumps occurring upon melting. By now, the melting curves for metals and the general form of the phase diagram have been defined at pressures of up to 50 kbar and the isothermal compressibility at pressures of up to 300 kbar. In the pressure range of 10–20 kbar, the properties of metals are determined on the basis of the absolute method of registration of the shock and isentropic compressibility of solid and porous samples [107, 110] and, at pressures of up to tens of megabars, by means of comparative measurements [11, 114–119].

Therefore, what remains uninvestigated is an extensive and practically important part of the phase diagram (Figs. 1.1, 3.1) characterized by great diversity and utter complexity of description of the physical processes occurring there. The following are realized in this region: a plasma non-ideal with respect to a wide spectrum of interparticle interactions; a dense heated metallic liquid whereby during its expansion, a "release" of degeneracy of the electron component, its recombination, takes place; "metal-dielectric" transition; and high-temperature evaporation of metal to the gas or plasma phase. The current information on these processes is very limited and exists in the form of semi-empirical estimates and a few measurement results. Suffice it to mention that, out of more than 80 metals in the periodic system, the critical point parameters have only been determined for three lowest-boiling ones [154], to say nothing of more detailed information on the form of the phase diagram for metals at high pressures and temperatures.

One can move into the region of lower densities and high pressures of plasma using the isentropic expansion technique [89, 155–157]. It is based on the generation of dense plasma under conditions of isentropic expansion of condensed material pre-compressed and irreversibly heated in a shock front.

To assess the possibilities of dynamic methods for isentropic generation of non-ideal plasma, Leontiev and Fortov performed [158] calculations of energy release leading, upon its disintegration, to melting and evaporation in the release wave. For such an assessment, one needs data on the critical parameters of metals, reference points on the thermodynamic surface of material. In view of the collectivization of valence electrons, their bond energies and, consequently, critical point parameters are extremely high [154] and, in most cases, inaccessible to conventional thermophysical experimental techniques. An exception is made for alkali metals and mercury, for which the registration of high-temperature boiling parameters makes it possible to verify the applicability, in their case, of the principle of thermodynamic similarity and, on this basis, to assess [154] the critical parameters for 80 metals that are still uninvestigated (cf., Appendix 1).

It follows from these estimates that the critical temperatures of metals in many cases prove to be comparable with their ionization potentials [107], which are considerably reduced in plasma due to strong interaction of charges with each other and with neutrals. Therefore, the metal vapors on the right side of the binodal are, apparently, in a thermally ionized state. Meanwhile, the high-temperature evaporation of metals corresponds to the transition, directly, to a highly non-ideal plasma state while bypassing, as distinct from the already investigated elements, the ionized gas region. This fact may affect the kinetics of high-temperature phase transitions [159], as well as drastically alter the usual form of the phase diagram of matter by causing the emergence of additional regions of phase separation and new exotic phase transitions (cf., Section 5.8).

The region of densities intermediate between solid and ideal gas, that is, characterized by the maximum uncertainty of theoretical predictions, prevents one from performing direct calculations of release isentropes under these conditions. Therefore, in order to assess the energy release leading to phase transitions under conditions of adiabatic expansion of metals, use was made of the entropy criterion [113] which allows for the condition of isentropic behavior of flow in the release wave. This criterion is based on a comparison of the critical point entropy and tabular values of entropy of phase transformations with the value of entropy of condensed phase of metals, in which the semi-empirical equations of state are valid [160] at high pressures and temperatures. The most detailed calculations have been performed [158] for aluminum, nickel, copper and lead, whereby perfect equations of state [161] with variable heat capacity are available to describe the shock compression of solid and porous samples and feature high-temperature asymptotics of ideal gas. The assessment results for copper and lead are given in Fig. 3.12 and those for other metals (using the equation of state from [162]) in Fig. 3.13. The states which occur upon impact against the target of an iron impactor accelerated to a velocity of the order of 5.9 km/s are marked in Fig. 3.12 and Table 3.2.

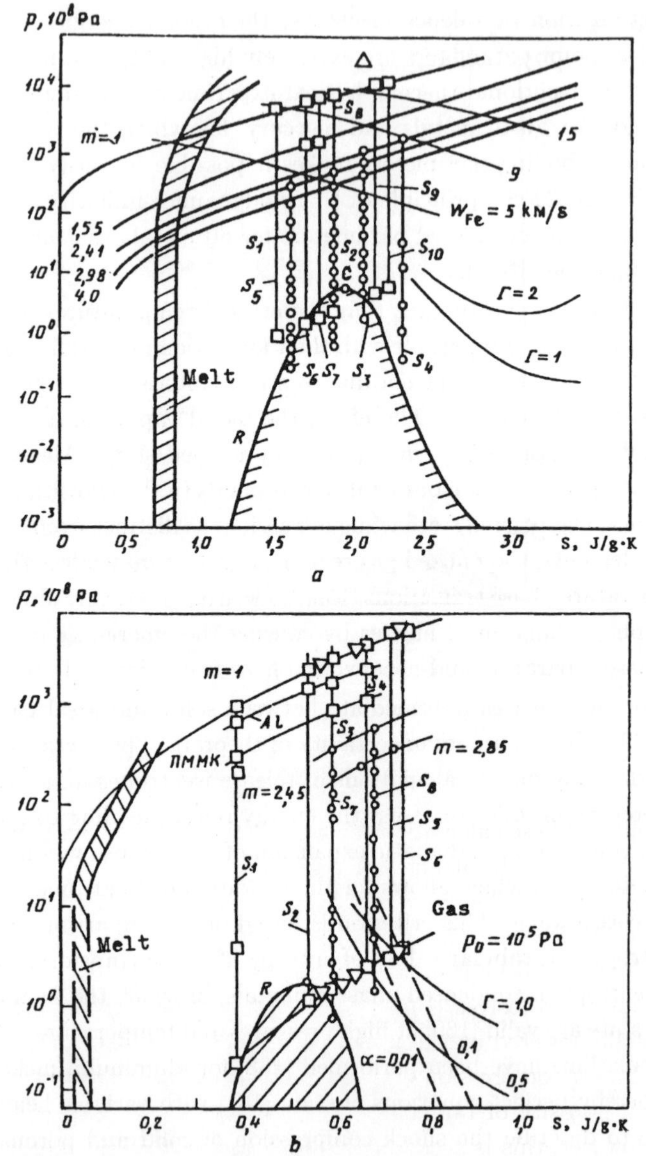

Fig. 3.12. Entropy diagrams for (a) copper and (b) bismuth: M and R, boundaries of melting and boiling, respectively; m, shock adiabats of different porosities; S, expansion isentropes; W_{Fe}, deceleration curves for iron impactors. The diagram in (b) gives the adiabats for materials in which bismuth was released. The points indicate the experimental data of [120, 157, 167–169].

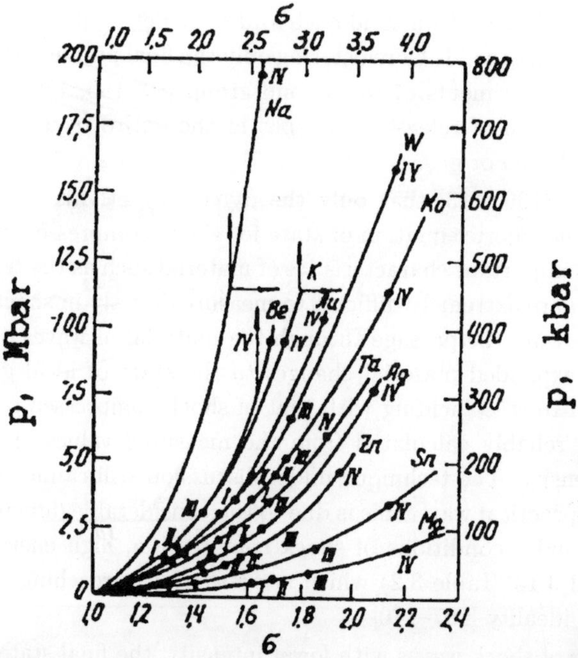

Fig. 3.13. Shock wave intensities causing melting (I, onset; II, end), boiling (III) and arrival at the critical point (IV) in isentropic release wave: $\sigma = \rho/\rho_0$, compression ratio of the material.

The results of these calculations show that the dynamic methods of plasma generation enable one to attain high energy states of metals in a wide range of parameters near the "liquid-vapor" equilibrium curve. However, in order to attain supercritical conditions in the release wave, shock waves of extremely high intensity need to be generated, which is just within the energy capabilities of chemical explosives [107]. The required shock wave intensities may be substantially lowered by the use of porous targets offering a more effective increase in shock compression entropy [113]. It is apparent that an increase in the initial porosity of targets at W = const leads to a decrease in shock pressure. However, this is accompanied by a rise in entropy, thereby enabling one to considerably expand the region of parameters accessible to dynamic experiment.

The few experiments involving the registration of plasma expansion under the effect of shock waves can be divided into two groups based on the nature of the physical information they yielded. The experiments of the first group

[157, 163–166] are based on Zeldovich's idea of determining the entropy of shock compression of solids by registering only the final parameters of expanded material. The experiments of the second group [89, 155, 156, 167] involve a detailed registration of release isentropes in the entire region of parameters from solid to plasma or gas.

It is known [106, 113] that only the dynamic methods make it possible to determine the caloric equation of state for shock-compressed medium. This equation lacks important characteristics of material such as the temperature or entropy, which are extremely difficult to measure directly in shock-compressed material. Following the passage through the material of waves of extreme intensities, the expanded material changes to the state of ideal gas or plasma [113] whose entropy, coinciding with that of shock-compressed condensed material, can be reliably calculated from the measured values of temperature, pressure or density. The technique under discussion still cannot be realized in this form in a practical way. This is due to the considerable difficulties involved in attaining, under conditions of shock compression, high energy states (cf., Figs. 3.12 and 3.13, Table 3.2) which are essential in reaching, upon release, the regions of ideality [167–170].

In the case of shock waves with lower intensity, the final states prove to be in the solid or liquid phase, which enabled [171] the use of the photoelectric technique of residual temperature registration. It made it possible for Fortov and Dremin [172] to find, on this basis, the entropy and temperature of copper at pressures of up to 1.9 Mbar. Scidmore and Morris [173] derived the entropy of sodium, strontium, barium and uranium by means of optical adsorption measurements of the evaporated metal fraction under the effect of short shock waves with $p \sim 0.2$–3 Mbar excited by thin (0.1–1 mm) impactors moving at velocities of 2–6 km/s. It follows from the entropy analysis [158] that shock waves with such moderate intensities may cause appreciable evaporation of metals, only as a result of their expansion to extremely low pressures, which compelled experiments in a vacuum of the order of 10^{-5} mm Hg [173] to be carried out.

Adsorption measurements performed under conditions of substantially non–one-dimensional flow enable one to derive the condensate fraction by relating it, on the basis of quality considerations on the kinetics of non-equilibrium process of evaporation and condensation, to the entropy of shock-compressed metal plasma [173]. In a series of subsequent studies, the general knowledge of evaporation and ionization in the release wave was used to analyze the high-frequency collision of solids and to develop fluorescent barium clouds in outer space for the purpose of studying the Earth magnetosphere [172]. The latter involved the explosions of metallic barium-lined cumulative charges in space.

Al'tshuler *et al.* [157] used the results of measurements of the expansion rate W for lead after the passage of shock waves with amplitudes of 40–300 GPa to analyze the vaporization of lead in release waves. The deviation from the rule of "velocity doubling", as registered by Al'tshuler *et al.* [157], is an extreme evidence of the vaporization effect in release waves.

Zhernokletov *et al.* [171] made an attempt to complement the traditional kinematic measurements with a registration of the temperature of isentropic expansion of lead in the neighborhood of its high-temperature boiling. The experiments involved adiabatic expansion of metal, shock-compressed to a pressure of about 600 GPa, into helium at different values of initial pressure, which remained transparent in the shock-compressed state and therefore transmitted the thermal radiation of expanded lead. This radiation was delivered, via

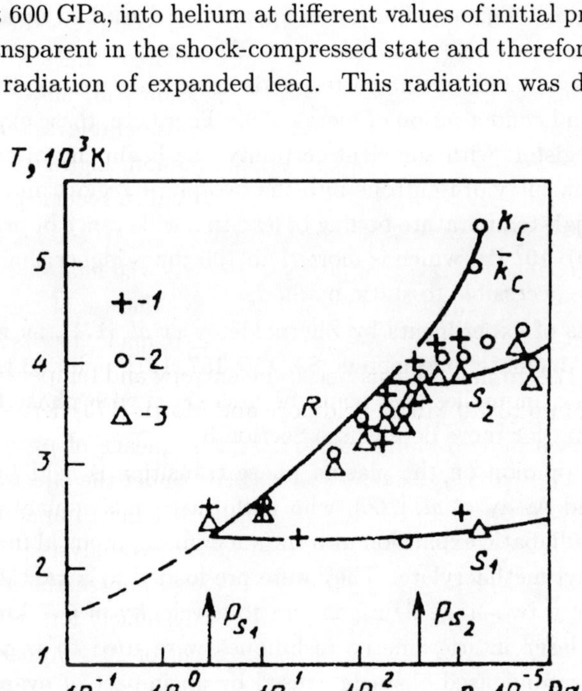

Fig. 3.14. Temperature measurements under conditions of adiabatic expansion of lead: the solid curves indicate the results of calculation, by the semi-empirical equation of state [57], of the boiling curve R, critical point K_c and release isentropes S; the broken curve indicates the static data on the boiling curve; $\bar{K}_0$, estimates of the critical point parameters from [158]. The symbols indicate the results of recording of temperature by photodiodes (1) and two-channel pyrometer: $\lambda_1 = 449$ nm (2); $\lambda_2 = 560$ nm (3). $p_{S_1} = 3 \cdot 10^5$ Pa and $p_{S_2} = 4.2 \cdot 10^7$ Pa, pressure on the isentropes upon their entry into the two-phase region.

optical-fiber communication lines, to fast silicon photodiodes and two-channel pyrometers, whereby the temperature could be determined using the brightness technique. The results of these measurements are given in Fig. 3.14. The transition of isentrope from the single-phase state (power dependence of T and p) to the region of phase mixture (exponential dependence) manifests itself in a sharp variation of the slope of the S and R curves. It is worthy to note that the special series of experiments to measure the temperature of states in the two-phase region, which is produced by expansion from the initial pressure of different values $p \sim 160$ GPa and $p \sim 240$ GPa, resulted, in accordance with Gibbs' phase rule, in identical temperature values. These serve as proof of the equilibrium of the process of adiabatic release in the two-phase region. This equilibrium is apparently disturbed at a pressure below $3 \cdot 10^6$ Pa, where the recorded values of temperature exceed those on the saturation curve. This result corresponds to calculations of the kinetics of evaporation and condensation of metals [59]. Therefore, these experiments enable one to register, with sufficient certainty, the beginning of evaporation of metal upon the entry of isentrope into the two-phase region, and to determine the curve of high-temperature boiling of lead in a wide range of pressure values, $p \sim (50\text{–}1{,}000) \cdot 10^5$ Pa, which is more than 100 times higher than those in the pressure range accessible to static methods.

The results of experiments by Zhernokletov *et al.* [171], as well as earlier recordings of kinematic parameters [89, 155–157, 167], point to the absence of thermodynamic anomalies that could be associated with phase transitions in a dense plasma (for more details, see Section 5.8).

The same opinion on the plasma phase transition is held by Kerley and Wise [165] and Assay *et al.* [166], who performed, analogously with [56], experiments in adiabatic expansion into the vacuum of porous aluminum, copper and polymethyl methacrylate. They were pre-loaded by a tantalum impactor accelerated by a two-stage light-gas gun to a velocity of 6–7 km/s. In these experiments, laser interferometry techniques were used to register the rate of motion of a condensed obstacle caused by an impact of evaporated metal against the latter. This rate of motion was compared against the results of calculation by a one-dimensional gas-dynamic computer code, in which various equations of state were entered.

The use of hydrodynamic recordings in the isentropic expansion method to obtain thermodynamic information assumes that the local thermodynamic equilibrium in the system is present. It corresponds to the smallness of characteristic times of phase transformations as compared to the typical time of experiment of the order of 10^{-7} s. It is interesting to note that in the case of isentropic expansion, one observes a unique (and unrealizable in most other cases) possibility [159] of penetrating the region of absolute instability of a single-phase system $[(\partial p/\partial V)_T > 0]$ by means of non-static variation of pressure. This is because in the critical point where $(\partial p/\partial V)_T = 0$, the quantity $(\partial p/\partial V)_S$ remains negative, which is the case inside the spinodal. The kinetics of removing the non-equilibrium state in this case are non-activative; they depend mainly on heat transfer and permit, under experimental conditions as shown by estimates, negligible supersaturations of not more than several degrees [159].

An analysis of the kinetics of disintegration of metallic metastable liquid has shown [159] that high values of critical temperatures (comparable with corresponding ionization potentials) lead to a marked thermal ionization of metal vapor on the binodal. In this case, the metastable phase contains a considerable number of charges which, while being effective condensation centers, lead to a phase relaxation time of the order of 10^{-9} s under experimental conditions. This is in agreement with the results of kinetic measurements upon evaporation of metals into vacuum [164–166] and expansion of shock-compressed lead into air [156] and helium [171]. According to the estimates [159], marked non-equilibrium can only be expected at pressures lowered to tens of bars. Apparently, this defines the lower limit of application of the isentropic expansion technique in thermodynamics. The measurement data support this viewpoint (Fig. 3.14) while pointing to the quasi-equilibrium behavior of high-temperature condensation of metals. To further clarify this point, a series of experiments were staged [156] involving the registration of the rate of expansion of metals into low-pressure air and an analysis of lead vapor impact against a copper shield. In the equilibrium case, a shock wave with $p = 0.73$ Mbar was excited in copper whereas an impact by superheated lead produced a pressure which is two times higher.

Mention should also be made of a number of other studies on the expansion of material following the passage of powerful shock waves [163, 172, 173]. Using the converging conic wave technique, de Beaumont and Leygonie [163] obtained two points for the expansion of uranium and copper plasma into air at atmospheric pressure. Based on these data, the semi-empirical equation of state has been corrected. The results relating to the reflection from soft obstacles of shock waves in porous copper are presented by Alekseev *et al.* [175], who were mainly interested in the problems of two-phase gas dynamics.

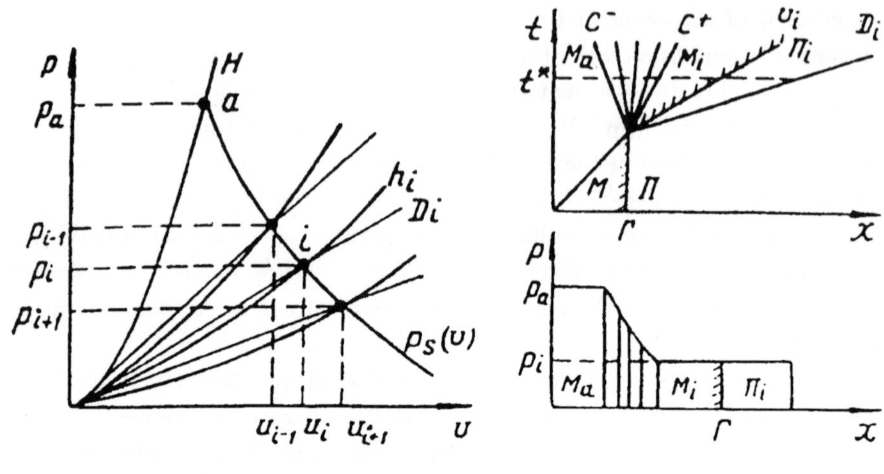

Fig. 3.15. Schematic diagram of experiments using the adiabatic expansion technique.

Another group of experiments [89, 155, 167, 169–171, 173], which involved the release adiabats, are based on registering both the final and intermediate states occurring upon isentropic expansion of plasma (Fig. 3.15). A shock wave propagates through the test substance M and causes the compression of the latter and its irreversible heating to an a state. The shock wave arrives at an interface Γ with a dynamically softer obstacle Π. It also generates a centered Riemann wave C^+C^- in which an adiabatic expansion of shock-compressed plasma from the a state to an i state takes place. Such expansion of metal generates a shock wave propagating at a velocity D_i in the obstacle. The registration of D_i enables one to find, from the known shock adiabat h_i, the pressure and mass velocity of the obstacle which, because of the condition of continuity [113] at the contact interface Γ, coincide with the respective characteristics of the expanding plasma.

By using obstacles of different dynamic rigidities and recording the respective values of p and u, one can continuously trace the variation of the expansion isentrope $p = p_s(U)$ from states on the Hugoniot to lower pressures and temperatures. The use of propulsion systems of different power allows one to vary the increment of entropy in the shock wave and, thereby, to use different isentropes overlapping the selected phase diagram region. The transition from the hydrodynamic variables p–u to thermodynamic variables p–V–E may be accomplished by calculating the Riemann integrals which express the conservation laws for the given type of self-similar flow.

Scidmore and Morris [173] employed this technique, while using linear and spherical explosion systems, to register three expansion isentropes for shock-compressed uranium. The experiments were performed mainly in the liquid phase while the most interesting neighbourhoods of the critical point, namely, two-phase and plasma regions, remained unstudied. This caused Scidmore and Morris [173] to restrict themselves to qualitative estimates in describing the isentrope behavior in the region of lower pressures.

A number of authors [89, 167–170] describe systematic investigations of adiabatic expansion of lead and copper used in the form of finely divided powder to enhance the dissipation effects. In doing so, special attention was given to the plasma region of parameters attained by using gases compressed to 1–50 bar as obstacles, while an increase in the density and pressure of metals was attained through the use of "soft" condensed obstacles such as light metals (aluminum, magnesium), Plexiglas, polyethylene, fluoroplastic and various foam plastics. The experimental scheme using linear explosion systems and gas obstacles is analogous to that shown in Fig. 3.11.

The velocity of propulsion of aluminum and steel impactors was 5–6.5 km/s. The approach velocity of impactors and the velocity of shock waves in shields, targets and "light" obstacles were measured using the electromagnetic and optical baseline methods. The error in electrocontact registration of wave velocity in condensed media was about 1%. The velocity of shock waves in gas obstacles (error of 1.5–2%) was measured by the optical baseline method using a fast photochronograph by means of recording the beginning of plasma luminescence, when a shock wave leaves the sample for gas, and a sharp increase in luminescence after the shock wave is reflected from a transparent obstacle installed at a pre-set distance.

In order to increase the propulsion velocity and, consequently, the amplitude shock wave pressure, layered cumulative propulsion systems have been developed [167, 169, 177] whose principle of operation is based on a mechanism analogous to the acceleration of a light body upon its elastic collision with a heavy one [178]. In order to optimize multi-stage propulsion systems, entropy and gas-dynamic calculations were performed, which described the thermal and gas-dynamic processes with due regard for melting, evaporation, elastoplastic effects and destruction.

As a result of these calculations and numerous experiments, two- and three-stage propulsion devices of various designs were developed. In one of these devices, a steel impactor with a thickness and diameter of 1 mm thick and 60 mm, respectively, was accelerated by the detonation products to 5–5.5 km/s over a speed base of 30 mm and struck against an organic-glass obstacle with a thickness of 1 mm. The outside of the latter was placed with a molybdenum impactor with a thickness of 0.1–0.2 mm and diameter of about 30 mm. As a result of wave interaction, this impactor would fly a distance of about 1.5 mm before striking against the target assembly, where plane shock waves of 10–12 mm in diameter could be produced in the latter. A three-stage propulsion system differs from a two-stage one in that a steel impactor with a thickness of 2.5 mm was accelerated in the first stage to load a layer of plastic explosive, which is about 5 mm thick. A steel impactor with a thickness of 1 mm was placed on the outside of the latter. The design of the outer stage and target assembly is the same as that found in a two-stage generator.

In doing so, the velocity range of molybdenum impactors is extended to 13 km/s as compared to 6 km/s for conventional propulsion. The characteristic time intervals measured in these experiments are approximately 10^{-8} s, which is an order of magnitude less than the characteristic registration time in typical explosion experiments [107].

Because of this, in experiments involving layered systems, the shock compressibility and adiabatic expansion of solid samples of aluminum, copper and bismuth were determined by the optical baseline method using step targets. For this purpose, recesses were made in the target on the side of the approaching impactor and on the outside, a transparent Plexiglas observation window was provided at a fixed distance from that surface. Agate-SF high-speed, high-sensitivity image converters operating in the photochronograph mode were used for registration, which made it possible to reduce the error of measurement of time intervals to 2% in the scanning range of approximately 2 ns/mm. Owing to this procedure, continuous and independent information can be obtained in each experiment at about the time of arrival of the impactor and the flow structure in the target, while monitoring the quasi–one-dimensionality and stationary position of the flow and allowing for a substantial degree of duplication of the kinematic information obtained.

With this experimental procedure, the approach velocity W of molybdenum impactor and the rate of shock wave propagation in the target over different speed bases are registered, as well as the rate of adiabatic expansion of metal into air at atmospheric pressure. Given these parameters, the "deceleration" method [106] enables one to find (using the shock adiabat for molybdenum) the pressure p and mass velocity of a shock-compressed target, as well as the states on the release isentrope. In doing so, every experimental point is obtained by averaging out the results of four to six experiments.

Along with experiments involving linear cumulative devices, experiments were performed using conical explosion generators of Mach shock waves [169], which utilize the effect of geometric cumulation under conditions of irregular (Mach) reflection of conically convergent shock waves [168]. As compared to the plane case, these devices are characterized by an additional concentration of energy during the shock wave convergence to the axis of symmetry and, at the same time, by an increased stability of flow as compared to the spherical convergence of shock waves [176].

Fig. 3.16. Schematic of explosion generator of Mach shock waves: *1*, detonator; *2* and *5*, lines of detonation arrangement; *3*, absorbing washer; *4*, textolite plate of detonation arrangement; *6*, conical explosive charge; *7*, conical liner projectile; *8*, conical target.

A typical generator of conical shock waves (Fig. 3.16) consisted [168, 169] of a solid metal cone with an angle of 45–60°. The shock wave was excited either by direct effect of the detonation products or by conical copper and aluminum liners with a thickness of about 3 mm, and accelerated by the detonation products to 3–4 km/s. The asynchronism of shock wave motion on the cone generator was at least 50 ns. The Mach shock wave, at a distance of 40–50 mm from the apex of copper cone, had a diameter of 7–8 mm, a velocity of about 12.1 km/s and a pressure of $p \sim 6 \cdot 10^5$ Pa.

Similar to layered systems, the kinematic parameters were registered using step targets. The optical signals were recorded using Imacon-640 (scanning of

50–100 ns/mm) and Kadr-2 (scanning of 10–20 ns/mm) high-speed image converters. In some experiments, the radiation was delivered to the photocathode of the image converter from the basal planes of targets via quartz light guides with a diameter of about 60 μm.

With this organization of experiments, one could register, in every experiment, the shock wave velocity in a copper cone (shield) over different bases in a target of test material, as well as the rate of its adiabatic expansion. In addition, the resultant photochronograms offer the possibility of monitoring the curvature and attenuation of shock waves. The subsequent use of the reflection method (using copper as the standard) helps find the remaining thermodynamic parameters. When registering the release isentrope, shock-compressed bismuth and copper were expanded into aluminum, Plexiglas and air at atmospheric pressure. Each experimental point was obtained by averaging out the results of four to 10 independent experiments with two to four recordings in every experiment. The error of individual measurement is estimated at 2–3% and the averaged error, at 1%.

A number of experiments were performed in generators utilizing, simultaneously, the effects of geometric and gradient cumulation. In these experiments, a conical Mach shock wave affected, via spacer of condensed explosive, a molybdenum (about 0.1 mm thick) or tungsten (about 0.25 mm thick) impactor, which accelerated over a base of 100–500 μm to the velocity of 16 km/s and collided with a step copper target. As a result of these experiments, a shock compression pressure of approximately 1,400 Pa ($T \sim 5$ eV) was attained, and isentropic scattering of copper into air at atmospheric pressure was registered. Note that these parameters of scattering are in reasonable agreement with the results of measurements by de Beaumont and Leygonie [163].

Figure 3.12 gives the $p - S$ diagram for copper and bismuth, and Figs. 3.17 and 3.18 give the $p - u$ diagram for lead and copper. In the case of bismuth, the experimental results cover a wide range of states from highly compressed matter on the shock adiabat with $p \sim 670$ GPa and $\rho \sim 2.6\rho_0$ to rarefied metal vapor with $\rho \sim 10^{-2}\rho_0$.

The experimental data presented in Figs. 3.17 and 3.18 demonstrate that the process of adiabatic expansion covers a very wide range of parameters (four orders of magnitude with respect to pressure and two orders of magnitude with respect to density): from a highly compressed metal plasma, where ions are disordered and electrons degenerate, to a quasi–non-ideal Boltzmann plasma and rarefied metallic vapor. As expansion proceeds, diverse little studied physical processes occur in the system: the electron degeneracy is removed;

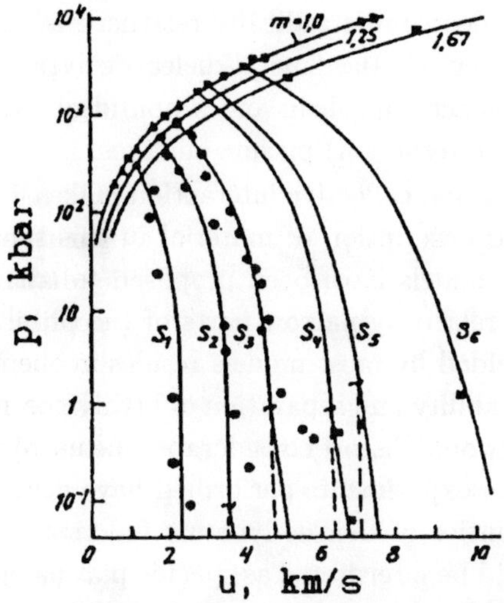

Fig. 3.17. $p-u$ Diagram for lead [157]: m, shock adiabats of different porosities; S, expansion isentropes. Dashed lines, metastable branches; dots, the experiment.

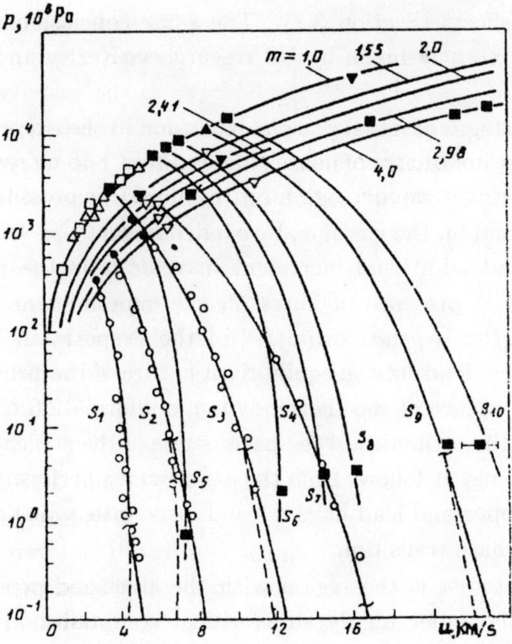

Fig. 3.18. $p-u$ Diagram for copper [169]. Designations are the same as Fig. 3.17.

the electron energy spectrum is radically restructured; partial recombination of dense plasma is effected; the "metal-dielectric" transition in a disordered electronic structure is accomplished; and a non-ideal (with respect to various types of interparticle interaction) plasma emerges.

The presence of strong collective interaction makes it difficult to provide a consistent theoretical description of material in this range of parameters. A number of heuristic models have been proposed to (cf., Chapter 5) describe individual effects in relatively narrow parts of the phase diagram. The main qualitative result yielded by most models resides in the possibility of the loss of thermodynamic stability and separation of highly non-ideal plasma into new exotic phases, which would distort considerably the usual form of the phase diagram for metals. The experiments performed have demonstrated the absence of marked discontinuities of thermodynamic functions or any hydrodynamic anomalies which could be interpreted as specific plasma phase transformations. Note that it is in the investigated range of parameters that the plasma phase transitions discussed in literature are most likely to occur. This is because an increase in temperatures and a decrease in density of Boltzmann plasma, as well as an increase in degenerate plasma pressure, lead to a relative decline in non-ideality effects (Section 3.1). The same conclusion on plasma phase transitions was recently made by US researchers Kerley and Wise [165] and Assay *et al.* [166].

The highest stages of metal plasma expansion in the experiments of [74–78] correspond to the realization of near-critical states. The entry of isentropes into the two-phase "liquid-vapor" region on the liquid phase side is accompanied by evaporation and on the gas side, by condensation. This leads to a variation of their slope and additional increment in velocity in the release wave (cf., Chapter 10). The presence of kinks on the experimental curves and their agreement with the *a priori* estimates of the evaporation effects [154, 159] provide further evidence of the equilibrium nature of the process of two-phase expansion. The nature of the thermodynamic information obtained is also in agreement with the estimates of the parameters of the critical point and phase boundary of boiling. It follows from the experiments performed that the phase diagrams for copper and lead have a usual form with a single critical point of "liquid-vapor" phase transition.

Of special interest is the region with the strongest expansion of metals, where a metal is under highly supercritical conditions. Figure 3.12 gives the values of the non-ideality parameter Γ and the degree of ionization $\alpha = N_e/(N_e + N_i)$ calculated from the chemical model of plasma [106]. The results of analogous treatment for copper demonstrate that the adiabatic expansion from the states on the shock adiabat with $p \sim 1,410$ GPa, $T \sim 5.25 \cdot 10^4$ K and $V \sim 0.0521$ cm^3/g leads to a weakly non-ideal plasma with the parameters $p \sim 0.73$ GPa, $T \sim 9,200$ K, $V \sim 0.3$ cm^3/g, $\Gamma \sim 0.1$ and $\alpha \sim 0.003$. This means that the entropy of a slightly ionized metallic vapor in these states may be reliably calculated by a quasi–ideal-gas approximation and, by virtue of the condition of isentropicity, compared with the values determined by the semi-empirical equations of state for highly compressed metal. A comparison with the equation of state of [89] produced close results.

One can see that the results, produced with modern explosion generators with powerful shock waves, made it possible, for the first time, to combine the regions of the phase diagram that correspond to radically different physical states [179].

The expansion isentropes relate the states with similar entropy of a superdense degenerate plasma on the shock adiabat with the near- or supercritical states of slightly ionized vapor. In addition, during the entry into the liquid-vapor two-phase region, they have the values of energy and volume that are in agreement with the parameters of the equilibrium line. This made it possible to realize, in experiment, Ya. B. Zel'dovich's idea [113] of deriving a thermodynamically complete equation of state from the results of mechanical measurements, namely, the entropy found for a highly expanded ideal vapor proves, by virtue of adiabaticity, to be equal to the entropy of shock compression. The equilibrium temperature of shock-compressed matter is then calculated by the thermodynamic identity of the second law of thermodynamics [172].

The experimental data on isentropic expansion served as a basis for the construction of semi-empirical equations of state [160, 179, 181] describing the entire available totality of static and dynamic data in the solid, liquid and plasma phases. They reproduce the melting and evaporation effects and feature, at superhigh pressures and temperature, correct asymptotic behavior with respect to the Thomas-Fermi and Debye-Hückel theories.

We shall discuss these problems in more detail in Chapter 5.

3.5. Generation of Superdense Plasma in Shock Waves

According to the existing concepts [106, 121], the description of plasma properties is radically simplified at extremely high pressures and densities when the inner electron shells of atoms and ions turn out to be "crushed" and a quasi-uniform distribution of electron density within the Wigner-Seitz cells is realized. In this case, a quasi-classical approximation to the self-consistent field method is valid (cf., Section 5.5) when the description is made in terms of the mean density of the electron component of plasma rather than in the quantum-mechanical language of wave functions and eigenvalues. These concepts form a basis for the Thomas-Fermi model. Its range of application depends on the smallness of exchange, correlation and shell effects [121]. The smallness of the corresponding latter criteria leads to a general estimate of the lower limit of application of the Thomas-Fermi model of $p_{\text{lim}} \sim e^2/a_0^4$. This corresponds to extremely high pressures which are approximately equal to or higher than 300 Mbar ($T = 0$ K). These exceed considerably the capabilities of experimental facilities based on the use of powerful condensed explosives. Therefore, an active search is currently under way for alternative methods of generation of superhigh plasma parameters. This would enable one to approach the region of quasi-classical description of matter with a view to estimating the real range of its application. Such methods include the use of powerful shock waves emerging in the short-range band of strong explosions [11, 114–119, 180–184], the use of coherent laser radiation [185–186] and relativistic electron and ion beams, electric explosion of thin metal foils, and the use of rail-gun electrodynamic accelerators. The last method is still being developed while the quantitative information on the thermodynamics of superdense plasma is provided by the measurements in strong-explosion experiments. For obvious reasons, the amount of experimental information in this case is rather small and, in the foreseeable future, will be hardly comparable with the scope of laboratory investigations. Therefore, it is practical to use the unique possibilities offered by such experiments for the solution to the key problems of the physics of high-energy processes. In recent years, the most representative example of such problems is provided by a study of the effect of electron shell structure of atoms on the thermodynamic properties of dense substances [118]. The shell effects show up in the oscillating behavior of thermodynamic functions (such as adiabats or isochores). This, in turn, may cause the existence of regions with $(\partial^2 p/\partial V^2)_S < 0$, which is responsible for the anomalous [113] behavior of substances in dynamic processes (the release occurs in a shock wave and compression is smooth; for more details, see Section 10.6). The inclusion of such properties of real substances in gas-dynamic calculations would call for a modification of the algorithms employed in modern mathematical computer codes [181]. This fact further rouses the interest in the shell effects.

At present, the vast majority of experiments with powerful shock waves are performed using a comparison method (the method of "reflections" [106]). It records the values of velocity of shock waves passing successively through the layers of materials being investigated, one of which is the standard. Clearly, such a method involves an uncertainty associated with the extrapolation procedure of constructing the shock adiabat for the standard in that very region of parameters, in which direct measurements must be performed. This uncertainty is absent from the experiments in recording the absolute shock compressibility of molybdenum [11] and aluminum [117] using penetrating physical fields (neutron and γ-radiation) for diagnostics.

Researchers in the US have suggested [11] a scheme to measure absolute shock compressibility in the high-pressure region. It is based on shifting the resonances of interaction of neutrons with nuclei of moving matter with respect to their position in the case of nuclei at rest (Doppler shift). The energy of fission of uranium nuclei by neutrons, formed during nuclear explosion, is used

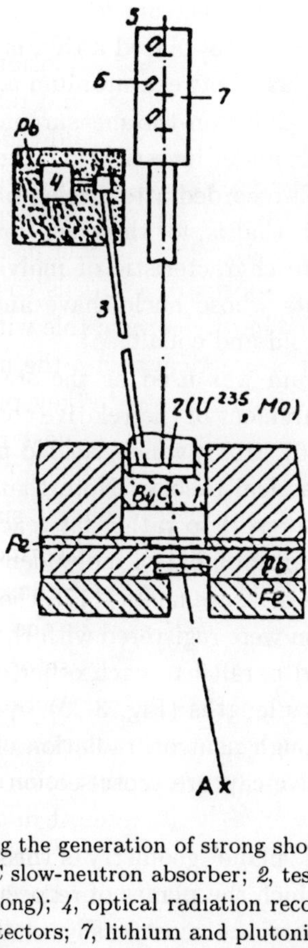

Fig. 3.19. Schematic illustrating the generation of strong shock waves by a nuclear explosive [11]: *A*, nuclear charge; *1*, B_4C slow-neutron absorber; *2*, test assembly of ^{235}U and molybdenum; *3*, light guides (12 m long); *4*, optical radiation recorders; *5*, time-of-flight neutron spectrometer; *6*, solid-state detectors; *7*, lithium and plutonium foils.

to produce high pressures. In the experiments performed by Ragan *et al.* [12], a block of ^{235}U, placed at a distance of 1.1 m from a nuclear charge behind a B_4C slow-neutron absorber, supported the molybdenum sample being investigated with light guides. The guides were provided for baseline registration of the shock front velocity (Fig. 3.19). The neutron flux generated upon detonation of the nuclear device A caused a rapid uniform heating of uranium to about 50 eV. This led to an expansion of uranium which generate a plane shock wave with an amplitude pressure of 20 Mbar in molybdenum. The second kinetic parameter of the velocity of shock-compressed molybdenum was found from the Doppler shift of resonance lines of neutron absorption in the 0.3–0.8 keV energy range. It was registered by a time-of-flight neutron spectrometer.

From the standpoint of experimental support of the method, the neutron flux in the resonance range of energy $10–10^3$ eV is essential. The number of such neutrons in the fission spectrum is very small. In order to increase their number, Prokhorov *et al.* [13] placed a thin layer of hydrogen-containing material, such as organic glass, between uranium and the sample.

The method discussed above on the measurement of mass velocity is not universal. For the materials under investigation, the cross-sections in the resonances must support a well-recorded attenuation of neutron flux by both stationary and moving matter, that is, for thickness comparable to the measuring bases. Such properties are characteristic of molybdenum, iron and copper. Besides, there are elements whose nuclei have anomalously large resonance cross-sections (tungsten, gold and cobalt).

Later [182], molybdenum was used as the standard property in experiments involving the measurement of the relative compressibility of uranium at a pressure of the order of 67 Mbar, which yielded pressures higher than those predicted by the quasi-classical theory. Note that in processing these data, the authors of [182] had to resort to rather far-reaching (from 20 to 50 Mbar) extrapolation of the standard adiabat of molybdenum.

In the case of absolute measurements [117], the phase and mass velocities of shock waves in aluminum were registered with the aid of reference layers and slit collimators L arranged parallel to each other, with the distance between them representing the gage lengths (Fig. 3.20). γ-radiation was recorded by scintillation detectors through neutron radiation of reference samples of material with a great radiative-capture cross-section (europium) placed in aluminum.

In the experiments, the "plane" geometry of the shock front, reference layers and collimating slits, in which the planes of respective surfaces are parallel to one another (Fig. 3.20), are the most easily realized. For this purpose, a

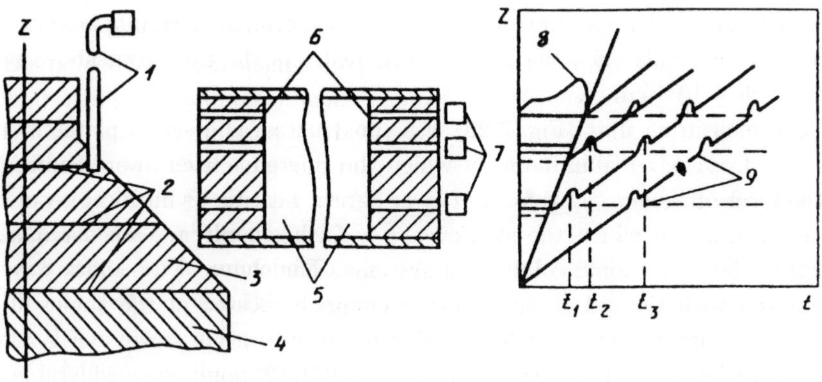

Fig. 3.20. Scheme of experiments [117] in recording the absolute shock compressibility of aluminum: *1*, optical channel; *2*, reference layers; *3*, material being investigated; *4*, channel where the shock wave is formed; *5*, collimating system; *6*, collimating slits; *7*, γ-radiation detectors; *8* and *9*, signals of radiation from stationary and moving reference points.

cylindrical channel was installed in the direction of wave propagation. It was manufactured from a material (such as magnesium and organic matter) whose density is less than the material in which the experimental facility is accommodated. It follows from the results of two-dimensional gas-dynamic calculations that the provision of such a channel ensures sufficient advance of the shock front relative to the front in the surrounding medium. The measuring unit is mounted at the cylinder end. The protection of the collimating system, which rules out the possibility of damage before completion of the recording of the moments when the reference layers pass the control positions, is accomplished by placing dense matter (lead, steel) in the direction of wave propagation.

In the experiments of Simonenko *et al.* [117], an intensive gamma source was obtained with pulsed irradiation by the neutrons of a substance whose nuclei had the radiative-capture cross-section exceeding those of the test material by a factor of approximately 10^3. The existing pulsed sources usually give rise to fast neutrons ($E \sim 1$ MeV). The radiative-capture cross-sections proceed efficiently at lower values of energy. Therefore, a neutron pulse must be ahead of the gas-dynamic motion being recorded by the time interval required for deceleration of neutrons in the test material to optimum values of energy. In a number of cases, europium may be used in the reference layers, for which the cross-section of the (n, γ)-reaction at $E = 10 \div 100$ eV is $q = (220\text{–}80)\ 10^{-24}$ cm^2.

The deceleration of neutrons in the sample leads to the heating of the test material. In general, this heating affects the shock compressibility of the material. To render an interpretation of the experimental results easier, it is necessary that this effect be weak, which restricts the flux from above by a value of $\Phi \leq 10^{17}$ cm^{-2}.

The measuring unit (Fig. 3.20) was constructed as a set of plates manufactured from AD-1 aluminum, in which the reference layers were embedded in the form of tablets. This form of layer agrees with the cylindrical geometry of the channel in which the shock wave is formed and can, when required, accommodate additional collimating systems. The shape of the shock wave is monitored with the aid of three optical channels located at the angles of an isosceles triangle on the diameter of 150 mm (one channel is shown in Fig. 3.20). The data of measurements by Simonenko *et al.* [117] point to considerable (in aluminum) errors in the quantum-statistical model at $p \sim 1.1$ TPa. These will be treated in more detail in Chapter 5.

The principal experiments in the ultrahigh-pressure region have been performed using the comparison method. The standard in experiments in dynamic compression of solid materials was provided by lead [114, 115] and iron [183]. Interpolation shock adiabats were constructed relating the ultrahigh-pressure ($p \geq 10$ TPa) region with a lower-pressure range of about 1 TPa, which is accessible for direct [106] experiments. In such experiments, the error of measurement of time intervals is 0.7–1.0%. The wave decay is at the level of 1–2%. By and large, it is an accurate estimate in determining the values of wave velocity on the contact boundary. These results may be used to verify the plasma models. However, they do not provide any answers to the question on the effect of the shell structure: the measurements were performed in the vicinity of the lower (with respect to pressure) boundary of the manifestation of shell effects. The transition from plane to spherical geometry of experiments in measuring Lepouter enables one to [14] raise the attainable pressure level to the exotically high values of about 400 TPa. However, such a transition results in a stronger decay of the shock wave in samples, and the indeterminacy of its inclusion once again made a major contribution to the total measuring error.

Unconventional experiments in studying the effect of the shell structure were performed by Model' and Narozhnyi [184]: the experimentally recorded time, required for the shock wave passage in samples of different materials, were compared with the calculated values. During the passage in samples, the pressure at the shock front dropped by a factor of three-four, that is, the simulation of the conditions of wave passage strongly calls for the knowledge of both the shock adiabat region and the release isentropes, which makes it difficult to interpret the results of these measurements.

The contribution by the shell effects was recorded most explicitly by Shatzman [15]. Shock waves with a clearly defined front shape were excited in test assemblies, and measurements were performed using a unified procedure for several values of pressure at the shock front. The gage lengths were selected such that the behavior of wave velocity inside the layers being investigated should be, according to calculations monotonic.

The moments of the shock front arrival at the reference surfaces were recorded by the optical glow of the air layers adjoining those surfaces, which was transmitted to a detector, such as a coaxial photocell, via light-guide channel with polished metal inner walls. In order to ensure the accuracy of measurements, three reference surfaces corresponding to the selected pressure were placed within the field of vision of a single photocell, namely, two surfaces of the standard and one surface of the sample. Two layers of the same material were provided for the experimental detection of wave decay. Every light-guide channel accommodated two detectors, with each recording a full signal. The complexity of the three-stage form of signal necessitated the use of oscillographic equipment.

Measurements were performed for the following pairs of materials: iron-aluminum, iron-lead, lead-iron, iron-water and iron-quartzite. The first material in each pair was placed first in the direction of wave propagation. The results are discussed in Chapter 5. However, we would like to note that Avrorin *et al.* [118] have reliably registered the shell effects on the shock adiabats in lead and aluminum.

The plasma effects are reflected most vividly in the experiments of Trunin *et al.* [116], who studied the compression of porous copper by powerful shock waves. These experiments involved the generation, behind the shock front, of a plasma with a specific internal energy of up to 0.75 MJ/cm^3, an electron concentration of $N_e \sim 10^{23}$ cm^{-3} and a pressure of up to 2 TPa (see Section 5.6). At the maximum temperature $T \sim (3\text{–}5) \cdot 10^5$ K, the electron degeneracy $N_e \lambda_e^3 \sim 1$ in a quintuply ionized plasma is removed while the Coulomb and short-range interaction remains strong, $\Gamma \sim 2$. Under these conditions, as shown in Section 5.6, the quasi-chemical plasma model, in view of the Coulomb interaction within the ring Debye approximation in the grand canonical ensemble of statistical mechanics, gives an acceptable description of dynamic experiment of [116], whereas the differences from the quasi-classical model may be as great as 20–30% with respect to density, and several times with respect to pressure.

3.6. Non-ideal Plasma Generation by Powerful Fluxes of Energy

Along with the customary techniques of shock wave generation (chemical and nuclear explosives, light-gas and powder guns), researchers are expressing an ever growing interest in unconventional methods based on the stimulation of matter by intense fluxes of directional energy. We refer to lasers, generators of pulsed electron and ion fluxes, electric-explosion and electrodynamic propulsion devices. The modern pulsed generators of directional energy, developed for the purposes of inertial controlled thermonuclear fusion, remote stimulation and technological applications, are capable of delivering energy to the test material on the kilojoule/megajoule scale at the megawatt power level. By focusing this energy into the submillimeter ranges of space, one is able to attain exotically high values of power density (10^{15}–10^{20} W/cm^3). This leads to extremely high values of pressure and temperature of a highly compressed plasma.

A broad spectrum of plasma states emerge under conditions of pulsed stimulation of condensed targets by intense fluxes of directional energy. It is difficult to identify a specific range of parameters for the thermodynamic description of those states [187, 188]. It is necessary to have the data on the properties of matter in an extremely wide region of the phase diagram, ranging from the highly compressed condensed state to ideal-gas, including the region taken up by strongly non-ideal plasma. We will explain this using the example of stimulation of a condensed target by intense laser radiation (Fig. 3.21) [185]. The laser radiation of frequency ω is absorbed in a heated plasma corona (at $\omega \sim \omega_p \sim \sqrt{4\pi N_e e^2/m}$) taken up by a high-temperature ideal plasma. The thermal energy is transferred to the region of relatively cold shock-compressed high-pressure plasma by electron thermal conductivity (shown by a broken line in Fig. 3.21) via non-ideal Boltzmann plasma embraced by complex hydrodynamic motion. In problems of controlled thermonuclear fusion with inertial confinement [186], the target shown in Fig. 3.21 is an external ablator of a multi-layer spherical thermonuclear micro-target. On its inside, a density of $\rho \sim 10^3\rho_0$ with a pressure $p \sim 10^{14}$ Pa and temperature $T \sim 10$ keV develop. The stimulation by intense fluxes of relativistic electrons and ions [85, 86] reminds one, qualitatively, of the situation corresponding to Fig. 3.21. The difference in this case is that the main energy of the beam is released in the high-density region $\rho \sim \rho_0$ (Fig. 3.21,b) and the overfly of the plasma corona is close to adiabatic.

The results of the analysis demonstrate that the effect of concentrated fluxes of directional energy on materials is a promising tool for the generation of powerful shock waves, which produce non-ideal plasma of extreme states.

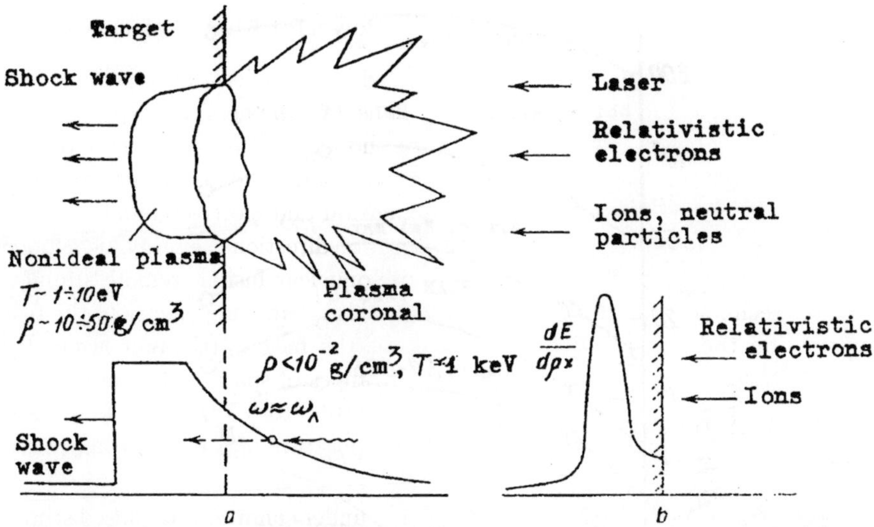

Fig. 3.21. Irradiation of a condensed target by fluxes of directional energy: (*a*) laser stimulation; (*b*) charged particles.

Most physical experiments involve the use of pulsed lasers [185], which offer the unique possibility of focusing coherent electromagnetic radiation on small ($\sim 10^{-4}$ cm^{-2}) surfaces. This leads to extremely high values of the local concentration of energy.

The specific powers currently attained and applied to targets are in the range of 10^{14}–10^{17} W/cm^2 and may be brought to 10^{21} W/cm^2 in the near future. The recoil impulse, which occurs as a result of the effect of such light fluxes, generates a powerful shock wave in the target. It can be used for compression and irreversible heating of dense plasma of materials being investigated. An analysis [189] of the hydrodynamic calculations of the intensity of shock waves, which emerge upon stimulation of various materials by existing and projected laser systems, demonstrates (Fig. 3.22) that in this case there is a real possibility of advancing into the ultramegabar range of pressures and investigating the properties of superdense plasma.

In designing laser targets for such experiments, a number of specific requirements arise [185, 189, 190] which are dictated by the physical singularities of the process and characteristics of diagnostic equipment. The target thickness depends on the absence of distorting release waves at the moment of laser pulse termination, as well as the small effect of "non-thermal" electrons occurring in

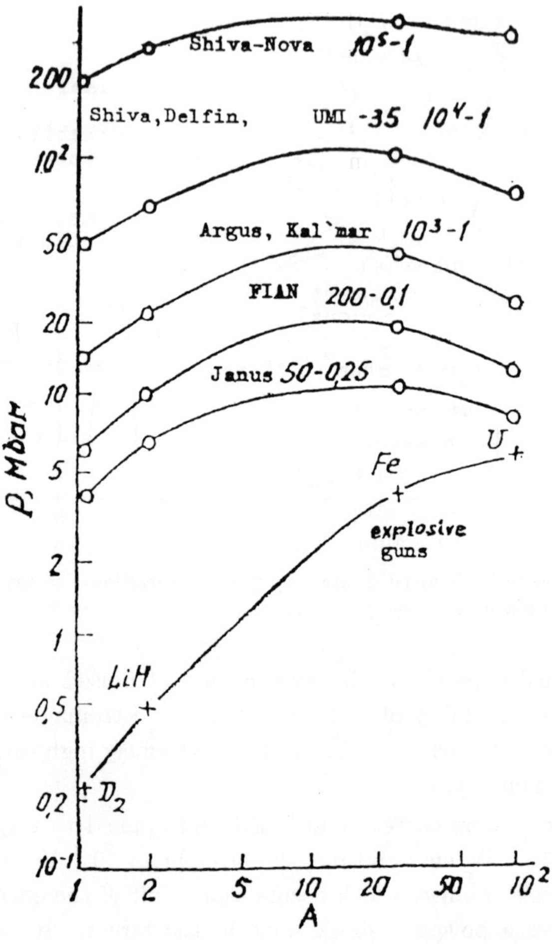

Fig. 3.22. Maximum plasma pressures [189] generated by different laser systems (the first numeral gives the energy in joules, the second numeral the pulse duration in 10^{-9} s); the bottom curve indicates the parameters accessible to the chemical explosive and light gun techniques; A, atomic number of the target element; (*a*) p, Mbar; (*b*) Shiva-Nova, $10^5 - 1$; (*c*) Shiva, Delfin, UMI-35, $10^4 - 1$; (*d*) Argus, Kal'mar, $10^3 - 1$; (*e*) FIAN, $200 - 0.1$; (*f*) Janus, $50 - 0.25$; (*g*) explosives, guns.

the laser radiation resonance-absorption band. The target diameter is selected such that it attains a fairly high radiation intensity and ensures the absence of side release waves. In addition, the target size should be sufficiently small in order to reduce the effect of surface currents from heated plasma. Otherwise, special shields should be used.

The initial experiments in the excitation of shock waves in hydrogen and Plexiglas involved the use of a low-power neodymium laser [194] with an energy of $E \sim 12$ J and pulse duration of $\tau \sim 5 \cdot 10^{-9}$ s. Because of the small size of the focal spot (40 μm), the shock waves decayed rapidly and degenerated into a sphere. To generate plane shock waves, Vessel and Solem [192] used a more powerful laser system ($E \sim 30$ J. $\tau \sim 0.3 \cdot 10^{-9}$ s). Measurements of the time of shock wave passage through a step-like aluminum sample enable one to register a shock front velocity of 13 km/s. This corresponds to a pressure of the order of 2 Mbar. These pressures were increased [193] by an order of magnitude (Fig. 3.23) using a laser of higher parameters ($E \sim 100$ J, $\tau \sim 0.3 \cdot 10^{-9}$ s). It produces a radiation intensity of $8 \cdot 10^{13}$–$3 \cdot 10^{14}$ W/cm^2 on the target. In this case, use was made of a small diameter target reduction. In the authors' opinion, the surface current effect and the agreement between theory and experiment were observed for the first time [195, 196]. The maximum pressures of 35 Mbar were attained in the Shiva laser facility [194] upon the irradiation of a composite aluminum/gold target by a light flux with a power density of the order of $3 \cdot 10^{15}$ W/cm^2. In subsequent studies [197], 10 beams of this facility (with an energy of about 1 kJ and a pulse duration of the order

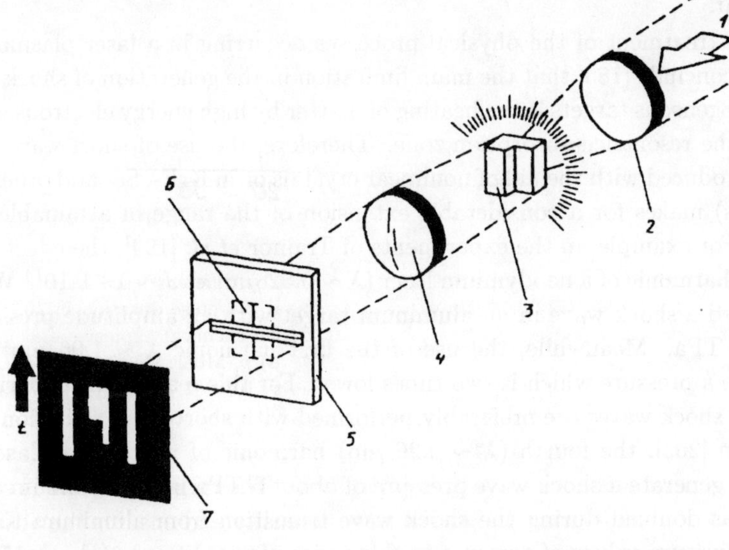

Fig. 3.23. Schematic illustrating the experiments in laser generation of shock waves [193]: *1*, laser radiation; *2*, focusing system; *3*, step-like target; *4*, image converter optics; *5*, image converter slit; *6*, image of target; *7*, photographic film.

of 3 ns) were used to accelerate a carbon foil 10 μm thick to approximately 100 km/s. Consequently, an impact against a carbon target produced a shock wave with an amplitude pressure of approximately 2 TPa. The same range of velocity of propulsion of plastic impactors was realized in experiments involving the use of iodine laser [198], with the characteristic values of power density ranging from 10^{12} to $6 \cdot 10^{15}$ W/cm^2.

To obtain information on the mass velocity of plasma, the authors of [192] used the "reflection" method in their experiments, which involve the shock wave transition from aluminum ($p \sim 3$ Mbar) to gold ($p \sim 6$ Mbar). The use of a CO_2 laser, with the [195] non-thermal electrons playing a minor role, yielded a shock wave intensity of 50 Mbar in aluminum. The value of power density J on a laminar target also ranged from $5 \cdot 10^{13}$ to $4 \cdot 10^{14}$ W/cm^2. A layer of gold, with a thickness of 10 μm, was included in the structure of this target to absorb the non-thermal electrons emerging from the aluminum ablation plasma.

An interesting example of the use of non-thermal electrons for isochoric heating of plasma is provided by Rennet *et al.* [199]. According to them, non-thermal electrons with an energy $E \sim 15$ keV can cause rapid heating and the explosion of a 3-μm upon expansion, layer of aluminum leading. This leads to the generation of a shock wave with an amplitude value of pressure of about 13 Mbar.

The treatment of the physical processes occurring in a laser plasma leads one to conclude [185] that the main limitation in the generation of shock waves in homogeneous targets is the heating of matter by high energy electrons emerging in the resonance-absorption zone. Therefore, the use of short-wave radiation (produced with the aid of nonlinear crystals or in KrF-, Se- and other laser systems) makes for a considerable extension of the range of attainable pressures. For example, in the experiments of Trainor *et al.* [194], the effect of the second harmonic of a neodymium laser ($\lambda \sim 0.52$ μm) at $J \sim 1 \div 1 \cdot 10^{14}$ W/cm^2 produced a shock wave in an aluminum target with an amplitude pressure of 1.0–1.2 TPa. Meanwhile, the use of the first harmonic $\lambda \sim 1.06$ μm would produce a pressure which is two times lower. For this reason, the experiments in laser shock waves are preferably performed with short-wave radiation [200–203]. In [203], the fourth ($\lambda \sim 0.26$ μm) harmonic of neodymium laser was used to generate a shock wave pressure of about 5 TPa in an aluminum range. This was doubled during the shock wave transition from aluminum to gold. The maximum values of pressure in this series of experiments (about 15 TPa) were obtained as a result of the impact of an aluminum foil with a thickness of 12 μm against the target.

Experiments with laser systems place severe demands on the diagnostic equipment: the time and space resolution should at least be as good as 10^{-11} s and 10^{-4} cm. Therefore, in the majority of laser experiments, it is only possible now to measure the shock velocity. The most diverse possibilities have been considered that would facilitate the registration of one more dynamic parameter: that of the plasma velocity. The possibilities include the "obstacle" technique [192, 203], flash radiography [197, 202], Doppler level shift and the like (for more details, see the review in [185]).

In order to study the structural properties of non-ideal plasma, Hall and Djaamet [202] employed the method of flash radiography of aluminum. The latter was heated to $T \sim 10^4$ K and triply compressed by colliding plane shock waves using the second harmonic of neodymium laser ($J \sim 2 \cdot 10^{13}$ W/cm^2). The source of external X-radiation was provided by a uranium target irradiated by a special laser. It was recorded using a fast micro-crystal X-ray spectrometer. The experiments revealed an oscillation in X-ray spectra. It was caused, in the opinion of Hall and Djaamet [202], by the emergence of the short-range order in the dense plasma, and is attributed to plasma phase transition.

Of considerable interest in experiments with laser shock waves is the study of the physical effects accompanying the arrival of shock waves to the free surface [113]. Ng *et al.* [200] used a fast image converter to analyze the emitted optical radiation of the rear side of an aluminum target after the passage of a shock wave with a pressure of about 0.03–1.2 TPa. The intensity of this radiation was then compared with the temperature of shock compression within the framework of too simplified a model. Then, for similar physical conditions, Ng *et al.* [201] measured the reflectivity of laser radiation with the wavelength $\lambda \sim 0.57$ μm from an adiabatically expanding plasma, which provided information about the high-frequency electrical conductivity of non-ideal plasma $\rho \leq 1$ g/cm^3.

The powerful pulse generators (of the order of 10^{14} W) of relativistic electrons and ions [197, 198], developed for the purposes of controlled fusion and solution of applied problems, permit are to focus high-intensity beams in targets which are several millimeters in diameter. The specific powers of the order of 10^{14}–10^{18} W/cm^2, applied in this manner, cause evaporation and expansion of the outer part of the target. This leads to the ablation generation of powerful shock waves.

Assuming the same (as in the case of laser stimulation) requirements [119], that is, the plasma flow in the target should be one-dimensional and quasi-stationary, and taking into account the characteristic ranges of electrons with energy in the megaelectronvolt range in metals are of the order of 0.1–1 mm, one can estimate the amplitudes of shock waves in plasma at a level of tens of megabars [187]. In the initial experiments performed to generate shock waves in metals, with the aid of relativistic electron beams [204–206], the maximum pressures do not exceed 3 Mbar. The rate of propagation and time profile of the shock wave were registered; they provide information about the strength properties of materials and the parameters of the equation of state.

A series of experiments with shock waves were staged in an effort to reveal the importance of anomalous deceleration ("magnetic stopping") of electron beam in condensed targets [205–207]. This anomalous deceleration, which is characteristic of intense relativistic electron beams, is attributed to the elongation of the electron trajectory in plasma as a result of the effect of intrinsic magnetic field of the beam. It shows up very clearly during the passage of the beam through thin metal foils [187]. In order to clarify this effect, Akkerman *et al.* [205], in measuring the group and phase velocities of shock waves, used diverse techniques which include monomode quartz light guides, two-step targets and targets with a complex configuration. Figure 3.24 gives the results of measurement of decay of the mass velocity of a dense aluminum plasma as the shock wave moves away from the zone of energy release. A comparison of these results with those for one- (curve *2*) and two-dimensional (curves *3*

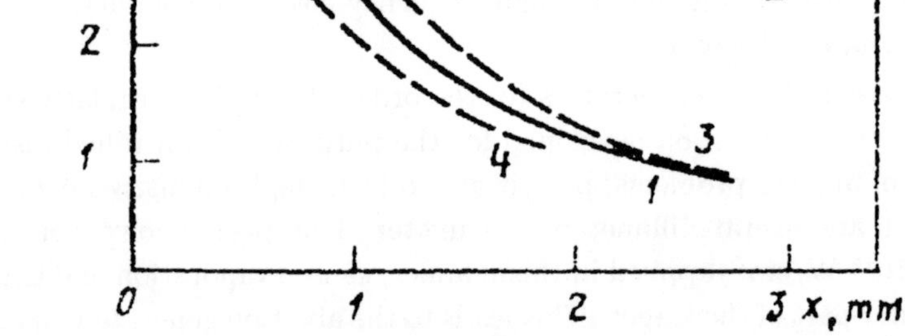

Fig. 3.24. Mass velocity of a shock wave generated by a high-current relativistic electron beam: *1*, experiment [205]; *2*, one-dimensional calculation; *3*, two-dimensional calculation with classical energy release (calculation using the Monte Carlo method); *4*, energy release with due regard to "magnetic stopping".

and *4*) hydrodynamic calculations point [205] to the validity of the classical (calculation using the Monte Carlo method is shown at *3*) mechanism of energy release, as well as to the insignificance (curve *4*) of "magnetic stopping" with reference to thick targets.

Interesting possibilities are offered by the method of ablation acceleration of foils under the effect of X-radiation, which occurs upon deceleration of the relativistic electron beam in the outer, metallic part of the target. In this manner, the rate of polyethylene envelope travel of the order of $5 \cdot 10^6$ cm/s was attained in a cone target. A neutron yield of $D - D$ reaction of about $3 \cdot 10^6$ neutrons per pulse was also registered.

Recently, along with the methods of pulsed controlled fusion excited by laser radiation or by charged particle fluxes, which have become traditional, ever increasing attention is paid to the scheme which utilizes an impact by macroscopic liners (with a mass of 0.1 g) accelerated to velocities of 10^7–10^8 cm/s [208]. Obviously, the high-velocity propulsion devices based on the electrodynamic acceleration methods, which are considered for such uses, can be used to generate superpower shock waves and to investigate the properties of plasma under extreme pressures and temperatures. In a linear magnetodynamic accelerator, a superconducting impactor is accelerated in a non-uniform magnetic field. The latter is generated by coils whose switching synchronized with the impactor travel. In one of the projects, the impactor acceleration is accomplished through a series of z-pinches collapsing toward the axis of symmetry [208].

At present, the rail-gun accelerator [209, 210] appears to be the most well-developed among a great number of electrodynamic accelerators. In this device (Fig. 3.25), an impactor *1* is accelerated by ponderomotive force $F = (1/2)I^2L$ upon the passage of electric current I in electrodes *3* (linear inductance L). In doing so, the accelerating force F should not exceed the ultimate strength of the impactor material ($I \leq 1$ MA). The optimal electric contact with the rails is provided by an electric arc *2* burning in the rear portion of the dielectric liner and urged against the latter by the magnetic field. The working capacity of such systems was demonstrated in experiments by Rayleigh and Marshall [209], who had supplied the rail-gun from an inductive storage and homopolar generator (energy of 500 MJ) to attain velocities of 6 km/s ($m \sim 3$ g), and by Kondratenko *et al.* [210] who had brought the propulsion velocity to 10–11 km/s with the use of a bank of capacitors (energy of about 0.8 MJ) and a linear magnetocumulative generator.

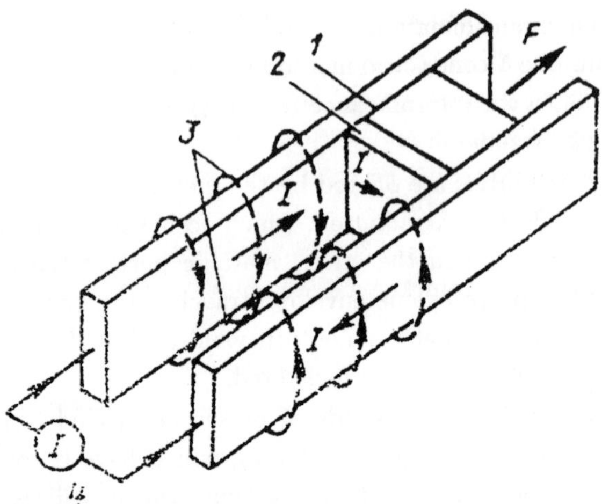

Fig. 3.25. Schematic of rail-gun generator [209, 210]: *1*, impactor; *2*, electric arc; *3*, current-carrying rails; *4*, power supply.

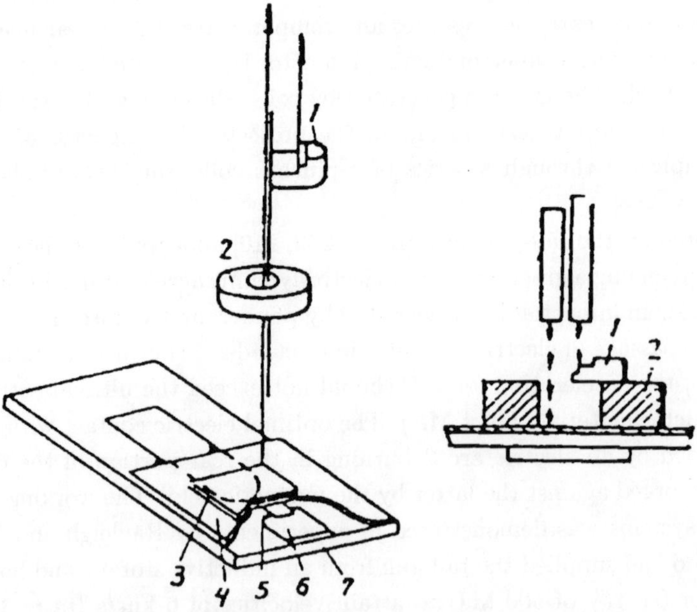

Fig. 3.26. Schematic of electric-explosion facility [211]: *1*, step-like target; *2*, channel for impactor travel; *3*, projectile plate; *4*, plastic insulation spacer; *5*, foil being exploded; *6*, current-carrying buses; *7*, massive substrate.

An interesting means of advancing through the pressure scale is offered by an electric explosion of conductors upon the discharge of a powerful bank of capacitors (Fig. 3.26) into the latter. According to [211], a bank of capacitors with a capacitance of 15.6 mF and voltage of 100 kV imparts to thin aluminum foil a specific energy exceeding by a factor of 10–100 the characteristic internal energy of condensed explosives. The foil expands to accelerate a plastic impactor, which has a thickness of 300 μm, to a velocity of 30 km/s and a tantalum impactor, which has a thickness of 30 μm, to 10 km/s. In this experiment, the shock front velocity (step-like target) is registered and the variation of the rate of impactor flight planned (laser Fabry-Perot interferometer). In accordance with the reflection method, these can provide quantitative information on the dynamic compressibility of plasma in the megabar range of pressures. The equations of state obtained for tantalum in the 1.9–7.8 Mbar pressure range are in a good agreement with data obtained in light-gas facilities.

Four

The Ionization Equilibrium and Thermodynamic Properties of Weakly Ionized Plasma

In a weakly ionized dense plasma, the electron concentration is not high and the non-ideality is due, primarily, to the interactions between charges and neutrals, and between neutral atoms and molecules. In materials of high polarizability, the ion-atom and electron-atom interactions become strong when the interatomic interactions are neglected.

4.1. Partly Ionized Plasma

4.1.1. *Classical Low-Temperature Plasma*

A degenerate low-temperature plasma, in which the inequality

$$\beta \mathrm{Ry} \gg 1 \,, \qquad \mathrm{Ry} = me^4/2\hbar^2 \,, \qquad \beta = 1/T \tag{4.1}$$

is valid, is a classical plasma because the inequality (4.1) implies that $e^2\beta \gg \lambda$. The quantity $e^2\beta$ is the mean scattering amplitude (Landau's length), which is equal to the square root of the mean effective scattering cross-section. Therefore, not only does the system conform to the classical statistics $N_e\lambda^3 \ll 1$, but the motion of individual particles in the system may also be described within the framework of classical mechanics.

From the standpoint of the problems treated in this book, partly ionized plasma poses the greatest interest. It is also referred to as the three-component plasma, bearing in mind that the atoms, free electrons and ions remained unbound. The emergence of atoms in the initial electron-ion system is due to the electron/ion interaction which, if $\beta\mathrm{Ry} \gg 1$ (Ry is the bonding energy of the electron-ion pair, the ionization energy of hydrogen atom), becomes strong even at relatively low values of density. This interaction results in the formation of bound electron-ion pairs, that is, atoms.

It is important that the electron and ion bound in an atom are spaced from each other to a distance of the order of Bohr's radius a_0, which is the least characteristic dimension in the system, $a_0 \ll N^{-1/3}$, $a_0 \ll e^2\beta$, $a_0 \ll \lambda$. Therefore, the electron and ion, which make up an atom, shield one another in such a way that, with good approximation, the interaction of such pairs with surrounding particles may be fully ignored.

4.1.2. *Three-Component Electron-Ion-Atomic Plasma*

The concentrations of components are related by the condition of ionization equilibrium. We will write this condition as Saha's equation and the first corrections to it. This may be done within the second virial approximation developed in the initial electron-ion system.

The second virial correction to Gibbs' thermodynamic potential of electron-ion system $\beta\Delta\Omega$ has the form [212, 213]

$$-\beta\Delta\Omega = \frac{1}{2}\sum_{i,j} \xi_i\xi_j \left(\frac{2\pi\hbar^2\beta}{m_{ij}}\right)^{3/2} \left[Z_{ij}(e^2) - Z_{ij}(0)\right],$$

$$Z_{ij} = \frac{1}{g_i g_j}\sum_{\nu} \exp(-E_{ij\nu}\beta)\,. \tag{4.2}$$

Here, ξ_i and ξ_j denote the fugacities of plasma particles. The fugacities of ions and electrons are

$$\xi_i = \exp(\beta\mu_i)(M/2\pi\hbar^2\beta)^{3/2} g_i\,,$$

$$\xi_e = \exp(\beta\mu_e)(m/2\pi\hbar^2\beta)^{3/2} g_e\,,$$

where μ_e, μ_i, g_e, g_i, m and M denote their chemical potentials, statistical weights, and masses; m_{ij} denotes reduced masses; ν denotes the quantum numbers of relative motion of the i- and j-particles; and $E_{ij\nu}$ denotes the respective energies of continuous and discrete spectra.

We retain in (4.2) the term that remains when $\beta\mathrm{Ry} \gg 1$, and write $\beta\Delta\Omega$ as

$$-\beta\Delta\Omega = \xi_i + \xi_e + \xi_i\xi_e(2\pi\hbar^2\beta/m_{ei})^{3/2}(g_a/g_i g_e)\exp(\beta E_1)\,. \tag{4.3}$$

Here, $m_{ei} = mM(M+m)^{-1}$ and g_a is the statistical weight of the state with the energy E_1, $E_1 = \mathrm{Ry}$. We will now proceed to the three-component description of plasma and introduce atoms into the treatment. For this purpose, we will define the activity of atoms using the following expression:

$$\xi_a = \exp(\beta\mu_a) \cdot [(M+m)/2\pi\hbar^2\beta]^{3/2} g_a \cdot \exp(\beta E_1)\,.$$

We will then use the condition of chemical equality $\mu_a = \mu_e + \mu_i$, that is, the chemical potential of new particle. Afterwards, we derive, instead of (4.3), the expression

$$-\beta\Delta\Omega = \sum_j \xi_j\,, \tag{4.4}$$

which describes a three-component ideal plasma ($j = e, i, a$). Hence the formulas $N_j = \xi_j[\partial(-\beta\Delta\Omega)/\partial\xi_j$ yield the particle density which is equal to $N_j = \xi_j$. This is followed by the equation of ionization equilibrium, Saha's formula:

$$\frac{N_e N_i}{N_a} = \frac{g_e g_i}{g_a}\left(\frac{m_{ie}}{2\pi\hbar^2\beta}\right)^{3/2} \exp(-\beta Ry)\,.$$

Because an atom and an ion (if this is not a proton) are multi-level systems, it is natural that both g_i and g_a be replaced by their internal partition functions of ion and atom Σ^+ and Σ. The ionization potential of a complex atom is different from Ry and is usually designated as I with the reduced mass as $m_{ei} \approx m$. Saha's equation has the final form

$$\frac{N_e N_i}{N_a} = \frac{2\Sigma^+}{\Sigma}\left(\frac{m}{2\pi\hbar^2\beta}\right)^{3/2} e^{-\beta I}\,, \tag{4.5}$$

where

$$\Sigma = \sum_k g_{ka} \exp(-E_{ka}\beta)\,, \qquad \Sigma^+ = \sum_k g_{ki} \exp(-E_{ki}\beta)\,.$$

The sums are taken over all bound states of atom (ion). g_{ka} and g_{ki} stand for the statistical weight of the k-th states of atom and ion, respectively, and E_{ka} and E_{ki} denote the bonding energy.

However, the three-component model in this form proves to be inadequate. First, the component of the second virial coefficient, which is still being ignored due to the interaction between the free electrons and ions, goes into infinity. Second, the partition functions over bound states diverge. This is due to the long-range nature of the Coulomb potential.

4.1.3. *Second Virial Coefficient and Partition Function of Atom*

We now discuss the divergence of the second virial coefficient due to the interactions of free electrons and ions over large distances:

$$T\sum_{k,j}\xi_k\xi_j\int r^2dr(\exp(-\beta e_ke_j/r)-1)\approx T\sum_{k,j}\int(e_ke_j\beta)^2dr/2\,,$$

where k, j assume the values of e, i. This divergence is eliminated if one takes into account the many-particle aspect of interaction. For this purpose, integration must be restricted to a distance which is less than the Debye length r_D. As a result, we have the Debye-Hueckel correction $\Delta(\beta\Omega)_D$:

$$\Delta(\beta\Omega)_D=(12\pi)^{-1}\left(4\pi\beta e^2\sum_j\xi_j\right)^{3/2}\,,\qquad j=e,i\,. \tag{4.6}$$

If (4.6) is included in the construction of the ionization equilibrium, the following correction [214] appears to the ionization potential in (4.5):

$$\Delta I=-2T\ln[1+\Gamma\varphi(\Gamma)/2]\,,$$

where $\Gamma=e^2/r_DT$ is the Debye parameter of interaction. The $\varphi(\Gamma)$ function is defined by the expression $\Gamma=2(1-\varphi^2)/\varphi^3$. Therefore, the ionization potential decreases.

The sum over bound states

$$\Sigma=\sum_{k=1}^{\infty}\exp(-\beta E_k)g_k$$

diverges for small values of bond energy which is like hydrogen: $E_k=-\mathrm{Ry}\,k^{-2}$ and $g_k=k^2$. This divergence, which occurs in a discrete spectrum, is compensated, in the case of correct inclusion of the electron/ion interaction, by a similar divergence in a continuous spectrum [212]. Rather than calculating $\sum_{k=1}^{\infty}g_k\exp(-\beta E_k)$, one must calculate the quantity

$$\xi_i\xi_e\left\{\left(\frac{2\pi\hbar^2\beta}{m}\right)^{3/2}\sum_k g_k\exp(-\beta E_k)-\int dr\left[1+\frac{\beta e^2}{r}+\frac{1}{2}\left(\frac{\beta e^2}{r}\right)^2\right]\right\}.$$

This quantity is derived if the terms which were previously included in expression (4.6) are subtracted from the electron-ion component of the virial addition (4.2). The first of those terms is related to the electroneutrality of plasma and was cancelled out in deriving (4.6). The second term had already made its contribution to (4.6). As a result of integration, a new (renormalized) partition function emerges, which will be designated Σ as before, with some additions to this sum. Hence

$$\Sigma=\sum_{k=1}^{\infty}g_k[\exp(-\beta E_k)-1+\beta E_k]\,. \tag{4.7}$$

One can see that Σ converges on the values of energy $|E_k|$, which is equivalent to thermal energy.

The problem of eliminating the divergence has long been discussed in the literature. A rigorous result was derived by Vedenov and Larkin [212]. A correct description of the solution to this problem may be found in [215, 216]. Expression (4.7) is valid on the condition that the second virial approximation is sufficient. Under these conditions, the values of T exceed considerably the characteristic values of interaction energy. Otherwise, no correct solutions to the problem exist at present.

The second virial coefficient of plasma is calculated in a number of studies (see [216] and the references cited therein). In addition to the main components treated above, this coefficient includes higher-density terms such as $N^2 \ln N$ (this term vanishes in symmetrical plasma), N^2, $N^{5/2} \ln N$ and $N^{5/2}$.

Quantum corrections appear in a term that is quadratic with respect to density. They are proportional to λ/r_D and are minor because $\lambda/r_\mathrm{D} \sim \gamma^{3/2}$ $(\beta\mathrm{Ry})^{-1/2}$.

4.2. Anomalous Properties of Metal Plasma

When the degree of ionization is low, the plasma non-ideality, even a very high one, may lead only to corrections to the thermodynamic properties of material. However, its effect on the composition of ionized components, on the free electron concentration, may be very strong. Therefore, a shift in the ionization equilibrium due to interparticle interactions show up most vividly in the electronic transport coefficients. The electrical conductivity, for instance, varies by orders of magnitude as a result of the variations in the number of carriers rather than in their mobility. Therefore, the anomalies in the electrical conductivity and thermal e.m.f., which exceed the limits of conventional concepts and are due to variations in the ionization equilibrium of non-ideal plasma, are discussed in Section 4.2.1. Afterwards, the basic theoretical principles are presented.

4.2.1. *Physical Properties of Metal Plasma*

Figure 4.1 plots the electrical conductivity of cesium in the entire range of pressures and temperatures.

Figure 4.2 shows the data on thermal e.m.f. Let us now discuss the salient features.

A plasma may be regarded as ideal at low pressures, such as at 0.1 bar $\ll p_c$. Therefore, at low temperatures, the electrical conductivity grows with

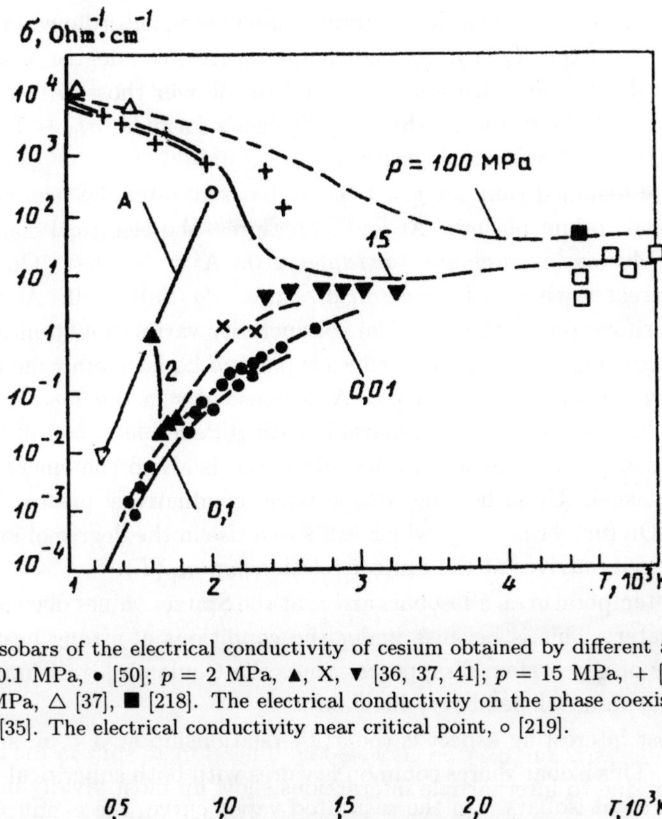

Fig. 4.1. Isobars of the electrical conductivity of cesium obtained by different authors: $p =$ 0.01 and 0.1 MPa, • [50]; $p =$ 2 MPa, ▲, X, ▼ [36, 37, 41]; $p =$ 15 MPa, + [218], □ [42]; $p =$ 100 MPa, △ [37], ■ [218]. The electrical conductivity on the phase coexistence curve: *A* [37], ▽ [35]. The electrical conductivity near critical point, ○ [219].

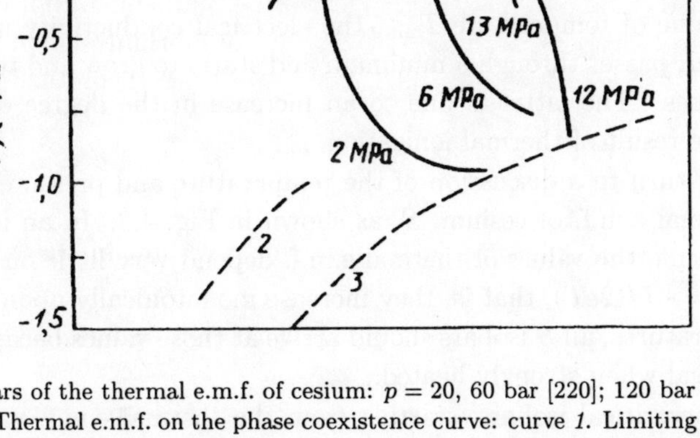

Fig. 4.2. Isobars of the thermal e.m.f. of cesium: $p =$ 20, 60 bar [220]; 120 bar [36]; $p =$ 130, 200 bar [40]. Thermal e.m.f. on the phase coexistence curve: curve *1*. Limiting relationships: Eqs. (6.37) and (6.38), curves *2* and *3*.

temperature because the electron concentration rises in accordance with Saha's equation, $N_e \sim \exp(-I/2T)$. At high temperatures, the degree of ionization increases and collisions with ions begin to prevail over those with atoms. In this case, the conductivity is defined by Spitzer's formula, $\sigma_{\rm sp} \sim T^{3/'2}/\ln\Lambda$, where $\ln\Lambda$ is the "Coulomb" logarithm.

It can be assumed that the $p = 1$ bar isobar separates the regions of ideal and non-ideal cesium plasma. At lower pressures, the electrical conductivity grows with decreasing pressure, $(d\sigma/dp)_T < 0$. At $p > 1$ bar. On the contrary, σ increases with an increase in pressure, $(d\sigma/dp)_T > 0$. At pressures above the critical point, the electrical conductivity varies monotonically upon heating or cooling. Starting from values typical of liquid metal, the electrical conductivity decreases upon heating. A decrease of σ in the $T \sim T_c$ region is especially sharp at pressures comparable with p_c, such as $p = 150$ bar. The "metal — non-metal" transition takes place and leads to the emergence of a non-ideal plasma. Upon heating, its electrical conductivity passes through a minimum. On further heating, which leads to a rise in the degree of ionization, the conductivity arrives at the Spitzer relationship $\sigma_{\rm sp}(T)$.

At high temperatures, all isobars arrive at the Spitzer values of conductivity sooner or later. This is because under the conditions of strong heating, the non-ideality of any system disappears. Since the Coulomb logarithm depends on p, the $\sigma_{\rm sp}$ values depend very little on p.

The most interesting aspect is the $\sigma(T)$ relationship at $p < p_c$, such as at $p = 20$ bar. This isobar shares common features with both subcritical ($p \to p_c$) and supercritical isobars. On the saturated vapor curve, the conductivity assumes anomalously high values (Fig. 2.10), which are three to four orders of magnitude higher than the results of conventional ideal-gas estimates. However, a slight increase in temperature leads to a sharp drop in conductivity. At some value of temperature $T_{\rm min}$, the electrical conductivity upon heating of the isobar, passes through a minimum and starts to grow and tends towards Spitzer values. The latter is due to an increase in the degree of ionization, which is the result of thermal ionization.

We now turn to a discussion of the temperature and pressure dependence of the thermal e.m.f. of cesium, S, as shown in Fig. 4.2. In an ideal, weakly ionized plasma, the values of thermal e.m.f. depend very little on p. They are close to $S = -I/(2eT)$, that is, they increase monotonically upon heating. At high temperatures, all S isobars should arrive at those values because a plasma becomes ideal when strongly heated.

The supercritical isobars, starting from the thermal e.m.f. values characteristic of liquid metal, fall sharply upon heating (Fig. 4.2). This is a result of the transition of the material from a metallic to non-ideal plasma state. In the non-ideal plasma region, the thermal e.m.f. values pass through a minimum to arrive, on further heating, at the $(-I/2eT)$ relationship.

Therefore, in measuring the electrical conductivity and thermal e.m.f. of non-ideal cesium plasma, results have been obtained which do not fit the conventional (from the standpoint of plasma physics) ideal-gas approximation limits. The "almost free" electron approximation used to great advantage in the liquid-metal state theory is also equally inapplicable.

4.2.2. *Decrease in the Ionization Potential*

The electrical conductivity on the supercritical isobars $(p > p_c)$ decreases by orders of magnitude with a small increase in temperature. Hence one can talk of the exponential density dependence of electrical conductivity (Fig. 4.3). It was interpreted by Vedenov [222] as a result of the growth in the electron concentration due to a decrease in the ionization potential. Alekseev and Vedenov [223] proposed the following theory.

The decrease in the ionization potential ΔI is the sum of two quantities, $\Delta I = \Delta I_1 + \Delta I_2$. First, the free electron energy decreases, due to strong polarization interaction with neutral atoms, by the quantity $\Delta I_1 = N\hbar^2(\pi q)^{1/2}/2m$.

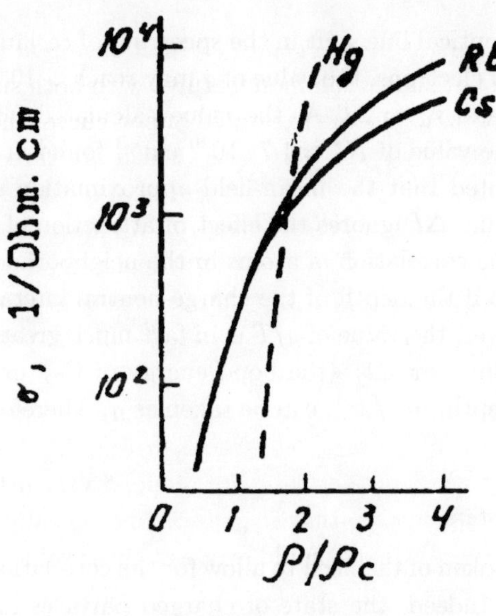

Fig. 4.3. Isotherms of the electrical conductivity of cesium, rubidium and mercury, $T = 1.025$ [12].

Here, q is the total electron-atom scattering cross-section and N is the atomic concentration. ΔI_1 is written in an optical approximation (1.12) while the electron scattering length for a slow electron is $|L| = (q/4\pi)^{1/2}$. Secondly, the energy of ion in gas is likewise reduced by a quantity $\Delta I_2 = 2\pi N\alpha e^2/r_a$, due to the polarization of the neighboring atoms (cf., Eq. (1.10)). Here, α is the atomic polarizability and r_a is the atomic radius. The decrease in ionization potential brings about an exponential growth in the electron concentration with density. Saha's equation in a weakly ionized gas, written with due regard for the decrease in the ionization potential,[1] gives

$$N_e \sim \exp[-I_m/(2T)], \qquad I_m = I - \Delta I_1 - \Delta I_2 . \tag{4.8}$$

If assuming that the electron mobility is regarded as varying slowly with density, the coefficient of electrical conductivity will then follow the variation of N_e to grow exponentially with density:

$$\sigma \sim \exp[(-I/2T) + (N/N^*)] . \tag{4.9}$$

Here, N^* is the density value from which the growth becomes appreciable. From the foregoing formulas, we obtain the following expression for N^*:

$$(N^*)^{-1} = (2mT/\hbar^2\sqrt{\pi q})^{-1} + (Tr_a/2\pi\alpha e^2)^{-1} . \tag{4.10}$$

An analysis of the optical line shift in the spectrum of cesium leads one to conclude that, for slow electrons, the value of q may reach $2 \cdot 10^{-13}$ cm^2 [224, 225]. Since $\alpha = 400a_0$ and $r_a = 2.6$ Å, the value calculated for N^* (Eq. (4.10)) agrees well with the value of $N^* \approx 1.7 \cdot 10^{20}$ cm^{-3} found in experiment.

It should be noted that the mean-field approximation used by the latter authors in calculating ΔI ignores the effect of attraction of atoms by ion and electron, that is, the correlation of atoms in the neighborhood of charged particles. This is valid if the depth of the charge-neutral interaction potential, q, is not great. However, the value of g/T is in fact much greater than unity. For the purpose of estimation, D_2^+ (the bond energy of Cs_2^+ or Hg_2^+ ions) or the polarization well depth, $\alpha e^2/2r_a^4$, can be taken as g. Therefore, the correlation is substantial.

4.2.3. *Charged Clusters*

We have already spoken of the need to allow for the correlation of atoms around charged particles. Indeed, the state of charged particles in a dense medium

[1] In Section 4.2.2, the ionization potential decrease is expressed in absolute values.

do experience radical changes. Thanks to attractive forces acting between the charged and neutral particles, charged clusters are formed. These heavy particles consist of an electron (ion) and a number of atoms; it is due to the presence of the electron (or ion) in them that the clusters are stabilized. The structure of clusters and the number of particles in them depend strongly on the degree of non-ideality. Let us follow the variation of the cluster structure in vapor upon isobaric heating. On the saturation line at $T = T_s$, one observes positively and negatively charged droplets. At a temperature somewhat higher than T_s, these are less dense formations which can no longer be regarded as liquid but which are still in dynamic equilibrium with vapor. At appreciably high temperatures, these are cluster (or, as they are sometimes referred to, complex) ions containing several atoms bound by polarization forces. Finally, in a weakly non-ideal plasma, the cluster ions dissociate and what remains are ordinary molecular ions. The existence of clusters may strongly affect the properties of material. For instance, in saturated cesium vapors, the thermal emission of electrons from droplets (while stabilizing the droplets) raises the degree of ionization and the electrical conductivity of the plasma by several orders of magnitude. This is responsible for the anomalously high electrical conductivity of high-temperature metallic vapors near the saturation line.

The clusters and like particles were observed in different temperature ranges in numerous gases and liquids; these phenomena were reviewed by Khrapak and Iakubov [4, 226]. Based on the review by Lagar'kov and Iakubov [227], we shall discuss below the ionization equilibrium in a low-temperature metal plasma.

We shall begin with the most simple case by considering a plasma consisting of electrons, atoms A, ordinary ions A^+ and molecular ions A_2^+. The electron density is defined by Saha's equation

$$N^+ N_e / N \sim \lambda_e^{-3} \exp(-I\beta) \tag{4.11}$$

and by the condition of electrical neutrality:

$$N^+ + N_2^+ = N_e \,. \tag{4.12}$$

The concentration of the molecular ion A_2^+ is given by the equation

$$N^+ N / N_2^+ \sim \lambda_a^{-3} \exp(-D_2^+ \beta) \,. \tag{4.13}$$

In Eqs. (4.11) to (4.13), N^+, N_e, N and N_2^+ denote the concentration of ordinary ions, electrons, atoms and molecular ions, respectively, λ_e and λ_a are the thermal wavelengths for electron and atom, and D_2^+ is the bond energy of atom in the ion A_2^+. Assuming that the degree of ionization is low, then on expressing N_2^+ and N^+ with the aid of Eqs. (4.11) and (4.12), we obtain

$$N_e^2/N = K_1 \exp(-I\beta)[1 + Nk_2 \exp(D_2^+\beta)], \tag{4.14}$$

where K_1 is the constant of ionization equilibrium $A \leftrightarrow A^+ + e$ and k_2 is the constant of the dissociation equilibrium $A_2^+ \leftrightarrow A^+ + A$. For cesium, $k_2^{-1} = 1.45 \cdot 10^{21}\, T^{-1/2}$ cm^{-3}, where T is expressed in electron-volts $D_2^+ \approx 0.7$ eV. It follows from Eq. (4.14) that the inclusion of the presence of molecular ions in the plasma leads to a shift in ionization equilibrium towards an increase in electron concentration. The higher the molecular ion dissociation energy, the greater the shift becomes. When $T = 1,400$ K and $N = 10^{20}$ cm^{-3} (near the saturation point for cesium vapors at $p = 20$ bar), the inclusion of Cs_2^+ ions gives an approximately fivefold increase of N_e. The electrical conductivity σ increases accordingly. An important part played by molecular ions in an alkali plasma was confirmed [228] by measurements of the electrical conductivity of sodium vapors near the saturation line.

Given high pressures and low temperatures, the inclusion of Cs_2^+ ion proves insufficient because, under these conditions, $Nk_2 \exp(D_2^+\beta) \gg 1$. This means that $N_2^+ \downarrow N^+$. One can naturally expect an appreciable number of Cs_3^+ ions to be present in the plasma. For the $A_3^+ \leftrightarrows A_2^+ + A$ equilibrium, we have the equation $N_3^+ = N_2^+ N k_3 \exp(D_2^+\beta)$, which is similar to Eq. (4.13). Taking these ions into account, we obtain for N_e^2, instead of Eq. (4.14),

$$N_e^2 = NK_1 \exp(-I\beta)[1 + Nk_2 \exp(D_2^+\beta) + N^2 k_2 k_3 \exp(D_2^+\beta + D_3^+\beta)]. \tag{4.15}$$

It is evident that as the formation of more and more complex ions (referred to as cluster ions) is taken into account, subsequent terms will appear in the square brackets in Eq. (4.8). Assuming that $D_3^+ = D_2^+$ and $k_3 = k_2/2$, the right side of Eq. (4.15) can be regarded as the beginning of a series whose sum is expressed by the exponent. Such summation yields [229]

$$N_e^2 = N_1 \exp[-I\beta + Nk_2 \exp(D_2^+\beta)]. \tag{4.16}$$

Apparently, the value of D_2^+ in Eq. (4.16) should be close to the heat of evaporation on the basis of a single atom, which is approximately equivalent to 0.77 eV for cesium. One of the first papers to contain a discussion of heavy cluster ions and summation, which is analogous to that performed above, was the paper by Magee and Funabashi [230].

The procedure of summation in Eq. (4.15) is heuristic. Indeed, even the third term in the series, calculated with due regard for the real values of D_3^+ and k_3, does not correspond to the third term of the exponential series. Nevertheless, Eq. (4.16) represents the major features of the phenomenon. It describes the presence, on the isobar, of a sharp temperature minimum of free electron concentration and, thereby, conductivity. Indeed, if $\sigma \sim N_e$,

$$\sigma \approx \sigma_0 \exp[-I\beta/2 + (Nk_2/2)\exp(D_2^+\beta)]\,, \tag{4.17}$$

where σ_0 is some constant. The position of the minimum electrical conductivity on the isobar corresponds approximately to the value of

$$T_{\min} \approx D_2^+ |\ln(Nk_2D_2^+/I)|^{-1}\,.$$

If $p = 20$ bar, $T_{\min} = 1,500$ K for cesium, which is in good agreement with experiment. In accordance with Eq. (4.10), the emergence of the minimum electrical conductivity is due to the competition between the dissociation of cluster ions, which causes a reduction in N_e, and the thermal ionization of atoms. This leads to a growth in N_e.

4.2.4. *Thermodynamics of Multi-particle Clusters*

As the non-ideality becomes stronger, the number of particles in the cluster increases. Therefore, we shall approach the formation of ion clusters from the standpoint of macroscopic thermodynamics. When an ion is placed in a gas with a mean atomic concentration N and temperature T, and if the density of particles attracted by the ion reaches a value that corresponds to the loss of thermodynamic stability, the vapor-liquid phase transition will occur in the neighborhood of the ion in the $r < R$ region. Let us now estimate the value of R. In a steady-state, the gradient of vapor pressure in the neighborhood of the ion must be balanced by the force of attraction of atoms by the ion located at the origin of co-ordinates

$$(M/\rho)\mathrm{grad}\, p = -dV/dr\,, \tag{4.18}$$

where $V(r)$ is the ion-atom interaction energy M is the atomic mass. Given the equation of state for vapor, one can integrate Eq. (4.18) to find the density distribution $N(r)$ in the neighborhood of the ion

$$M\int_p^{p(r)} \rho^{-1}(p,T)dp = -V(r)\,. \tag{4.19}$$

In an ideal gas, Eq. (4.19) gives an obvious result of $N(r) = N\exp[-\beta V(r)]$. At great distances from the ion, the potential is a polarization, $V(r) = -\alpha e^2/2r^4$. The atomic density $N(r)$ increases as the ion is approached.

Near the saturation line and spinodal, the equation of state differs from that for ideal gas. Primarily, the interatomic attraction is appreciable. We shall therefore assume that $p = NT - aN^2$, where a is the parameter characterizing the attraction in van der Waals' equation. In this case, Eq. (4.19) gives, after integration,

$$N(r) = N\exp\{-\beta V(r) + 2aN\beta[N(r)/N - 1]\}\,. \qquad (4.20)$$

With $N = 10^{20}$ cm^{-3} and $T = 1,400$ K, a distance from the ion which is equal to $R = 17a_0$, the density $N(R)$ is obtained. It exceeds the mean density by a factor of two. This value of density is close to the loss of thermodynamic stability. Whilst doing so, the number of particles in a drop of calculated radius is close to 20.

This gives rise to the question of whether one can rightly regard the formation of several dozens of atoms as macroscopic. One can assume that as the number of atoms in a complex increases, the valence electrons will be located on molecular orbitals covering the entire complex. They will also form a band whose width increases as another atom is joined. Clearly, it is when the distance between the levels is $\delta\varepsilon \to T$ that one can talk of a drop with the macroscopic properties being close to metallic. The distance between levels can be evaluated by dividing the atomic ionization potential I by the number of states, $\delta\varepsilon \approx I(2Rk_{\rm F})^{-1}$, where $k_{\rm F}$ is the Fermi wave vector and R is the radius of the complex. As usual, the wave vector $k_{\rm F} \approx r_s^{-1}$, where r_s is the interatomic distance. Upon substituting the values $R = 17a_0$ and $r_s = 6a_0$ in $\delta\varepsilon$, we see that $\delta\varepsilon/T$ is markedly less than unity. Therefore, one can assume that the properties of an ion cluster with a number of particles exceeding 10 will be close to metallic.

The foregoing assessment of the conditions of emergence of ion clusters is independent of the sign of the ion charge. If equal numbers of positive and negative ion clusters were formed, no anomalous growth of free electron concentration would be observed. In fact, the "electron-atom" interaction is weaker than the "ion-atom" interaction (although their polarization asymptotics are identical). This is reflected by the fact that the bond energy of positive ion D_2^+ is appreciably higher than that of negative ion. For cesium, $D_2^+ \approx 0.7$ eV and $E = 0.47$ eV, and for mercury the bond energy of ion is $D_2^+ \approx 1$ eV while the mercury atom does not bind the electron at all. Therefore, it turns out that there are considerably more positive ion clusters in a plasma than the negative ones. Consequently, a large number of free electrons is present in the plasma. The electronic transport coefficients are proportional to the free electron concentration and, therefore, are very sensitive to the concentration ratio between the positive and negative clusters.

In order to understand how the emergence of ion clusters shifts the ionization equilibrium, we now consider a simple model to follow Khrapak and Iakubov [4]. This model permits the ion-atom interactions and ignores the interatomic and electron-atom interactions. If, assuming that F_0 is the free energy of a mixture of ideal gases of electrons, atoms and ions, the free energy of the system F (with due regard for interaction) is lower than F_0 by the quantity TN^+m, where m is the number of ions in a cluster. This quantity is close to the excess number of atoms around the ion

$$m = N \int d\mathbf{r}[\exp(-\beta V) - 1]\,.$$

Therefore,

$$F = F_0 - TN^+N \int d\mathbf{r}[\exp(-\beta V) - 1]\,.$$

By varying F in a regular manner with respect to the concentrations of particles entering the ionization reaction, and allowing for quasi-neutrality, we obtain the equation of ionization equilibrium:

$$N_e^2 = NK_1 \exp\left\{-I\beta + N \int d\mathbf{r}[\exp(-\beta V) - 1]\right\}\,. \tag{4.21}$$

Equation (4.21) includes the exponential density dependence of electron concentration, which describes the increase in N_e upon compression. Of great importance is the presence of a strong temperature dependence of the "decrease in ionization potential", $\Delta I = NT \int dr[\exp(-\beta V) - 1]$, in the exponent. This enables one to describe the non-monotonic dependence of σ upon temperature on the isobar. The potential $V(r)$ can be characterized by two parameters: the sphere of action of the forces between the ion and atom, v_0, and the potential well depth, g. The depth of potential $V(r)$ proves rather great. At low temperatures, when $I\beta \gg 1$ and $\exp(g\beta) \gg 1$ (such are the temperatures of $T \leq T_c$, where T_c is the critical temperature), the exponential relationships in Eq. (4.21) are very sharp. Therefore, the principal relationships are reproduced by the simple formula

$$\sigma = \sigma_0 \exp[-I\beta/2 + (N\mathrm{v}_0/2)\exp(g\beta)]\,, \tag{4.22}$$

where σ_0 is some constant. This relationship describes the isobars of electrical conductivity given in Fig. 4.1. At low pressures, the densities N are low (hence the second term in the exponent can be neglected) and the conductivity grows monotonically upon heating. At pressures of the order of $p \sim p_c$, the non-ideality leads to the emergence of a minimum conductivity on the isobar. The value of temperature at the minimum is

$$T_{\min} = g/[\lg(I/gN\mathrm{v}_0)]\,. \tag{4.23}$$

As pressure grows, the position of the minimum shifts towards higher temperatures.

As in the case above, assume that $g = 0.7$ eV for cesium. The action radius of the "ion-atom" forces will thus be assumed to be equal to $10a_0$ (which is close to both the polarization radius $(ae^2\beta/2)^{1/4}$ and the internuclear distance of the Cs_2^+ ion). Hence Eq. (4.23) yields $T_{\min} = 1,700$ K for the 20-bar isobar and $T_{\min} = 2,500$ K for the 150-bar isobar. These values correspond rather well (given the roughness of approximation) to the experimental values.

4.2.5. *"Vapor-Liquid" Phase Transition and Metallization*

It follows, from a consideration of the diagrams (Fig. 1.3) demonstrating the region of existence of non-ideal plasma of alkali metals and mercury, that this region adjoins the high-temperature portion of the two-phase region.

The question of whether the "metal-dielectric" transition is a phase transition, and how the metallization is related to the "liquid-vapor" phase transition, has been discussed by Zel'dovich and Landau [232]. Special note was made of the coincidence of the thermodynamic critical point with the "metal-dielectric" transition or of the loss of metallicity in the liquid phase region. Apparently, the latter was revealed in mercury experiments [31, 32]. The temperature coefficient of electrical conductivity loses the value that is characteristic of metal at a density of $\rho \approx 9\ \mathrm{g \cdot cm^{-3}}$, which exceeds considerably the critical density (Fig. 1.4). In cesium, the metal-dielectric transition is apparently accomplished in the critical density region. This is indicated by measurements of electrical conductivity [38, 223] and data on magnetic susceptibility [233].

Figure 4.3 demonstrates the electrical conductivity of cesium and mercury at temperatures slightly above the critical point. There is a qualitative difference between the values of electrical conductivity for mercury and cesium. At $\rho \approx \rho_c$, the electrical conductivity of mercury ($\sigma = 10^{-1}\ \mathrm{Ohm^{-1} \cdot cm^{-1}}$) is three orders of magnitude less than that of cesium ($\sigma = 200\ \mathrm{Ohm^{-1} \cdot cm^{-1}}$).

The value of the latter corresponds to the level of minimum metallic conduction, which is in accordance with Mott's theory [234], while that of the former is too low. This (along with the data on the temperature coefficient of conductivity, thermal e.m.f. and so forth) enabled Kikoin and Senchenkov [31] and Hensel [57] to make the following conclusion. In mercury, the phase transition is similar to the transition in non-metallic media in the neighborhood of the critical point, that is, liquid mercury at high density exists in the metallic state and at lower density in the low-conducting state. For cesium (as well as for potassium and rubidium), the "liquid-gas" and "metal–non-metal" transitions coincide.

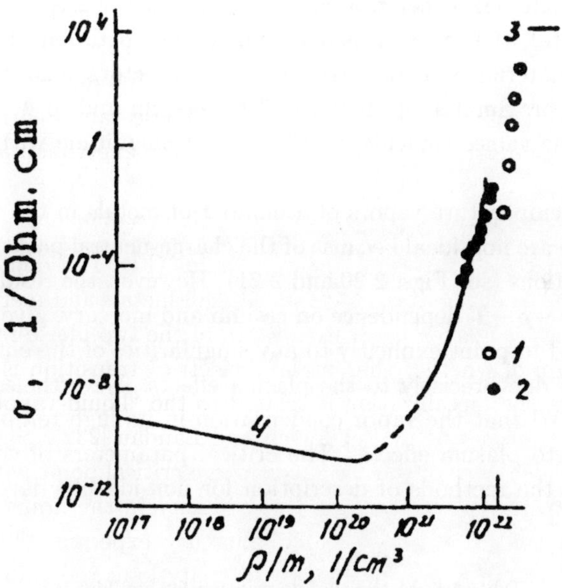

Fig. 4.4. Isotherm of electrical conductivity for mercury, $T = 1,973$ K [4]. Measurement results: *1*, [31]; *2*, [32]; *3*, level of electrical conductivity for metal; *4*, electrical conductivity for ideal plasma.

We now discuss the metallization of metallic vapor plasma. The supercritical isotherm of electrical conductivity for mercury, shown in Fig. 4.4, gives a clear indication of the behavior of the "dielectric-metal" transition. The transition region is stretched out with respect to density by an order of magnitude. This indicates that the first stage of metallization is due to the presence of long-range forces in the plasma, which show up even at moderate densities.

Such an interaction refers to the polarization interaction of ions and electrons with atoms. Many years ago, Herzfeld proposed the criterion of metallization, which was discussed by Frenkel [235]. The ionization potential (the bond energy of electron and atom) in a dense medium, $I_m = I/\varepsilon$, decreases according to the microscopic theory because of the increase in the dielectric permeability ε. In accordance with the Clausius-Mossotti equation,

$$\varepsilon = 1 + 4\pi N\alpha(1 - 4\pi\alpha N/3)^{-1}. \tag{4.24}$$

If the density increase approaches the value of $3/(4\pi\alpha)$, the dielectric permeability will increase indefinitely. Corresponding to the case of $N > 3/(4\pi\alpha)$ is spontaneous ionization, that is, the transition of electrons to free or, rather, collectivized state. In other words, if $N \geq 3/(4\pi\alpha)$ (α is the polarizability in gaseous state), the material is a metal. If the condition of $N \leq 3/(4\pi\alpha)$ is valid, the material is a dielectric or semiconductor. The value of $3/4\pi\alpha$ amounts to approximately $4 \cdot 10^{21}$ cm^{-3} for cesium and to $4 \cdot 10^{22}$ cm^{-3} for mercury. These values coincide by an order of magnitude with the observed values.

The high-temperature vapors of a number of metals in the vicinity of the saturation line are non-ideal because of the charge-neutral particle and charge-charge interactions (see Figs. 2.20 and 2.21). However, the results of measurements of the $p - \rho - T$ dependence on cesium and mercury, given in Figs. 2.20 and 2.21, failed to point explicitly to any singularities of the equation of state that would be due precisely to the plasma effects. Nevertheless, it was suggested [236, 237] that the vapor condensation in its high-temperature region is due largely to plasma effects. The critical parameters of cesium were estimated using the methods of description for non-ideality developed in [216] (Table 4.1).

Table 4.1. Methods of description for non-ideality.

	T_c, K	p_c, MPa	N_{ec}, 10^{21} cm^{-3}
Experiment	1,925	9.25	1.7·
Calculation [238]	2,200	22.0	1.0·
Calculation [239]	2,600	180.0	3.0·

Source: [216].

Apparently, these assumptions about the nature of phase transition in metals are valid. One can therefore hardly expect that, at higher temperatures, plasma interactions will once again bring about phase separation. An interesting possibility of a new "plasma" phase transition was discussed in [240] (see Chapter 5).

4.3. Lowering of Ionization Potential and Cluster Ions in a Weakly Non-ideal Plasma

In a moderately dense plasma, the interaction of electrons and ions with atoms and molecules brings about two effects: the lowering of the ionization potential, and the formation of molecular ions which are followed by heavier cluster ions. At moderate temperatures, these effects show up most clearly in a metal plasma. The data on molecular ions of alkali metals are given below. Ionization equilibrium in a multi-component mixture has been treated.

4.3.1. *Interaction Between Charged Particles and Neutrals*

The initial corrections to the free energy of the system, F, due to electron-atom and ion-atom interactions, can be calculated given $B_{ae}(T)$ and $B_{ai}(T)$, the virial coefficients of these interactions,

$$F = F_0 - TN_eNB_{ae} - TN^+NB_{ai}\,.$$

First, we consider the electron-atom interaction. The second virial coefficient, $B_{ae}(T)$, should be calculated from the Beth-Uhlenbeck formula (cf., [1]),

$$\begin{aligned} B &= (2\sqrt{\pi}\lambda_e)^3\left[\sum_n \exp(-\beta E_n) + \frac{1}{\pi}\int_0^\infty \sum_\ell (2\ell+1)\frac{d\delta_\ell}{dE}\exp(-\beta E)dE\right] \\ &= B^d + B^c\,, \end{aligned} \tag{4.25}$$

where $\lambda_e = \hbar(2mT)^{-1/2}$ is the electron thermal wavelength, δ_ℓ is the partial phase of electron-atom scattering, and E_n is the energy of electron-atom bound states. The sums are taken over all states of negative ion and scattering moments. A set of scattering phases enable one to determine the effect of interaction in continuous spectrum. Meanwhile, the bond energies provide the possibility of finding the second virial coefficient of interaction in discrete spectrums. Therefore, $B = B^c + B^d$. Let us first discuss the "electron-atom" interaction in the continuous spectrum.

Within the framework of the scattering theory in an effective radius approximation, the electron-atom interaction is characterized by only two parameters: the scattering length L and atomic polarizability α [241]. If k is the electron wave number, then

$$\delta_0(k) = -Lk - \pi\alpha k^2/(3a_0)\,;$$

$$\delta_\ell(k) = \pi\alpha k^2[(2\ell+1)(2\ell+3)(2\ell-1)a_0]^{-1}, \qquad \ell \geq 1\,.$$

By substituting $\delta_\ell(k)$ in Eq. (4.25) and performing the integration and summation over ℓ, we obtain the virial coefficient of interaction in the continuous spectrum, $B^c = -4\pi L\lambda_e^2$ (in a medium of molecules possessing a dipole moment, $B^c = 2q(T)\hbar(2mT)^{-1/2}$, where q is the transport cross-section of electron-molecule scattering [242]). This value should be substituted in the expression for plasma free energy F, $F = F_0 - TN_eNB^c$, where F_0 is the free energy of ideal plasma. From this expression, Saha's equation, which has the potential for atomic ionization in a medium, follows:

$$I_m = I + \Delta I\,, \qquad \Delta I = 2\pi\hbar^2 LN/m\,.$$

Should attractive forces prevail between atoms and electron, then $L < 0$ and $\Delta I < 0$, that is, the ionization potential in the medium I_m proves less than the ionization potential of isolated atom, I. The conditions of applicability of the effective radius theory are

$$|L| \ll \lambda_e\,, \qquad \alpha/a_0 \ll \lambda_e^2\,.$$

This is the low-temperature region.

Table 4.2. Lengths L of electron scattering by inert gas atoms, a_0.

He	Ne	Ar	Kr	Xe	Reference
1.16	0.45	−1.63	−3.8	−6.8	[243]
1.12	0.14	−1.40	−3.1	−5.7	[244]

The lengths of electron scattering by inert gas atoms are listed in Table 4.2 and by atoms of alkali metals in Table 4.3. Given the scattering length L of the order of a_0 and $T = 2{,}000$ K, $\Delta I \geq T$ only at $N \geq 7\cdot 10^{21}$ cm^{-3}. This may be a dense plasma of inert gas with a readily ionizable seed such as alkali. If $L = -10a_0$, $|\beta\Delta I| \geq 1$ even at $N \geq 10^{20}$ cm^{-3} and $T = 2{,}000$ K. Great (from the standpoint of their absolute value) scattering lengths are characteristic of the atoms of alkali metals. However, the effective radius theory proves inapplicable in their case because of a number of reasons. Therefore, one needs a set of experimental values of δ_ℓ to perform calculations. However, such a set

Table 4.3. Singlet, L_+, and triplet, L_-, lengths of electron scattering by atoms of alkali metals [245], a_0.

Quantity	Li	Na	K	Cs
L_+	3.65	4.23	0.45	−4.04
L_-	−5.66	−5.91	−15	−25.3

Source: [245].

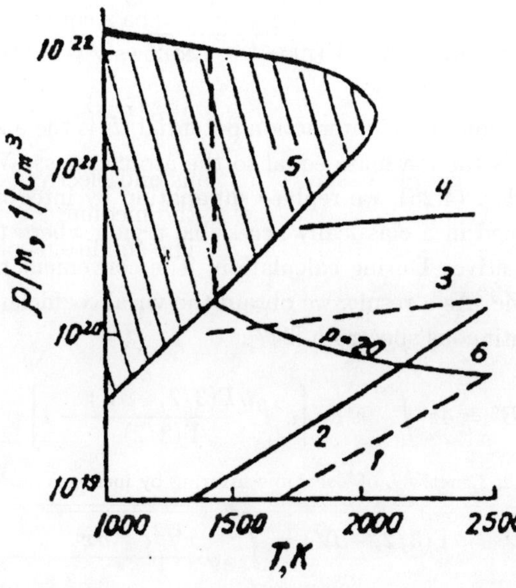

Fig. 4.5. Density-temperature diagram for cesium [4]: *1–4*, $NB^c = 1$ and $NB^d = 1$ for the electron-atom (*1*, *2*) and ion-atom (*3*, *4*) interactions; *5*, two-phase region; *6*, 20-bar isobar.

is not available. We shall take advantage of the calculated values of [245] and [246].

The $B^dN = 1$ and $B^cN = 1$ curves are plotted in Fig. 4.5, which represents the "density-temperature" diagram for cesium. These curves delimit the regions of strong and weak non-ideality, which is due to the "electron-atom" interaction in discrete and continuous spectra.

It follows from Table and Fig. 4.5 that at low temperatures, $B^d \downarrow B^c$ and, with an increase in temperature, the difference between them becomes smaller. It has been shown above that the contribution to free energy by the term

TN_eNB^c can be interpreted as a shift in the continuous spectrum boundary, that is, as a reduction of the ionization potential. In turn, the interaction contribution in discrete spectrum is described by introducing negative ions into the plasma composition. The effect of the reduction in the ionization potential manifests itself as a minor correction and becomes important at high densities.

The "ion-atom" interaction in continuous spectrum can be regarded as classical. Therefore, based on the general expression (4.25), we employ a quasi-classical approximation for scattering phases to obtain the virial coefficient B^c:

$$\delta_\ell(E) = \sqrt{M/\hbar^2}\int d\mathbf{r}\Big[\sqrt{E - V(r) - \hbar^2\ell^2/Mr^2} - \sqrt{E - \hbar^2\ell^2/Mr^2}\,, \quad (4.26)$$

where $V(r)$ is the "ion-atom" interaction potential, E is the energy of colliding particles, and M is the ion mass equal to the atomic mass. When Eq. (4.26) is substituted in Eq. (4.25), we replace summation by integration. The integration is performed in a classically accessible region where the radicands in Eq. (4.26) are positive. During calculation, it is convenient to first integrate with respect to one. As a result, we obtain the virial coefficient B^c due to the interaction in continuous spectrum [247]:

$$B^c = 4\pi\int_0^\infty r^2 dr\left[e^{-\beta V}\frac{\Gamma(3/2,-\beta V)}{\Gamma(3/2)} - 1\right], \quad (4.27)$$

where

$$\Gamma(3/2,-\beta V) = \int_{-\beta V}^\infty x^{1/2}e^{-x}dx$$

is the incomplete gamma function, $\Gamma(3/2) = \pi^{1/2}/2$.

The expression obtained resembles the classical virial coefficient

$$B = 4\pi\int_0^\infty r dr[\exp(-\beta V) - 1]\,,$$

but differs from it by the factor with gamma functions. This difference results from an isolation of the contribution by bound states, that is, states of diatomic molecular ions which make a contribution to B^d. The expression for B^d is reduced to the constant of dissociation equilibrium of the reaction $A_2^+ \leftrightarrow A^+ + A$. If D_2^+ is the dissociation energy of the ion A_2^+, then

$$B^d = (2\sqrt{\pi\lambda_a})^3(\Sigma_2^+/\Sigma\Sigma^+)\exp(D_2^+\beta)\,,$$

where Σ_2^+, Σ^+, Σ denote the internal statistical sums of the ions A_2^+ and A^+, the atom A and $\lambda_a = \hbar(MT)^{-1/2}$. The value of B^d can be calculated based on the structure of A_2^+.

The value of B^c was obtained by Likal'ter [214]. Since small distances in the integral (4.27) make no big contributions, one can use the polarization potential $V(r) = -\alpha e^2/2r^4$ for $V(r)$, where α is the atomic polarizability. The calculations then lead one to the expression

$$B^c = 4\pi C(\alpha e^2 \beta/2)^{3/4}\,, \qquad C \approx 1.61\,.$$

The $B^c N = 1$ and $B^d N = 1$ curves are plotted in Fig. 4.5. This was done earlier in the case of "electron-atom" interaction. Following an analysis of Fig. 4.5, the non-ideality of cesium vapors occurs earlier in an "ion-atom" interaction than in an "electron-atom" interaction, both being in the continuous and discrete spectra.

Thus, in a moderately dense plasma, the non-ideality due to "charge-neutral" interaction is reflected by the emergence of negative and positive molecular ions in the plasma composition, and by the reduction of the ionization potential ΔI:

$$\Delta I = 2\pi\hbar^2 LN/m - 4\pi CNT(\alpha e^2\beta/2)^{3/4}\,. \tag{4.28}$$

4.3.2. *Molecular and Cluster Ions*

We now turn to a consideration of the region where $NB^d > 1$. This means that the concentration of molecular ions A_2^+ exceeds that of atomic ions A^+. In other words, the interaction between ions and atoms becomes strong. Allowance should be made for the heavier ions A_3^+. With an increase in density or decrease in temperature, one can expect the concentrations of heavier ions to grow.

The information on bond energies and the structure of molecular and cluster ions has accumulated in recent years [227, 248]. Table 4.4 lists the parameters of alkali diatomic ions A_2^+ and tri-atomic ions A_3^+, dissociation energies D_2^+ and D_3^+ (for the $A_2^+ \to A^+ + A$ and $A_3^+ \to A_2^+ + A$ processes), equilibrium internuclear distances R_e and vibrational frequencies ω_e, as recommended by Gogoleva *et al.* [49]. Table 4.5 lists the same quantities for inert gas ions based on the data by a number of authors.

Table 4.6 correlates D_2 (dissociation energies of molecules A_2), D_2^+, D_3^+ and q (vaporization heats of metals on a single atom basis). The most important

Table 4.4. Parameters of alkali ions A_2^+, A_3^+.

Parameters	Li_2^+	Na_2^+	K_2^+	Rb_2^+	Cs_2^+
D_2^+, eV	1.28	1.02	0.84	0.77	0.66
R_e, a_0	5.91	6.69	7.9	8.5	9.1
ω_e, cm^{-1}	273	119	69	43	31
Parameters	Li_3^+	Na_3^+	K_3^+	Rb_3^+	Cs_3^+
D_3^+, eV	1.37	1.12	0.95	0.88	0.79
R_e, a_0	5.67	6.48	7.94	8.76	9.40
ω_e, cm^{-1}	342	149	86.7	53.7	39.4
	239	104	60.6	37.5	27.5

Source: [49].

Table 4.5. Parameters of inert gas ions A_2^+, A_3^+.

Parameters	He_2^+	Ar_2^+	Kr_2^+	Xe_2^+
D_2^+, eV	2.23	1.33	1.15	1.03
$R_e \cdot 10^{-8}$ cm	1.08	2.50	2.75	3.22
ω_e, cm^{-1}	1,627	300	160	123
Parameters	He_3^+	Ar_3^+	Kr_3^+	Xe_3^+
D_3^+, eV	0.17	0.20	0.27	0.27
R_e 10^{-8} cm	–	2.67	2.97	3.47

Table 4.6. Dissociation energies and vaporization heat q, eV.

Atom	D_2	D_2^+	D_3^+	q
Li	1.03	1.28	1.37	1.69
Na	0.73	1.02	1.12	1.13
K	0.514	0.84	0.95	0.87
Rb	0.49	0.77	0.88	0.78
Cs	0.45	0.66	0.79	0.74

fact is that D_2^+ is so much greater than D such that, for instance, at $T = 1,500$ K the N_2^+/N^+ ratio for sodium and cesium proves to be an order of magnitude greater than N_2/N. It is known that the concentrations of trimers and dimers in alkali metal vapors are low. In spite of this, trimer and dimer ions can play an important role in ionization equilibrium.

The validity of the relationship $D_2^+ > D$ also means that the ionization energy, $I(A_2)$, of molecule proves to be less than that of atom. Indeed, it follows from the condition of conservation of energy in the Born-Haber cycle, $\mathrm{A_2 \to A_2^+ + e \to A^+ + A + e \to A + A \to A_2}$, that

$$I(\mathrm{A_2}) = I(\mathrm{A}) - (D_2^+ - D_2)\,.$$

Another important circumstance which follows from Table 4.6 is the closeness of the values D_2^+, D_3^+ and q. The closeness of these values lead to the belief that the value of the separation of atom from heavy ion is almost constant. Therefore, the values of D_m^+ for heavier ions are known to a first approximation.

Very important information on heavy A_m^+ ions was yielded by experiments in which the A_m^+ ions emerged upon molecular beam passage through an ionization chamber (measurements in sodium, potassium, cesium and mercury).

Figure 4.6 presents a schematic of the facility by Forster *et al.* [249]. A source of metal vapors was provided by a heater whose temperature reached

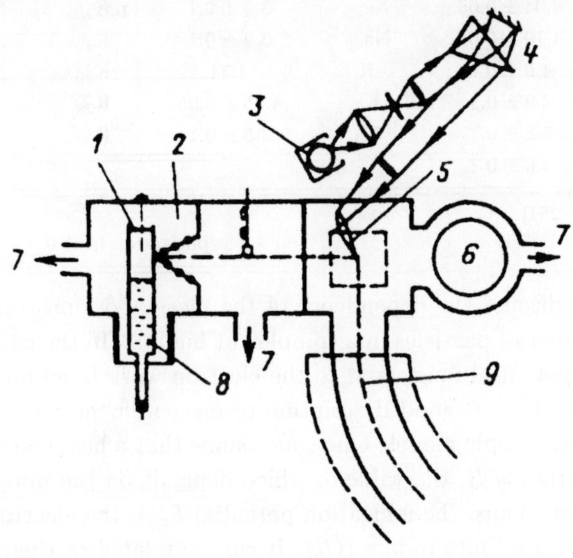

Fig. 4.6. Schematic of the facility by Forster *et al.* [249]: *1*, boiler; *2*, cone heater; *3*, xenon lamp; *4*, monochromator; *5*, ionization chamber; *6*, trap; *7*, evacuation; *8*, liquid metal; *9*, magnetic field region.

1,000°C. The molecular beam was developed as a result of molecular effusion through a hole with a small diameter (of the order of $5 \cdot 10^{-3}$ cm). The beam was collimated with the aid of diaphragms before entering the ionization chamber. A high-pressure xenon lamp served as a UV-radiation source. By varying the wavelength of radiation emanating from a monochromator, one could measure the photoionization thresholds. The emerging ions were drawn off to a mass-spectrometer. The results are listed in Table 4.7. Analogous data were also obtained for clusters of lead, iron, nickel and aluminum (see the list of references in [252]).

Table 4.7. The values of photoionization thresholds for sodium and potassium clusters, eV.

Cluster	I_m	Cluster	I_m	Cluster	I_m
Na	5.15	Na_{13}	3.6 ± 0.1	K_4	3.6 ± 0.1
Na_2	4.934 ± 0.01	Na_{14}	3.5 ± 0.1	K_5	3.4 ± 0.1
Na_3	3.97 ± 0.05	Na_{15}	3.7 ± 0.1	K_6	3.44 ± 0.10
Na_4	4.27 ± 0.05	Na_{19}	3.6 ± 0.1	K_7	3.40 ± 0.10
Na_5	4.5 ± 0.05	Na_{35}	3.4 ± 0.1	K_8	3.49 ± 0.10
Na_6	4.12 ± 0.05	Na_{45}	3.3 ± 0.1	K_9	3.40 ± 0.10
Na_7	4.04 ± 0.05	Na_{55}	3.3 ± 0.1	K_{10}	3.27 ± 0.10
Na_8	4.10 ± 0.05	Na_{65}	3.2 ± 0.1	K_{15}	3.21 ± 0.10
Na_9	4.0 ± 0.1	K	4.34	K_{20}	3.23 ± 0.10
Na_{10}	3.9 ± 0.1	K_2	4.05 ± 0.05	K_{28}	3.05 ± 0.10
Na_{11}	3.8 ± 0.1	K_3	3.3 ± 0.1	K_{34}	3.01 ± 0.10
Na_{12}	3.6 ± 0.1				

Source: [249–251].

Let us now discuss the dependence of the measured ionization potentials, I_m, on the number of particles in a complex at high m. In the case of $m \to \infty$, the ionization potential must tend to the electron work function for a macroscopic metal sample. A good description of measurement results is provided by a simple macroscopic model. One can assume that a heavy ion is a metallic droplet with a radius R, the value of which depends on the number of atoms in the droplet m. Thus, the ionization potential I_m is the electron work function for drop with a finite radius $I(R)$. It can be related to the electron work function for plane surface (in the $R \to \infty$), $I(\infty)$, by the following relationship:

$$I(R) = I(\infty) + e^2/2R. \tag{4.29}$$

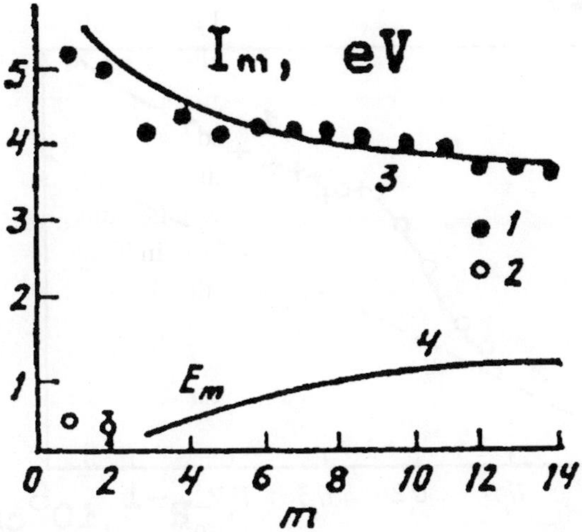

Fig. 4.7. I_m and E_m of sodium complexes Na_m, eV: *1*, experiment; *2*, Na^- and Na_2^-; *3* and *4*, theory (4.29), (4.30).

The experimental and calculated $I_m(m)$ relationships for sodium are plotted in Fig. 4.7. For this, use was made of the relationship between the drop radius and the number of particles in the drop $R = R_m m^{1/3}$, $R_m = R_{\text{inf}} \exp[-2k\gamma/(3R_m m^{1/3})]$, where R_{inf} is the atomic radius in a macroscopic metal sample, and k and γ denote the macroscopic compressibility and surface tension, respectively. The inclusion of surface tension somewhat reduces the value of R_m for small clusters, which corresponds to reality.

The results point to the good stability of positively charged cluster ions and provide some information on their structure, which apparently approaches the liquid-metal structure. At low m, considerable fluctuations are observed in the m-dependence of I_m. These are well-described by the results of fairly complicated quantum-mechanical calculations performed by a number of authors.

While the situation appears very simple in the case of alkali and a number of other metals, a different situation arises with mercury. As demonstrated by the measurements in [252] (Fig. 4.8), clusters of mercury are only metallized at $m \geq 70$. In this case, as in a number of other situations, this is due to the fact that mercury is among non-common metals (see Section 4.5).

Considerably less experimental information is available on negative alkali complexes A_m^-. However, in a number of experiments in gas discharge plasma, quite heavy complexes were reliably identified, such as complexes of K_m^- up to

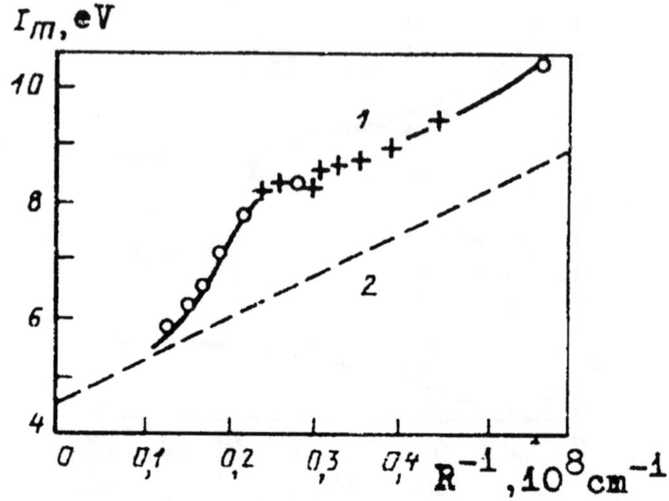

Fig. 4.8. The ionization energy of mercury clusters, I_m, as a function of the cluster radius R [252]: *1*, plotted by experimental points; *2*, calculation by formula (4.29).

$m = 7$ [253]. It is known that the binding energies of electron with an atom and a diatomic molecule (E_1 and E_2) are substantially smaller than that of a positive atom with the same particles. For example, the affinity energies for cesium are $E_1 = 0.47$ eV and $E_2 \approx 0.3$ eV [64] and, as ions Hg^- and Hg_2^- are unstable. Table 4.8 lists the parameters of ions A^-, A_2^- and A_3^-, as recommended by Gogoleva *et al.* [49]. It presents the energies of electron affinity for atom and molecule (E_1 and E_2); the D (A_2^-) dissociation energies of $A_2^- \leftrightarrow A^- + A$; D_3^+ (A_3^-); the dissociation energies of $A_3^- \leftrightarrow A_2^- + A$; and the internuclear distances and vibrational frequencies.

Table 4.8. Parameters of negative alkali ions.

Parameters	Li	Na	K	Bb	Cs
E_1, eV	0.620	0.548	0.501	0.486	0.470
E_2, eV	0.55	0.49	0.45	0.43	0.42
$D_e(A_2^-)$, eV	0.96	0.66	0.47	0.43	0.40
$R_e(A_2^-)$, a_0	6.1	7.2	8.9	9.8	10.5
$\nu(A_2^-)$, cm^{-1}	213	92	63	38	28
$D_e(A_3^-)$, eV	0.90	0.59	0.49	0.40	0.38
$R_e(A_3^-)$, a_0	11.2	13.2	16.3	18.2	19.3

Source: [49].

For energies of electron affinity for a heavy complex, one can write an expression which is analogous to Eq. (4.29):

$$E(R) = E(\infty) - e^2/2R\,. \tag{4.30}$$

Obviously, $E(\infty) = I(\infty) = A_e$, where A_e is the electron work function for metal. This relationship is plotted in Fig. 4.7.

Therefore, an electron is bound by neutral particles in a much weaker way than a positive ion. This suggests that in alkali metal vapors, free electrons play an important role in the balance of charged particles by providing a high electrical conductivity of plasma.

4.3.3. *Ionization Equilibrium in Alkali Metal Plasma*

It is well-known that metal vapors contain few diatomic molecules and a negligible fraction of heavier neutral complexes. An inverse relationship may show up in the composition of ionized components. If one considers the relationship between positive and negative ions, the A_m^+ concentration may exceed the A_m^- concentration considerably. Such asymmetry is due to the difference in bond energies.

In order to determine the composition of plasma formed by electrons (e), atoms (A), diatomic molecules (A_2), di- and tri-atomic positively charged ions (A_2^+, A_3^+), and diatomic negatively charged atoms (A_2^-), consider the following plasmochemical processes:

$$\begin{aligned}
e + \mathrm{A}^+ &\Leftrightarrow \mathrm{A} & N_e N^+/N &= K_1 \\
e + \mathrm{A} &\Leftrightarrow \mathrm{A}^- & N_e N/N^- &= K_2\,, \\
\mathrm{A} + \mathrm{A} &\Leftrightarrow \mathrm{A}_2 & N^2/N_2 &= K_3\,, \\
\mathrm{A} + \mathrm{A}^+ &\Leftrightarrow \mathrm{A}_2^+ & N N^+/N_2^+ &= K_4\,, \\
\mathrm{A}_2 + \mathrm{A}^+ &\Leftrightarrow \mathrm{A}_3^+ & N_2 N^+/N_3^+ &= K_5\,, \\
\mathrm{A}_2 + e &\Leftrightarrow \mathrm{A}_2^- & N_2 N_e/N_2^- &= K_6\,,
\end{aligned} \tag{4.31}$$

where K_i are the chemical equilibrium constants. Equation (4.31), along with the conditions of conservation of charge and full number of particles, enables one to determine the concentration of any component. By simple manipulations, we derive

$$N_e^2 = K_1 N \frac{1 + N/K_4 + N^2/(K_3 K_5)}{1 + N/K_2 + N^2(K_3 K_6)} \,. \tag{4.32}$$

It is easily seen that the great fraction in Eq. (4.32) is reflective of the effect of molecular ions on N_e. Individual components in the fraction numerator (denominator) correspond to the contribution by some or other positive (negative) ions. Obviously, the inclusion of cluster ion A_4^+ would lead to the emergence of a subsequent term of the series in the numerator, while the inclusion of A_3^- ion would lead to the emergence of a subsequent term of the series in the denominator, and so forth.

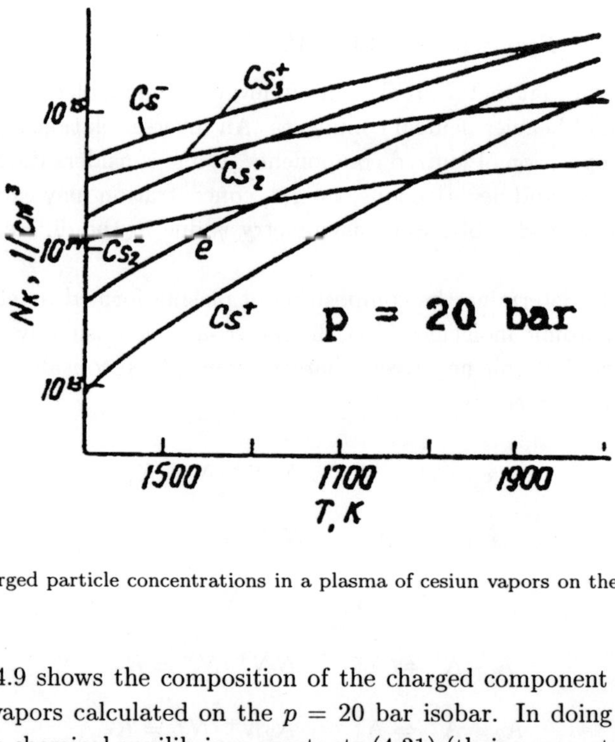

Fig. 4.9. Charged particle concentrations in a plasma of cesiun vapors on the 20-bar isobar [254].

Figure 4.9 shows the composition of the charged component of a plasma of cesium vapors calculated on the $p = 20$ bar isobar. In doing so, use was made of the chemical equilibrium constants (4.31) (their parameters are listed in Table 4.9). The temperature dependencies of the equilibrium constants K_i take the form of

$$K_i = c_i T^{\eta_i} \exp(-E_i/T) \,.$$

The peculiarities referred to in the beginning of this section show up in Fig. 4.9. As the temperature decreases, heavy ions play an increasingly

Table 4.9. Constants required to perform calculations according to Eq. (4.32).

		Na		Cs	
i	η_i	c_i, 10^{21} cm$^{-3}\cdot$ eV$^{-\eta_i}$	E_i, eV	c_i, 10^{21} cm$^{-}\cdot$ eV$^{-\eta_i}$	E_i, eV
1	3/2	3	5.14	3	3.89
2	3/2	12	0.54	12	0.47
3	−1/2	29	0.73	5	0.45
4	−1/2	4.22	1.14	1.45	0.85
5	1	2.5	1.63	0.95	1.2
6	3/2	4.9	0.25	0.45	0.30

Source: [254].

important role. The A^+ ion, which prevails at $T \geq 2,200$ K among the positive ions, is replaced by the A_2^+ ion. It, in turn, gives way to the A_3^+ ion. The A^- ion prevails at $T \leq 2,000$ K among negatively charged components. However, the A_2^- ion remains inconspicuous and there is no reason to expect the emergence, on further cooling, of appreciable concentrations of A_3^- ions.

4.4. Droplet Model of Non-ideal Plasma of Metal Vapors. Anomalously High Electrical Conductivity

4.4.1. *The Proposed Droplet Model of Non-ideal Plasma*

The vapors of alkali metals on the saturation line and in the vicinity of the latter represent a strongly non-ideal plasma. At a distance that is not too close to the critical point, this plasma may be regarded as weakly ionized. In this case, the non-ideality is mainly caused by the strong interaction between charged and neutral particles. This interaction promotes the formation of heavy charged complexes, that is, clusters, in plasma. The plasma of metal vapors may be treated as a mixture of gases of electrons and clusters of different charges and sizes. The inclusion of the effect of clusters on the degree of plasma ionization, made within the framework of different models in [229, 254–257], enables one to understand the reason for the emergence of an anomalously high electrical conductivity of cesium and mercury vapors in the vicinity of the saturation line. The Saha and Lorentz formulas give the value of the coefficient of electrical conductivity $\sigma = 3\cdot 10^{-5}\cdot$ (Ohm $\cdot$ cm)$^{-1}$ for $p = 2\cdot$ MPa and $T = 1,400$ K. The value of $\sigma = 10\cdot$(Ohm cm)$^{-1}$ obtained in experiment is five orders of magnitude higher than these estimates, which are common with non-ideal plasma.

One commonality of the studies [229, 254–257] is the assumption of a sharp increase in the concentration of charged polyatomic clusters when the saturation line is approached. In doing so, the concentration of positively charged clusters exceeds considerably that of negatively charged ones. The electroneutrality of plasma is made possible by a corresponding increase in the electron number density. This leads to a rise in the coefficient of electrical conductivity of plasma. Differences in theoretical models occur when one tries to clarify the properties of clusters.

It was proposed in [255] that clusters be treated as small droplets. The advantages of this model include the possibility of using the characteristics of metals, such as surface tension and electron work function, to determine the properties of clusters.

The phenomena of variation in the aggregate state of matter, which is caused by electrostriction, in the neighborhood of charged particles are well-known in physics [235, 258, 259]. Charged droplets may form in vapor and icicles, in liquid. On the one hand, the conditions of charge stabilization of a droplet in a plasma of metal vapors are less preferable. This is because the temperature is high. On the other hand, the very high values of polarizability enhance electrostriction.

The elementary droplet model of [255] enables one to set forth the basic ideas and make rough estimates. We will assume that a plasma consists of atoms, positively charged droplets and electrons. The concentration of these particles is N, N^+ and N_e, and the total number of all particles is $\tilde{N} = N + N^+ + N_e$. The thermodynamic potential of the system has the following form:

$$\Phi = N \cdot \varphi_{\mathrm{G}} + N^+(g \cdot \varphi_{\mathrm{L}} + 4\pi\gamma R^2 + W + e^2/(2R)) + \sum_k N_k \cdot \ln(N_k/\tilde{N})\,. \quad (4.33)$$

Here, φ_{L} and φ_{G} are the thermodynamic potentials of liquid and vapor per atom, so that the quantity $gT\ln(p_s/p) = g(\varphi_{\mathrm{L}} - \varphi_{\mathrm{G}})$ is the work of formation of a neutral droplet of radius R; $g = (4\pi/3) \cdot R^3 \cdot N_{\mathrm{L}}$ is the number of particles in the droplet; N_{L} is the concentration of particles in the liquid; and p_s is the saturation pressure. Moreover, $4\pi\gamma R^2$ is the surface energy (γ is the surface tension); $W + e^2/(2R)$ is the electron work function from a droplet of radius R. In the last entropy term in (4.33), the summation is made over all sorts of particles. Such a simple approach, when we are not interested in droplets with radii other than the most likely radius R, and do not describe the internal degrees of freedom of the droplets, conforms to Frenkel's theory of pre-transition phenomena [235].

We minimize, on the radius, the work required to produce a single charged droplet, $[gT \cdot \ln(p_s/p) + 4\pi\varphi \cdot R^2 + W + e^2/(2R)]$, in order to derive the Kelvin equation for determining the radius of the most likely droplet,

$$TN_L \cdot \ln(p/p_s) = 2\gamma/R - e^2/(8\pi \cdot R^4).$$

In saturated vapor, R is equal to the "electrocapillary" radius

$$R = [e^2/(16\pi\gamma)]^{1/3}. \tag{4.34}$$

This enables an estimate of the parameters of ion droplets along the curve of saturated cesium vapors (Table 4.10). Of course, the emerging droplets are too small for the macroscopic description to be sufficiently correct (corrections for curvature to γ improve matters only slightly). However, according to the nucleation theory, it is usually held that droplets containing more than 10 particles are macroscopic.

Table 4.10. Parameters of ion droplets in saturated cesium vapors.

Parameter	T, K			
	1290	1430	1600	1690
p_s, 0.1 MPa	10.8	20.6	40.3	51.3
N_s, 10^{22} cm^{-3}	0.95	1.6	3.0	3.9
γ, g/s^2	23.9	17.5	11.7	8.9
R, a_0	16.4	17.6	19.2	20.5
g	1.5	18	21	23
N_e, 10^{16} cm^{-3}	0.13	1.5	11	36

We will further treat only saturated vapor $p = p_s$. We will derive the equation of ionization equilibrium, $gA \Leftrightarrow A_g^- + e$, and vary (4.33) to include the stoichiometry $g\delta N = -\delta N^+ = -\delta N_e$. In view of the fact that the plasma is weakly ionized ($N_e = N^+ \ll N$), we finally derive

$$N_e = N \cdot \exp[-(W + 4\pi\gamma R^2 + e^2/R)/2T] = N \cdot \exp[-(W + 3e^2/4R)/(2T)]. \tag{4.35}$$

For a number of reasons, the pre-exponential factor in (4.35) has been very poorly determined. Therefore, we will discuss the exponent which produces the main effect. The exponent contains W, the work function of electron from metal, which is known to decrease as the metal temperature rises (as the density decreases). This correlation was investigated in [260]. We extrapolate the derived values in view of the fact that the work function must reach zero at the critical temperature T_c. It turns out that one can write

$$W(T) = W_0(T_c - T)/T_c\,, \tag{4.36}$$

where $W_0 = 1.8$ eV is the tabular (for cesium) value of the work function.

The results of the calculations for N_e are listed in Table 4.11. The interaction effects led to an increase in N_e by orders of magnitude as compared to the ideal-gas approximation. However, such a simple analysis is too crude. The drop model was improved considerably in [261] and [262]. The presentation that follows is based mainly on the results of [261].

Table 4.11. The "surface energy" of atom α, vaporization heat q, and ratio α/q at triple point.

Parameter	Li	Na	K	Pb	Cs	Fe	Cu
α, eV	0.98	0.68	0.60	0.54	0.52	2.9	2.2
q, eV	1.6	1.1	0.93	0.85	0.82	3.5	4.4
α/q	0.60	0.63	0.64	0.65	0.63	0.69	0.64

Source: [261].

4.4.2. *Ionization Equilibrium*

We will treat the ionization equilibrium in a multi-component mixture of ideal gases of electrons and complexes. It consists of g atoms and a charge Z. For such a mixture, the thermodynamic potential Φ is

$$\begin{aligned}\Phi &= \sum_{g,Z} \Phi_g^Z + \Phi_e = \sum_{g,Z} N_g^Z(T\ln p_g^Z + \chi_g^Z) + N_e(T\ln p_e + \chi_e)\\ &= \sum_{g,Z} N_g^Z(T\ln p + \chi_g^Z + T\ln(N_g^Z/F)) + N_e(T\ln p + \chi_e + T\ln(N_e/F))\,,\end{aligned} \tag{4.37}$$

where

$$F = \sum_{g,Z} N_g^Z + N_e = pV/T\,;$$

$$\chi = f - T\ln T + T\ln V\,, \qquad f = -T\ln\Sigma\,. \tag{4.38}$$

In (4.37) and (4.38), N_g^Z, N_e, p_g^Z and p_e denote the number of particles in a volume V and the partial pressure of the respective components; p is the total pressure; and Z is the partition function over internal and translational degrees of freedom of the respective components.

An electron possesses only translational degrees of freedom,

$$\sum_e = \exp(-f_e/T) = 2V\lambda_e^{-3}\exp(-W/T) \equiv VN_{es}\,, \tag{4.39}$$

where W is the electron work function from a flat surface, λ_e is the thermal wavelength of electron, and N_{es} is the equilibrium number density of electrons in the vicinity of the metal surface. We will derive the correlation between the concentrations ($n = N/V$) of charged and neutral components after treating the equilibrium $A_g^Z \Leftrightarrow A_g^{Z+1} + e$, which yields

$$n_g^Z = n_g^0(n_{es}/n_e)^Z \exp\{-[f_g^Z - f_g^0]/T\,. \tag{4.40}$$

We will now treat another reaction, that of condensation of atoms leading to a rise of $A_{g-1}^Z + A_1^0 \Leftrightarrow A_g^Z$. This gives

$$N_g^0 = \exp[-(f_g^0 - g\mu_{\rm G})/T]\,. \tag{4.41}$$

The chemical potential of gas $\mu_{\rm G}$ is defined by the expression

$$\mu_{\rm G} \equiv \mu_1^0 = T\ln(n_1^0\lambda_1^3) = f_1^0 + T\ln n_1^0\,, \tag{4.42}$$

where λ_1 is the thermal wavelength of atom. One usually tries to express f_g^Z or f_g^0 in terms of the macroscopic characteristics of the disperse phase. Two assumptions are made. First, the presence of charge has no effect on the internal properties of particle. Second, the dependence of f_g^Z on Z is defined by the electrostatic energy of self-action of excess charge, which is only true of large particles, the surface fraction of whose atoms is much less than the volume fraction. Hence

$$f_g^Z = f_g^0 + Z^2/(2\varepsilon R) = f_g^0 + \beta Z^2/2g^{1/3}\,; \tag{4.43}$$

$$\beta = (4\pi n_L/3\varepsilon)^{1/3}\,, \tag{4.44}$$

where $n_{\rm L}$ is the number density of atoms in a liquid and ε is the dielectric permeability of vapor. The substitution of this expression into (4.40) gives

$$n_g^Z = n_g^0(n_{es}/n_e)^Z \exp(-\beta Z^2/2g^{1/3}T)\,. \tag{4.45}$$

The maximum concentration is observed in positively charged complexes with

$$Z^* = R \cdot T \cdot (n_{es}/n_e)\,. \tag{4.46}$$

The charge Z^*, under conditions in the vicinity of the saturation line for alkali metals, is close to unity.

We will identify, in f_g^0, the contributions made by the translational $f_{g,\mathrm{tr}}$, degrees of freedom of the complex as a whole,

$$f_{g,\mathrm{tr}} = T \ln(\lambda_g^3/V)\,, \qquad f_g^0 = f_{g,\mathrm{tr}} + E_g^0\,,$$

where $\lambda_g = g^{1/2}\lambda_1$ is the thermal wavelength of a g-atom complex. The quantity E_g^0 includes the kinetic energy and the energy of interaction of atoms in the complex. It is evident that, in the limit of high values of g, E_g^0 is proportional to g. The dependence of E_g^0 on g may be sought for in the form of expansion in inverse powers of the complex radius,

$$E_g^0 = \mu_L \cdot g + \alpha \cdot g^{2/3} + \xi \cdot g^{1/3}\,, \tag{4.47}$$

where μ_L is the chemical potential of atoms in a liquid. The second term in Eq. (4.47) corresponds to the free surface energy of the complex, and the third term may be treated as the inclusion of the curvature dependence of the surface tension.

In a large number of studies, E_g^0 is calculated from the "first principles". Figure 4.10 gives the values of E_g^0 borrowed from papers cited in [261]. They relate to the Ag complexes formed by atoms of different elements (Li, Na, K, Fe and Cu), and are marked in Fig. 4.10 using different symbols. One can see that, with good accuracy, the quantity $E_g^0/(g, \mu_\mathrm{L})$ is independent of atoms in the complex. This fact is in keeping with the conclusions of Langmuir and Frenkel [235] on the correlation between the sublimation energy and specific surface energy. This justifies the choice of expression (4.47) for E_g^0. Indeed, according to Frenkel, at fairly low temperature (when $\mu_\mathrm{L} \approx -q$, where q is the sublimation energy),

$$E_g^0/(g \cdot \mu_\mathrm{L}) = 1 - \alpha/(q \cdot g^{1/3})\,, \qquad \alpha/q \approx (\eta - \eta')/\eta$$

where $\alpha = 4\pi \cdot \gamma \cdot (3/4\pi \cdot n_L)^{2/3} = 4\pi \cdot \eta\gamma \cdot r_s^2$ is the "surface energy" of a single atom and η and η' are the numbers of the nearest neighbors of an atom deep inside the liquid and on the surface, respectively. It is clear that $\alpha/q \approx 1/2 < 1$. It follows from Table 4.11 that the ratio α/q, taken at triple points of metals, is close to 0.64.

Let us now revisit to Eq. (4.41). The first two terms of the expansion of E_g^0 in the power $g^{1/3}$ are retained. In view of the fact that $\mu_\mathrm{L} - \mu_\mathrm{G} =$

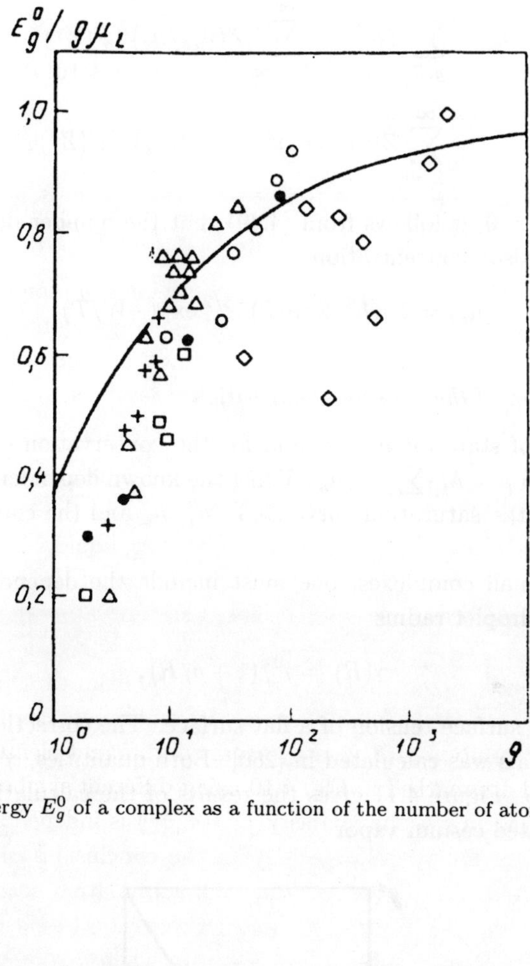

Fig. 4.10. The energy E_g^0 of a complex as a function of the number of atoms in this complex [261].

$T \cdot \ln(p/p_s)$, the "classical" expression for the concentration of uncharged droplets is derived:

$$N_g^0 = AN_1^0 \exp[-(-gT\ln(p/p_s) + \alpha g^{2/3})/T]\,. \tag{4.48}$$

The pre-exponential factor A, which is usually assumed to be equal to unity, will be determined with the aid of the experimental data on the equation of state for saturated vapor.

The condition of plasma neutrality and expression (4.45) for the concentration of charged droplets yield the equations for determining the electron number density:

$$n_e = \sum_{g,Z} Z n_g^Z = \sum_{Z=-\infty}^{\infty} Z(n_{es}/n_e)^Z F(Z^2)$$

$$= \sum_{Z=0}^{\infty} Z[(n_{es}/n_e)^Z - (n_e/n_{es})^Z] F(Z^2) . \qquad (4.49)$$

Because $F(Z^2) > 0$, it follows from (4.49) that the number density n_e is less than the Richardson concentration:

$$n_{es} = 2 \cdot (h^2/2\pi m T)^{-3/2} \exp(-W/T) . \qquad (4.50)$$

4.4.3. *Calculation of the Plasma Composition*

The equations of state for matter and for the conservation of mass are $p = T \cdot \sum_{g,Z} n_g^Z$, and $\rho = M \cdot \sum_{g,Z} g n_g^Z$. Using the known dependencies of pressure and density on the saturation curve [263], n_1^0, n_e and the constant A can be determined.

For fairly small complexes, one must include the dependence of surface tension on the droplet radius

$$\gamma(R) = \gamma^0/(1 + \delta/R) ,$$

where γ^0 is the surface tension of a flat surface. The correction for curvature δ for alkali metals was calculated in [260]. Both quantities, γ^0 and δ, depend on temperature. Figure 4.11 gives the results of the calculation for quantity $g^*(T)$ in saturated cesium vapor.

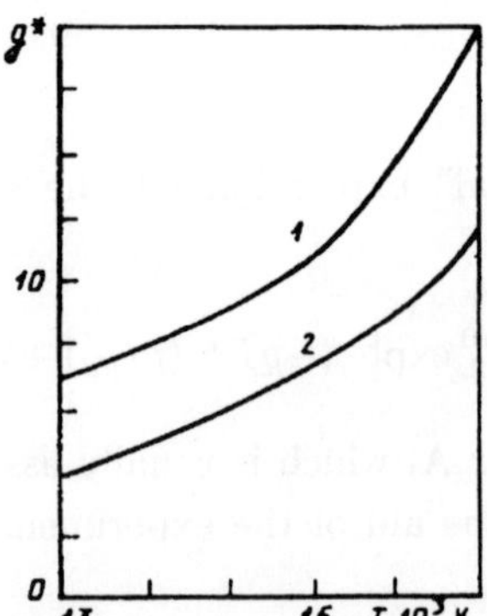

Fig. 4.11. The temperature dependence of g^* in saturated cesium vapor [261]: *1*, in view of δ; *2*, with $\delta = 0$.

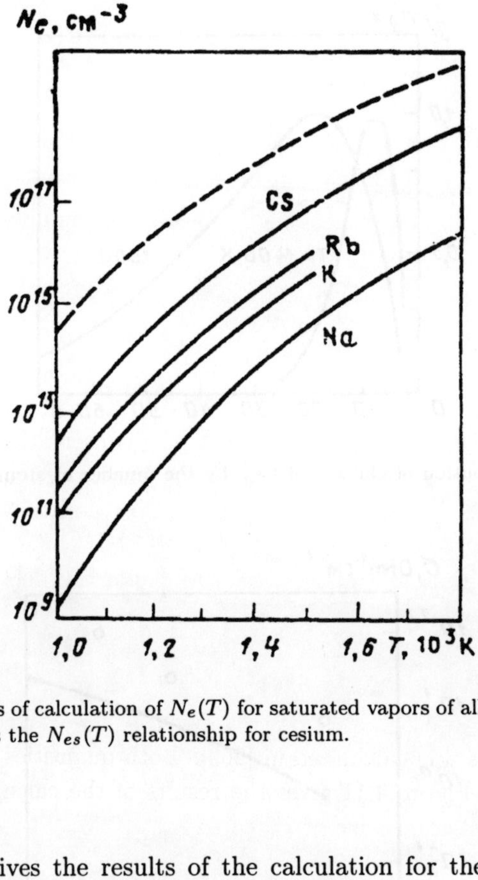

Fig. 4.12. The results of calculation of $N_e(T)$ for saturated vapors of alkali metals [261]. The broken line indicates the $N_{es}(T)$ relationship for cesium.

Figure 4.12 gives the results of the calculation for the electron number density in the plasma of Na, K, Rb and Cs vapors.

The validity of the droplet model is mainly defined by the values of $g^*(T)$ and by the half-width of distribution $n_g^{+1}(T)$. The normalized distribution of charged complexes for cesium is shown in Fig. 4.13 by the number of atoms in them, g, for $T = 1,400$ K and 1,800 K. As the temperature rises, the distribution of droplets by the number of particles becomes wider and the positions of maxima shift towards greater values of g.

The magnitude of A proved to be temperature-dependent. For cesium at $T = 1400$ K, $A = 5.3 \cdot 10^{-2}$ and at $T = 1,800$ K, $A = 3.3 \cdot 10^{-3}$. Against the background of a very sharp variation of $n_e(T)$, such a dependence, as well as the magnitude of A proper, appears acceptable.

Figure 4.14 gives the results of the calculation and measurements for the coefficient of electrical conductivity of saturated cesium vapors. The

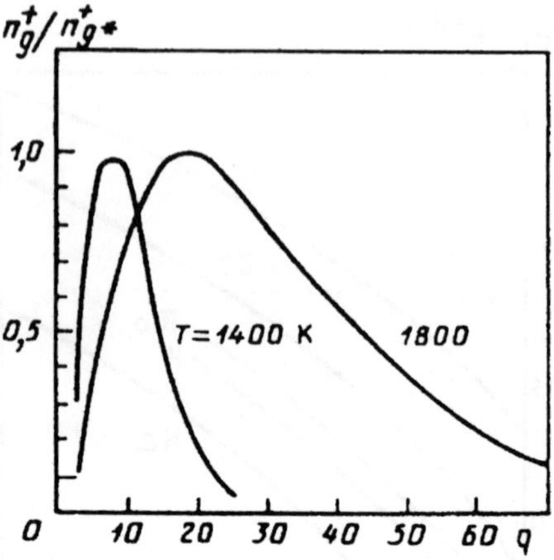

Fig. 4.13. The distribution of clusters of Cs_g^+ by the number of atoms bound in them, g [261].

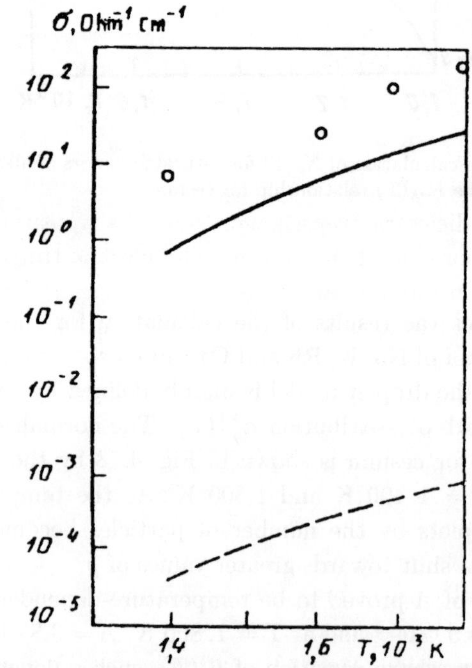

Fig. 4.14. The coefficient of electrical conductivity of saturated cesium vapors [261]: the solid curve indicates the droplet model; the broken curve indicates ideal plasma; and the points indicate the experimental data.

coefficient of electrical conductivity was calculated by the Lorentz formula $\sigma = e^2 n_e / \tilde{q} m \tilde{V} n_1^0$, where $\tilde{V} = \sqrt{8T/(\pi m)}$; and $\tilde{q}$ is the cross-section of electron-atom scattering. In highly polarizable media, the density effects may bring about a strong increase in mobility. This is due to the fact that, under conditions of high density, the long-range (polarization) components of electron-atom potentials overlap. The resultant potential field is smoothed out and the electron is scattered by the short-range components of potentials (see Chapter 6). Therefore, the cross-section $\tilde{q}$ in [255, 261] is of an order of magnitude less than the tabular value.

The qualitative pattern is analogous to the remaining alkali metals. Pogosov and Khrapak [261] do not describe the transition from anomalous to normal electrical conductivity. If one proceeds from the results of measurements in [264], this transition may occur at $T \approx 1,200$ K. Zhukhovitskii [262] has suggested yet another version of the droplet model, in which an attempt is made to produce a unified description of the equation of state for vapors, electrical conductivity and its transition from anomalous to normal.

4.5. Metallization of Plasma

4.5.1. *Mott's Transition*

A continuous decrease in the density of metals during their heating causes, in the long run, a sharp decrease in electrical conductivity. Reference is usually made to the metal-dielectric transition or to Mott's transition [234]. Mott and Hubbard have demonstrated that the metal-dielectric transition is caused by the electron-electron interaction, which leads to band splitting. As a result, the empty and filled bands were separated by an energy gap.

The initial Mott estimation of the parameters of dielectric-metal transition is very simple. The spectrum of bound states of shielded potential $V(r) = -(e^2/r) \cdot \exp(-r/r_0)$ is investigated, and it is demonstrated that the bound states disappear completely when r_0 decreases to a value close to

$$r_0^M = 0.84 a_0 \,. \tag{4.51}$$

Depending on the absence or presence of electron degeneracy, r_0 is the Debye-Hueckel or Thomas-Fermi screening radius.

Unlike phase transition, Mott's transition in metal occurs at high temperature. Therefore, it turns out smeared and has no critical point. Under the conditions of expansion of metal, the coefficient of electrical conductivity drops sharply but continuously. Mott has introduced the concept of the minimum coefficient of electrical conductivity of metal, which is estimated from the condition of equality of the electron path to its wavelength. It is customary to assume that this value is approximately 200 $\text{Ohm}^{-1}\ \text{cm}^{-1}$.

We will now discuss the transition from a different perspective, namely, the metallization of plasma as a result of compression. The increase in the coefficient of electrical conductivity, upon compression, is due to a decrease in the ionization potential. The rate of this decrease rises during compression. It is often suggested that non-thermal ionization or ionization by pressure occurs. These assumptions, however, remain valid as long as one can talk about individual atoms, the electrons bound in them, and about free electrons. At higher values of density, the shells of the ground atomic states overlap. The description of such a system calls for a different approach [265, 266].

4.5.2. *Quasi-Atomic Model*

Atoms of metal come so close together that the classically accessible regions of motion of valence electrons start overlapping (the radius of such a region is e^2/I). In doing so, the classical transition of electron from one atom to another proves possible. The quasi-atomic state is formed when the density reaches a value corresponding to the percolation threshold of

$$\xi_0 = (4\pi/3) \cdot (e^2/I)^3 N_i \approx 1/3\,.$$

It is then that the electron may cross the entire metal sample and makes a transition from one ion to another. Each quasi-atom takes up a spherical cell with a volume of N_i^{-1}. A valence electron spreads out over its own cell and, to some extent, over neighboring cells. Each ion is mainly shielded by its own electron and, partly, by extraneous electrons as well. This constitutes the essential difference from the liquid-metal state, when the entire plasma volume is accessible to the free motion of electron. Broadly speaking, this is realized when close packing of allowed regions is attained, that is, at $\xi_0 = 0.74$. Therefore, the region being described is narrow with respect to density (approximately from one-half to two values of the critical density of metal N_c), but important events occur in this region which define the critical state of matter.

Because $\beta I \gg 1$, it follows from the foregoing that the range of validity of the theory corresponds to the condition of strong non-ideality, $\gamma \gg 1$. These are the conditions in the neighborhood of the critical points of a number of metals. In cesium, for example, total ionization is attained, $N_e = N_c$, so that $\gamma_c = e^2 N_c^{1/3} \beta \approx 10$.

The plasma energy is made up of the interaction energy ΔE, the critical energy of ions, and the electron excitation energy,

$$E = \Delta E + c_\nu nT = -(1/2)nI - \alpha \cdot e^2 n^{4/3} + (3/2)nT\,, \tag{4.52}$$

where $n = N_e + N_i$. The isolation of the term $(-nI/2)$ from ΔE is specific precisely to the quasi-atomic model. This suggests the need for renormalization of the Madelung constant α. It is usually calculated for a system of ions against the background of uniformly smeared electrons and amounts to 0.57. In the case at hand, the value of α must be smaller. It takes into account the electrostatic interaction between an ion and the periphery of the cell containing the electric charge brought about by "extraneous" electrons. According to the estimate of [267], the Madelung constant in (4.52) is $\alpha \approx 1/4$.

The foregoing enables one to write the expression for free energy and derive the equation of state:

$$p = n \cdot T - \alpha \cdot e^2 \cdot n^{4/3}/3\,. \tag{4.53}$$

This equation reveals thermodynamic instability. The finite compressibility of the liquid phase is included in (4.53) in Van der Waals' approximation:

$$p = n \cdot T(1 - nb)^{-1} - \alpha \cdot e^2 \cdot n^{4/3}/3\,. \tag{4.54}$$

The parameter b corresponds to the minimum volume per particle and is defined by the quantity I such that $b \sim (e^2/I)^3$.

4.5.3. *Phase Transition in Metals*

The degrees of ionization of a number of metals in the neighborhood of the critical points are so high that plasma interactions are predominant. Therefore, a number of researchers suggested that the phase transition in its high-temperature region may be due to plasma interactions, that is, a "plasma" phase transition. More generally, the question of plasma phase transition is discussed in Section 5.8.

Equation (4.54) describes, qualitatively, the phase transition in metals and helps to estimate the critical parameters,

$$n_c = (1/7) \cdot b\,; \quad T_c = (48/7^{7/3}) \cdot (a/b^{1/3})\,; \quad p_c = 7^{-7/3}(a/b^{4/3})\,, \tag{4.55}$$

where $a = \alpha e^2/3$. The parameter a, which describes attraction, is universal for all monovalent metals. It is only parameter b that is specific. This is an important singularity of the equation of state (4.54), which distinguishes it from Van der Waals' equation. The similarity relations that relate the critical parameters [267] follow from Eq. (4.55):

$$n_c \sim I^3 , \qquad T_c \sim I , \qquad p_c \sim I^4 .$$

Likal'ter [267] generalized the theory to metals with Z valence electrons, which yielded more general similarity relations:

$$n_c \sim Z \cdot I^3 , \qquad T_c \sim Z \cdot I , \qquad p_c \sim Z^2 \cdot I^4 . \tag{4.56}$$

Relations (4.56) helps to establish a number of simple correlations between critical parameters. In particular, we eliminate valency and find the correlation between the critical pressure and temperature:

$$p_c \sim T_c^2 I^2 . \tag{4.57}$$

The similarity relations make it possible to express the critical parameters of a number of metals: by maintaining the metallic state in the critical point in terms of the parameters measured for one of them. It is natural to use the critical parameters of cesium for this purpose. In view of the fact that cesium is monovalent, and using a bar to indicate the parameters related to the analogous parameters of cesium, we have

$$\bar{T}_c = Z\bar{I} , \qquad \bar{p}_c = Z^2 \bar{I}^4 . \tag{4.58}$$

We use the experimentally obtained values of the critical factors of compressibility and of the fraction of allowed volume of cesium [268] to derive the compressibility at the critical point:

$$z_c = p_c/(N_c T_c) = 0.11 \cdot (Z + 1) .$$

The range of validity of the theory is discussed in [266]. As revealed above, this range is not wide. Nevertheless, verification shows that a number of metals satisfy the condition of metallic state at the critical point. These include all alkali metals; copper and silver of the noble metals; beryllium of the elements of Group II; and all elements of the main subgroup of Group III, including aluminum. The transition metals include all elements found in the iron group, some elements in the palladium and platinum groups, including molybdenum, and uranium. However, there is no sharp dividing line between the different types of states. In a number of metals, which are not on this list, the interaction at the critical point is not a purely Coulomb interaction, but makes a transition to Van der Waals interaction due to the mutual polarization of atoms. This transition is terminated in the case of uncommon metals, such as mercury, arsenic, bismuth and tellurium.

The critical temperature and pressure of some simple and transition metals are calculated by formulas (4.57) and (4.58) on the basis of the critical parameters of cesium [270], namely, $T_c = 2,013$ K and $p_c = 11.35$ MPa (Table 4.12).

Table 4.12. Critical parameters of metals.

Element and Valence	I, eV	T_c, 10^3 K	p_c, 0.1 GPa	Refs.
Li(1)	5.39	2.8	0.42	[267]
		3.2 ± 0.6	0.69 ± 0.14	[271]
Na(1)	5.14	2.65	0.345	[267]
		2.6 ± 0.35	0.35 ± 0.07	[271]
		2.5	0.255	
K(1)	4.34	2.25	0.175	[267]
		2.22 ± 0.33	0.16 ± 0.03	[271]
		2.20 ± 0.03	0.155 ± 0.015	[270]
Cu(2)	7.73	8.0	7.05	[267]
		8.4	7.45	[154]
		7.6	8.3	[268]
		7.85	9.05	[89]
B(3)	8.3	13.0	21.0	[267]
		8.2	9.6	[154]
Al(3)	5.99	9.3	5.7	[267]
		8.0	4.5	[154]
		7.1	5.5	[268]
		7.5	6.0	[269]
Cr(3)	6.77	10.5	9.35	[267]
		9.6	9.7	[154]
Ni(3)	7.63	11.5	15.0	[267]
		10.5	9.1	[154]
		9.6	11.0	[268]
U(3)	6.05	9.4	6.0	[267]
		11.5	6.1	[154]
		13.0	8.5	[268]
		12.5	5.0	[82]

Source: [267].

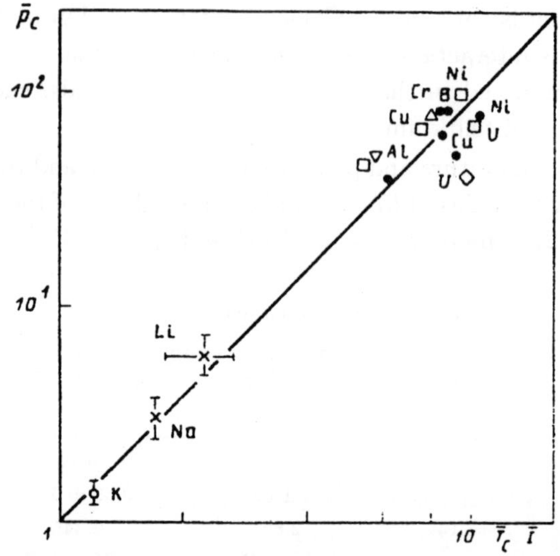

Fig. 4.15. The relative critical pressure of metals as a function of the relative critical temperature [267]: the points indicate the data of a number of researchers (see list of references in [267]) related to the critical parameters of cesium.

Figure 4.15 shows the relative critical pressure as a function of the relative critical temperature. These quantities correspond to formula (4.56) if the values of T_c and p_c are related to the respective values known for cesium. The estimates of the critical parameters, available from the literature and made on the basis of available experimental data and semi-empirical relations, correlate well with the dependencies given above. One can conclude that metallic states of this type with Coulomb interaction at the critical point are fairly characteristic of metals.

Five

The Thermodynamics of Plasma with Developed Ionization

5.1. One-Component Plasma Against the Neutralizing Background of Compensating Charge

We shall begin with a description of the thermodynamics of strongly ionized plasma by discussing a model system such as the one-component plasma. A one-component plasma (OCP) represents a system of point ions placed in a homogeneous medium of charges of the opposite sign. Due to a favorable combination of numerical and analytical investigation techniques, OCP has been comprehensively studied in a wide range of its parameters [272–285]. Below, we shall mainly follow these authors. We shall restrict ourselves to a discussion of OCP which obeys the classical statistics. Such a model serves as a good approximation for the plasma at ultrahigh pressure realized in the center of white dwarfs and heavy planets of the Jupiter type (Table 5.1). In these cases, matter is ionized under the effect of pressure and degenerate electrons have sufficient kinetic energy, $\varepsilon_k \approx \varepsilon_{\mathrm{F}} = (3\pi^2)^{2/3} hN_e^{2/3}/(4\pi m)$, to produce an almost uniform background distribution of density. In view of the small mass of electrons, the kinetic energy of electron gas is $\varepsilon_{\mathrm{F}} \gg kT$ at high density ($r_s \to 0$), and its pressure is much higher than that of an ion subsystem. In fact, two systems are realized in this case: the Coulomb system of point nuclei described by the Boltzmann statistics (interaction parameter $\gamma = e^2/(kT\bar{r})$, and quantum electron fluid (parameter $r_s = [3/(4\pi \cdot N_e)]^{1/3} \cdot (4\pi^2 me^2/h^2)$). The weak interaction between these components reduces, to an insignificant polarization, the increase in electron density in the neighborhood of nuclei, and most attention is concentrated on the analysis of Coulomb internuclear interaction. The OCP model is the most simple, non-trivial model of plasma. This is because the form of interaction potential does not cause any doubts, and the absence of quantum effects enables one to eliminate, from treatment, the formation of bound states (of molecules and atoms [287, 288] and ions) and the effect of degeneracy and interference [289–293]. Therefore, this model of plasma has been exhaustively studied with the aid of analytical and numerical methods in a wide range of values of the non-ideality parameter.

Table 5.1.

	Jupiter	White Dwarf	Neutron Star
Z	1(H)	6(C)	26(Fe)
N, cm^{-3}	$6 \cdot 10^{24}$	$5 \cdot 10^{30}$	10^{32}
T, K	10^4	10^8	10^8
γ	50	10–200	870
$\bar{r}, a_0$	0.65	$0.4 \cdot 10^{-2}$	$0.8 \cdot 10^{-3}$

5.1.1. *Monte Carlo Method*

The principal results, with respect to the one-component model, have been obtained using the Monte Carlo method [286]. It enables one to perform computer experiments with a plasma of any density. The method proceeds from the first laws of statistical physics and is based on direct computer calculations of mean thermodynamic values:

$$\langle F\rangle = Q^{-1}(V,N,T)\int\cdots\int F(\mathbf{q})\exp\{-\beta U_N(\mathbf{q})\}d^N\mathbf{q}\,, \tag{5.1}$$

where $Q(V,N,T) = \int\cdots\int\exp\{-\beta U_N(\mathbf{q})\}d^N\mathbf{q}$ is the configuration integral defining the equilibrium properties of the thermodynamic system and $\mathbf{q} \equiv \mathbf{q}_1,\ldots,\mathbf{q}_N$ are the co-ordinates of particles. In doing so, the interparticle interaction potential is assumed to be pre-assigned and, in most concrete calculations, two-particle:

$$U_N(\mathbf{q}) = \sum_{a,b,i<j}\Phi_{ab}(\mathbf{r}_{ij})\,, \qquad \mathbf{r}_{ij} = \mathbf{q}_i - \mathbf{q}_j\,. \tag{5.2}$$

The Monte Carlo method consists, in this case with the aid of pseudorandom-number generator of the computer, of the generation of a Markov chain which represents a set of space configurations, $A_1, A_2 \ldots A_N$, with a possibility of $A_i \to A_j$ transition between configurations:

$$\omega_{ij} = \exp\{-\beta[U(A_j) - U(A_i)\}\,. \tag{5.3}$$

In doing so, the thermodynamic functions are averaged along the obtained Markov chain, which is equivalent to averaging over the canonical ensemble.

In view of the limited capabilities of the computer, direct calculations are performed for a single cubic elementary cell containing N particles. In order to allow for the interaction of these particles with those of the neighboring cells, use is made of the Ewald's procedure and periodic boundary conditions.

In their pioneering study involving the use of the Monte Carlo method in one-component plasma, Brush and Sahlin [272] calculated the values of energy, heat capacity c_V, and binary correlation function $g(r)$ as a mean over $\alpha \sim 10^5$ configurations, with the number of particles in a cell being $N \sim 108$. In doing so, a wide range of parameters, $\gamma = 0.1$–125, was covered.

With the dense plasma of Jupiter in mind, DeWitt and Hubbard [278] used the method of linear response theory to perform numerical calculations of hydrogen plasma, as well as the plasma with a mixture of light elements, with due regard for polarization of background ($r_s \neq 0$).

The calculations by Brush and Sahlin [272] describe OCP in a liquid state in the $0.05 \leq \gamma \leq 100$ range (from 32 to 500 particles in a cell). The calculations by Hansen [276] cover the $1 \leq \gamma \leq 160$ range (from 16 to 250 particles in a cell). The results of these studies are in good agreement, except for the region of the highest values of γ which is described more accurately by Hansen [276]. Pollock [277] used OCP in a crystalline state in the $140 \leq \gamma \leq 300$ range (128 and 250 particles on a cell). A number of studies are devoted to the concrete OCP of Jupiter [280, 281].

5.1.2. *Results of Calculation*

The binary correlation function $g(r)$ is presented in Fig. 5.1. Figure 5.2 demonstrates the behavior of the static structure factor:

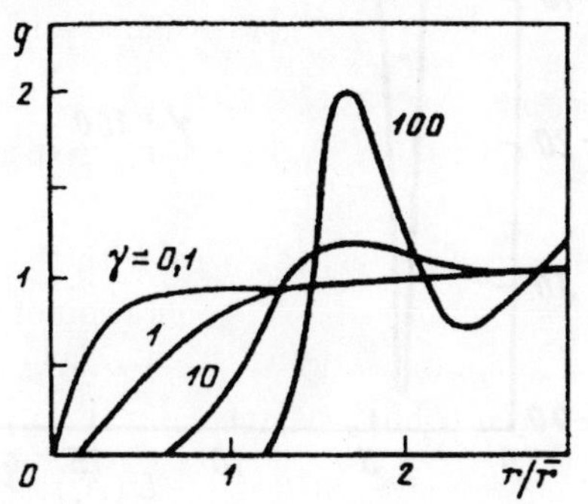

Fig. 5.1. The binary correlation function of OCP distribution for different values of the non-ideality parameter γ [275].

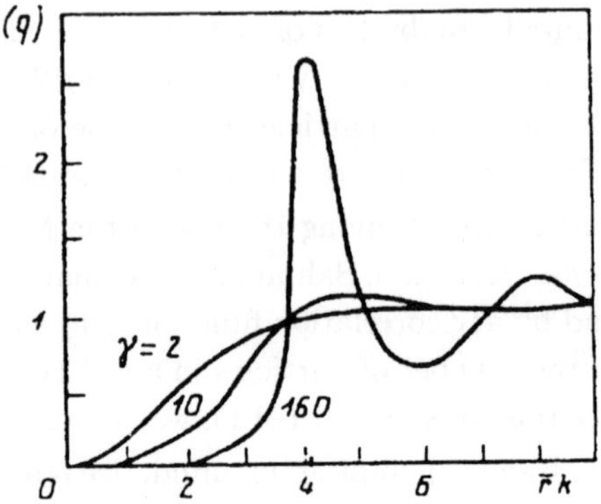

Fig. 5.2. The structure factor $S(q)$ of OCP for different values of the non-ideality parameter γ [275].

$$S(q) = 1 + N \int d\mathbf{r}[g(r) - 1] \exp(-i\mathbf{q}\mathbf{r})\,. \tag{5.4}$$

The static dielectric permeability $\varepsilon(q, 0)$ is given in Fig. 5.3:

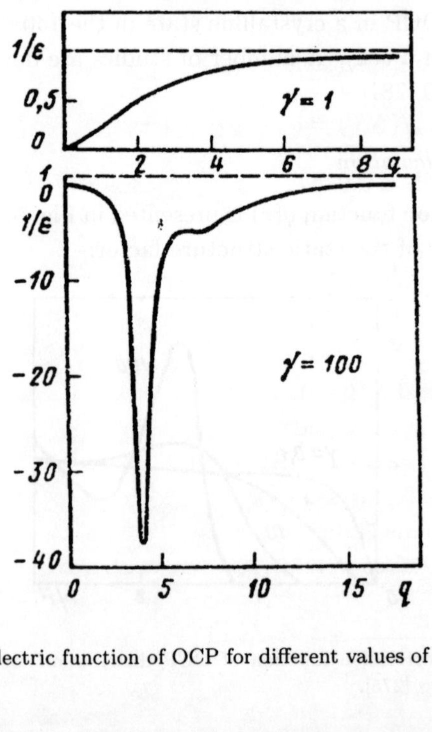

Fig. 5.3. The static dielectric function of OCP for different values of the non-ideality parameter γ [275].

$$\varepsilon(q,0) = [1 - q_{\rm D}^2/q^2 S(q)]^{-1}, \qquad q_{\rm D}^2 = 4\pi N(Ze)^2/T. \tag{5.5}$$

With small values of γ, these functions vary monotonically. At $\gamma \ll 1$, they are described by the linearized Debye approximation:

$$S_{\rm D}(q) = q^2(q^2 + q_{\rm D}^2)^{-1}, \qquad \varepsilon_{\rm D}(q,0) = (q^2 + q_{\rm D}^2)/q^2. \tag{5.6}$$

The pair correlation function is monotone up to $\gamma \sim 2.5$ and is in qualitative agreement with the Debye approximation:

$$\begin{aligned} g(r) &= 1 - \Gamma(r_{\rm D}/r)\exp(-r/r_{\rm D}), \\ \Gamma &= (eZ)^2/r_{\rm D}T, \qquad r_{\rm D} = (4\pi Z^2 e^2 N/T)^{-1/2}. \end{aligned} \tag{5.7}$$

At $\gamma \le 2.5$, $g(r)$ oscillations occur. These point to the emergence of a short-range order: the system changes from the ideal-gas to a liquid state. With an increase of in γ, the oscillations increase and an effective hard core is formed.

Given $g(r)$, one can calculate the internal energy u:

$$u = 3NT/2 + u_{\rm ex}, \qquad u_{\rm ex}/(NkT) = (N/2T)\int d\mathbf{r}(Z^2e^2/r)[g(r) - 1]. \tag{5.8}$$

The results of the Monte Carlo calculations for $1 \le \gamma \le 160$ were approximated by the following expression [281]:

$$u_{\rm ex}/(NT) = a\gamma + b\gamma^{1/4} + c\gamma^{-1/4} + d, \tag{5.9}$$

where $a = -0.89752$, $b = 0.94544$, $c = 0.17954$ and $d = -0.80094$. By integrating Eq. (5.9) over γ, one can obtain the free energy of OCP, F:

$$\begin{aligned} f = F/(NT) &= a\gamma + 4(b\gamma^{1/4} - c\gamma^{-1/4}) \\ &\quad + (d+3)\ln\gamma - (a + 4b - 4c + 1.135). \end{aligned} \tag{5.10}$$

The referencing of f in Eq. (5.10) to the point $\gamma = 1$ was done by Slattery *et al.* [281] while using the expressions for $u_{\rm ex}$ valid at $\gamma \le 1$.

Note that in accordance with the virial theorem $pV = u/3$, a negative value of u leads to the negative pressure of the ion component. The total pressure is certainly positive. This is due to the high value of the electron background pressure. The isothermal compressibility behaves analogously:

$$\frac{1}{NT}\frac{1}{\chi_T} = 1 + \frac{u_{\rm ex}(\gamma)}{3} + \frac{\gamma}{9}\frac{d}{d\gamma}u_{\rm ex}(\gamma). \tag{5.11}$$

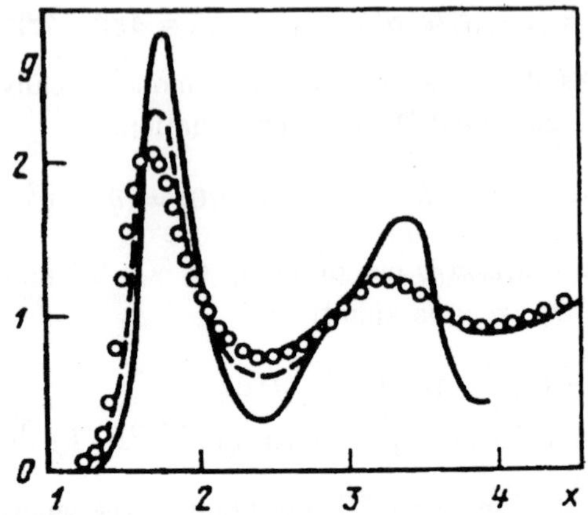

Fig. 5.4. Binary correlation function $g(x)$ for $\gamma = 160$. Solid curve, solid phase [277]; broken curve, liquid phase [276]; dots, HNC approximation [280].

The results of the calculation for solidified OCP (Fig. 5.4) were given for a *BCC* lattice, which yields the lowest energy. Slattery *et al.* [281] approximated the excess energy at $160 \leq \gamma \leq 300$ by the expression

$$u_{\text{ex}}/NT = a_{BCC}\gamma + (3/2)b\gamma^{-2} = 0.895929\gamma + 1.5 + 2800/\gamma^2\,. \qquad (5.12)$$

For a physical interpretation of the individual terms in the approximation expressions of the type of (5.9) and (5.10), simplified models [282, 285] were used. These are discussed below.

5.1.3. *Ion-Sphere Model*

The ion-sphere model, or cell model, breaks the system up into spheres with a radius $\bar{r}$. Each sphere accommodates an ion surrounded by a homogeneous cloud of neutralizing electrons. The negative charge density is $(-3Ze/4\pi\bar{r}^3)$. Within this model, one can easily calculate the mean excess energy

$$u_{\text{ex}}/(NT) = -0.9\gamma + I.5\,. \qquad (5.13)$$

Note, first, that Eq. (5.13) satisfies one of the global inequalities imposed on OCP by the Coulomb potential singularity. The following inequality [282] must be satisfied:

$$u_{\text{ex}}/NT \geq 0.9\gamma\,.$$

Equation (5.13) should be compared with Eq. (5.10), as well as with Eq. (5.12). In the latter, a_{BCC} is Madelung's constant for bcc lattice.

Therefore, the dominant term in Eq. (5.10) is static. The remaining terms make up the so-called thermal part of u_{ex}. Its contribution to u_{ex}, p, χ at $\gamma \geq 100$ proves to be less than 3%.

As we can see, the leading term in expressions (5.9) and (5.10) is close to the Madelung term for the bcc lattice of Coulomb charges. The OCP goes in the limit of strong interaction in the lattice. Use is also made of the short-range effect of interparticle interaction, which is characteristic of cell models.

In order to clarify the role of this short-range effect and to describe the second (proportional to $\gamma^{1/4}$) term in (5.9), DeWitt and Rosenfeld [294] employed the hard-sphere model popular in the theory of liquids. By treating the hard spheres as "zero" approximation models (H_0, F_0) in describing OCP (H, F), using the Gibbs–Bogoliubov inequality

$$F \leq F_0 + \langle H - H_0 \rangle ,$$

and employing the solutions of the Percus–Yevick equations, we derive, as a result of the variational procedure (see Section 5.4.3 for more details), the expression

$$u_{ex}/NkT = -(9/10)\gamma + [(8/9)\gamma]^{1/4} - 1/2 + (7/80)(18/\gamma)^{1/4} - .$$

In fact, this relation reproduces the first term in the ion-sphere model (5.13) and explains the appearance of the second term in (5.9), which may be interpreted as a manifestation of the short-range effects in a highly compressed plasma.

5.1.4. *Wigner Crystallization*

The Wigner crystallization was first treated in [295] for a degenerate plasma, where it was shown that the electron gas, against a background of compensating charge, must be arranged in an ordered manner in the case of fairly low concentration. Indeed, during the expansion of plasma, the Coulomb energy $\varepsilon_0 \sim e^2 N^{1/3} \sim e^2/r_s (r_s = r_0/a_0 = ([(4/3)\pi N]^{1/3}/(h^2/4\pi^2 m_0 e^2))$, which stabilizes the lattice, decreases slower than the kinetic energy that destroys the lattice, the scale of the latter being the Fermi energy $\varepsilon_k \sim \varepsilon_F \sim h^2 N^{2/3}/8\pi^2 m \sim 3.68/r_s^2$. Therefore, with fairly small values of density, the kinetic energy $\varepsilon_k \sim N^{2/3}$ becomes less than the potential energy $\varepsilon_0 \sim N^{1/3}$, and is incapable of destroying the ordered structure of electrons which emerged because of their repulsion.

The problem of determining the conditions of the Wigner crystal formation is treated in numerous studies, which are analyzed in [296]. We will briefly describe the results of this analysis. A simple comparison of the potential and kinetic energy of electrons gives the limits of crystallization:

$$r_s^c \sim 1.1 - 3.7; \qquad N_c^{1/3} \cdot a_{\mathrm{B}} \sim 0.56 - 0.17\,,$$

where the lower estimate for r_s^c was derived by disregarding the exchange and correlation, and the upper estimate was derived by including them in the Hartree–Fock approximation.

Based on the Lindemann melting criterion, one derives the estimate

$$r_s^c \sim 20, \qquad N_c^{1/3} \cdot a_{\mathrm{B}} \sim 3.1 \cdot 10^{-2}\,,$$

which is very sensitive to the details of this empirical criterion.

Assuming that the melting of crystal corresponds to the disappearance of bound states of electron, one can derive the stability limit:

$$50 \leq r_s^c \leq 100; \qquad 6.2 \cdot 10^{-3} \leq N_c^{1/3} \cdot a_{\mathrm{B}} \leq 1.2 \cdot 10^{-2}\,.$$

The Monte Carlo calculations for paramagnetic and ferromagnetic Fermi liquids result in the critical values

$$r_s^c \sim 75 \pm 5; \qquad N_c^{1/3} \cdot a_{\mathrm{B}} \sim (8.3 \pm 0.5) \cdot 10^{-3}\,;$$

$$r_s^c \sim 100 \pm 20; \qquad N_c \cdot a_{\mathrm{B}} \sim (6.2 \pm 1.3) \cdot 10^{-3}\,.$$

It is assumed [295] that the latter values of r_s are the most reliable ones.

For a non-degenerate plasma, the estimates of the melting parameters are obtained by comparing the free energy of the gas and the crystal phases. From Eq. (5.12), the expression for free energy density [281] is derived:

$$f(\gamma) = -0.895929 + 9\gamma/2 - 1.8856 - 1490/\gamma^2\,. \tag{5.14}$$

The free energy curves for liquid and solid phases intersect at $\gamma_m = 165$ [281]. With this value of non-ideality parameter, the Wigner crystallization of OCP occurs. The value of γ_m is very sensitive to the details of the calculation procedure. Different authors give, in their opinion, the most probable values of γ_m, which straddle the range of 155 and 171.

Wigner crystallization was experimentally observed in a two-dimensional system of electrons forced against the surface of liquid helium. A classical electron crystal occurred at $\gamma_m \approx 140$. This value of γ_m is close to the calculated value of 125 ± 15 [283]. In a two-dimensional system, $\gamma = (Ze)^2(\pi N)^{1/2}/T$.

The quasi-periodic structure of an ion cloud with $N_i \approx 10^8$ cm^{-3} and $T \sim 10^{-2}$ K in magnetic field was interpreted by Gilbert *et al.* [297] as a manifestation of the Wigner crystallization. From the same perspective, Hall *et al.* [298] treated the oscillation of the binary function of plasma, measured in laser-driven shock waves and a characteristic of an ordered structure.

It is possible that at $\gamma > \gamma_m$, supercooled OCP is in an amorphous vitrified state [285]. Essentially, this state is a disordered system of monocrystals with a size of several lattice spacings. It is interesting to note that such may be the state of matter in the interior of white dwarfs because, according to some estimates, the parameter γ in them may even exceed γ_m.

5.1.5. *Integral Equations*

In the theory of liquid, integral equations relate the binary correlation function to the two-particle interaction potential $V(r)$ [287, 288]. The Percus-Yevick equation, which had shown itself to be advantageous in describing the interaction of the systems properties with the short-range potential [289], proved unfit for OCP [290]. Apparently, the hypernetted chain approximation (HNC) is most suited to plasma. This is seen in the qualitative analysis of the HNC levels, as well as the good agreement between the results obtained by the Monte Carlo method and the results of numerical integration of hypernetted chain equations [280, 290]. These equations have the form

$$g(r) = 1 + h(r) = \exp[-V(r)/T + h(r) - c(r)]\,, \tag{5.15}$$

$$h(r) = c(r) + N \int d\mathbf{r} c(|\mathbf{r} - \mathbf{r}'|)h(r')\,, \tag{5.16}$$

where $h(r)$ and $c(r)$ are the complete and direct correlation functions, respectively, $V(r) = Z^2e^2/r$. The solution is reached through the following procedure. Starting from some initial $h(r)$, one calculates its Fourier transform $h(k)$, followed by $c(r)$:

$$c(r) = \int \frac{d\mathbf{k}}{(2\pi)^3} \exp(i\mathbf{k}\mathbf{r}) \frac{h(k)}{1 + h(k)}\,. \tag{5.17}$$

Upon substituting Eq. (5.17) in Eq. (5.16), one derives the subsequent interpretation of $h(r)$ and so forth.

Springer *et al.* [290] found $g(r)$ and u_{ex} in the $0.05 \le \gamma \le 50$ range. Ng [280] did that with high accuracy for $20 \le \gamma \le 7{,}000$. The binary correlation functions in the HNC approximation describe well those obtained using the Monte Carlo method (cf., Fig. 5.4).

Numerical HNC results were approximated by DeWitt [291] using the following expression:

$$u_{\rm ex}/(TN) = a\gamma + b\gamma^{1/2} + c\ln\gamma + d\,, \tag{5.18}$$

where $a = -0.900470$, $b = 0.2688263$, $c = 0.0719925$ and $d = 0.0537919$. Because of the predominance of the static term, Eqs. (5.9) and (5.18) coincide with an accuracy higher than 1%. As for the thermal parts, they differ both literally and numerically (45% at $\gamma = 150$). With the highest values of γ, the strict constraint (5.13) is not satisfied.

In spite of the considerable advantages of the HNC approximation in general, it proves unsatisfactory in situations when the thermal part of $u_{\rm ex}$ comes to the forefront. Such is the case, for instance, in calculating excess heat capacity:

$$(c_v/N) - 1 = -\gamma^2 d(u_{\rm ex}/TN\gamma)/d\gamma\,. \tag{5.19}$$

Here, the error reaches 20%. The shortcomings of the HNC approximation are due to the neglect of the contribution made by ladder diagrams in calculating $g(r)$. The semi-empirical inclusion of this contribution, made by Rosenfeld and Ashcroft [291], proved successful. It was performed in such a manner as to satisfy the sum rule for compressibility violated by the HNC approximation. As a result, the thermodynamic quantities and correlation functions were found to be very close to those obtained using the Monte Carlo method.

5.1.6. *Polarization of Compensating Background*

The degenerate electronic gas, which forms the compensating background, polarizes due to the non-uniformity of ion charge distribution. An electron cloud forms around each ion, which modifies the ion-ion potential. In a Fourier representation,

$$V(r) = 4\pi(Ze)^2 q^{-2}\varepsilon_e^{-1}(q)\,, \tag{5.20}$$

where $\varepsilon_e(q)$ is the dielectric function of a system of electrons. The static dielectric permeability $\varepsilon_e(q)$ is known in various approximations. This problem was reviewed by Gorobchenko and Maksimov [292]:

$$\varepsilon_e(q) = 1 - \frac{4\pi(Ze)^2\chi_0(q)q^{-2}}{1 + 4\pi(Ze)^2 G(q)\chi_0(q)q^{-2}}\,, \tag{5.21}$$

where $\chi_0(q)$ is the static polarizability of free electronic gas (Lindhard function), and $G(q)$ is the local field correction allowing for the exchange and correlation effect [294]. In the limit of high electron density ($r_s \to 0$), the exchange and correlation can be neglected and $\varepsilon_e(q)$ is given by a random phase approximation (RPA):

$$\begin{aligned}\varepsilon_e(q) &= 1 - \frac{4\pi(Ze)^2\chi_0(q)}{q^2}\\ &= 1 + \frac{1}{(q\lambda_{\rm TF})^2}\left\{\frac{1}{2} + \frac{q_{\rm F}}{2q}\left[1 - \frac{q^2}{4q_{\rm F}^2}\right]\left[1 - \frac{q^2}{4q_{\rm F}^2}\right]\ln\left|\frac{q+2q_{\rm F}}{q-2q_{\rm F}}\right|\right\}. \quad (5.22)\end{aligned}$$

Here, $\lambda_{\rm TF} = \bar{r}(\pi/12Z)^{1/3}$ is the Thomas-Fermi screening distance and $q_{\rm F}$ is the Fermi momentum, $q_{\rm F}^{-1} = \bar{r}(4/(9\pi Z))^{1/3}$. Indeed, Eq. (5.22) is quite valid while $r_s \le 1$.

To the roughest Thomas–Fermi approximation, the interion potential is shielded as follows:

$$V(r) = (Ze)^2\frac{1}{r}\exp\left(-\frac{r}{\lambda_{TF}}\right). \quad (5.23)$$

If $\lambda_{\rm TF} \gg \bar{r}$, screening is then not important, that is, the background polarization can be neglected. In the opposite case, the results of the calculation depend on two, rather than one, parameters, namely, γ and r_s [279]. In the $r_s \le 0.5$ range, the results of the Monte Carlo calculations are approximated as follows [275]:

$$\begin{aligned}u_{\rm ex}/(NT) = &-(0.8946 + 0.0543r_s)\gamma + (0.8165 - 0.1853r_s)\gamma^{1/4}\\ &-(0.5012 - 0.0659r_s). \quad (5.24)\end{aligned}$$

At $r_s > 1$, the results of the calculation are very sensitive to the form of $G(q)$.

It turns out unexpectedly that the background polarization effect on the thermodynamic properties of plasma is relatively small, even if the screening is no longer weak ($r_s \approx 1$). It is mainly the static part of the energy that varies. The inclusion of correlations varies it by several per cent at $r_s \sim 0.5$. The plasma pressure varies even less, because the polarization correction, $\sim r_s \cdot \Gamma = \beta \cdot e^2/a_0$, does not depend on the system volume.

5.1.7. *Charge Density Waves*

In OCP, charge density waves emerge in view of the fact that the inter-electron interactions reader the state with the density profile, which is oscillating in space, more advantageous as compared to its uniform distribution. This leads to a modulation of the density of positive charge of the background, which provides the condition of electroneutrality [296]. This effect is realized [275] with the emergence of a "soft mode" instability associated with the formation of the Wigner crystal [295]. Therefore, the state of charge density waves may be treated as an intermediate phase between the Wigner crystal and common metal with a uniform charge distribution.

The condition for the emergence of charge density waves is the vanishing of the dielectric permeability,

$$\varepsilon(q,0) = 0\,,$$

which is related to the structure factor $S(q)$ by the fluctuation-dissipation theorem (5.5). Figure 5.3 demonstrates that, according to the Monte Carlo calculations, such states may emerge in OCP under conditions of considerable plasma non-ideality: $\gamma \geq 160$.

The real possibility for the emergence of charge density waves in a three-dimensional electron gas depends largely on the degree of deformability of the ion lattice and its closeness to the model of deformable "jellium". Apparently [296], alkali metals are the most suitable three-dimensional objects in the search for charge density waves. This is because the weakness of ion-ion interaction promotes the deformability of ion lattice [297–299]. At present, the question of charge density waves in three-dimensional metals remains open, while in the case of quasi-one-dimensional structures (such as chalcogenides of transition metals TaS_2, $NbSe_3$ and the like) the existence of those waves is reliably registered, for instance, by electron-microscopic observations.

5.1.8. *Sum Rules*

The numerical results obtained using the Monte Carlo method serve as a comparison for the "experimental" data. It enables one to assess the range of application of different analytical approximations. Direct comparison shows, for instance, that the methods developed for a weakly non-ideal plasma are fairly valid upon extrapolation to $\gamma \approx 1$ [293]. It is important that in the case of such an extrapolation, general restraints should be met which are valid for the OCP of any non-ideality. These include the following relationships [293, 200]:

(i) The condition of positiveness of the binary correlation function:

$$g(r) \geq 0\,. \tag{5.25}$$

(ii) The condition of exponentiality, which restricts the value of $g(r)$ at

$$g(r) \sim \exp[-\beta V(r)]\,, \qquad r \to 0\,. \tag{5.26}$$

(iii) The condition of local electroneutrality in a Coulomb system:

$$N \int d\mathbf{r}[g(r) - 1] = -1\,. \tag{5.27}$$

(iv) The condition of screening obtained by Stillinger and Lovett [301]:

$$N \int d\mathbf{r} r^2 [g(r) - 1] = -6k_{\mathrm{D}}^{-2}, \qquad k_{\mathrm{D}} = (4\pi Z^2 e^2 N/T)^{1/2}\,. \tag{5.28}$$

We now consider, in more details, the condition of exponentiality, which we shall rewrite in a more definite manner:

$$g(r) = \exp[-V(r)/T + a + br^2 + \ldots]\,, \qquad r \to 0\,. \tag{5.29}$$

The constant $a > 0$ [302]. This is especially important from the standpoint of the theory of rates of nuclear reactions in the interior of stars.

A great deal of effort has been spent to determine the constants in the exponent (5.29). The point is that the behavior of $g(r)$, at low values of r, governs the lowering of the Coulomb barrier whose magnitude and shape define the rate of nuclear reactions taking place in the interior of stars. The reaction rate depends on the probability of two nuclei closing with each other to a distance of the order of 10^{-13}–10^{-12} cm. In a strongly non-ideal plasma, the screening effects increase $g(r)$ at low values of r. Consequently, they increase the rates of nuclear reaction [302–304]. The lowering of the barrier is defined by the expression

$$\exp(a) = \lim\{g(r)\exp[V(r)/T]\}\,. \tag{5.30}$$

This effect is also important from the standpoint of laser fusion.

5.1.9. *Asymptotic Expressions and Their Extrapolation*

The equilibrium properties of OCP are defined by the free energy

$$\beta F_1 = \ln \int \cdots \int \frac{\prod_{i=1}^N d^3\mathbf{r}_i}{V^N} \exp\left\{(-\beta)\left[\sum_{i<j}^N \frac{(Ze)^2}{|\mathbf{r}_i - \mathbf{r}_j|}\right.\right.$$

$$\left.\left. - \frac{N}{V}\sum_{i=1}^N \int d^3r \frac{Ze^2}{|\mathbf{r}_i - \mathbf{r}|} + \frac{N^2}{2V^2}\int\int d^3\mathbf{r} d^3\mathbf{r}' e^2 |\mathbf{r} - \mathbf{r}'|\right]\right\}; \beta = kT\,.$$

The analytical estimates for βF_1 can only be obtained in the limit of weak ($\gamma \ll 1$) and moderate ($0.3 \leq \gamma \leq 1$) non-ideality. The technique of estimating

βF_1 on the basis of Mayer's diagrams for Coulomb systems, which is well-developed for rapidly decreasing (faster than $\sim r^{-3}$) interaction potentials, is inapplicable because of the divergence of the respective integrals. In order to eliminate these divergences, it is necessary to perform regrouping and selective summation of a series of the perturbation theory. This leads to the result [305]

$$\beta F_1 = N\{(-1/3)(\sqrt{3}\cdot\gamma^{3/2}) + S_2(\gamma) + S_3(\gamma) + \ldots\},$$

where the first term corresponds to a summation of the "ring" diagrams (Debye approximation). The terms S_2 and S_3, which are of a higher order with respect to γ, are non-exponential functions of density and effectively describe the interaction of groups of particles in terms of the shielded potential

$$\beta\cdot V(r) = (\beta\cdot e^2/r)\cdot\exp(-r/\lambda_{\mathrm{TF}}).$$

The expression for S_2,

$$S_2 = \frac{N}{2}\int d^3r \oint\left[e^{-\beta V} - 1 + \beta V - \frac{1}{2}(\beta V)^2 \ldots\right],$$

has, in the $\gamma \to 0$ limit, the order $O(\gamma^3 \ln\gamma)$. Instead of using cumbersome analytical expressions for S_2 and S_3, Rogers and DeWitt [306] performed a numerical calculation of the respective integrals with a subsequent approximation of the results. The relation for βF_1 leads, in a standard manner, to

$$\frac{\beta V}{N} = -\frac{1}{2}(\sqrt{3}\gamma^{3/2}) + \frac{3}{2}\gamma\frac{d}{d\gamma}[S_2(\gamma) + S_3(\gamma) + \ldots];$$

$$\frac{c_v}{NkT} = -\frac{1}{4}(\sqrt{3}\gamma^{3/2}) - \frac{3}{2}\gamma^2\frac{d^2}{d\gamma^2}[S_2(\gamma) + S_3(\gamma) + \ldots].$$

In the absence of electron shielding ($r_s = 0$), and in accordance with the virial theorem,

$$\frac{p - p_0}{N} = \frac{1}{3}\frac{\beta U}{N}, \qquad p_0 = N\cdot k\cdot T.$$

In the limit of weak non-ideality ($\gamma \ll 1$), the dominant term of these expressions has the order $O(\gamma^{3/2})$, which corresponds to the well-known Debye–Hueckel approximation

$$\frac{\beta F_1}{N} = \frac{1}{3}\varepsilon; \qquad \beta\frac{U}{N} = -\frac{1}{2}\varepsilon;$$

$$\frac{c_v}{NkT} = \frac{1}{4}\varepsilon; \qquad \varepsilon = \sqrt{3}\gamma^{3/2}.$$

However, as the non-ideality increases ($\gamma \to \infty$), the term $O(\gamma^{3/2})$ is compensated by the contributions of other integrals in the βF_1 expansion. The dominant terms in this limit are of the order of γ, which clearly follows from the results of the Monte Carlo calculations. The transition from a dependence, which is proportional to $O(\gamma^{3/2})$, to a weaker dependence that is proportional to γ occurs in the $0.3 \le \gamma \le 0.75$ range. Therefore, the value of $\gamma \sim 0.75$ is taken to be the lower limit of the range of existence of a strongly non-ideal plasma. Note that the lower energy estimates have been formulated for OCP [307]:

$$\frac{\beta U}{N} = -\frac{1}{2}\sqrt{3}\gamma^{3/2} \qquad \text{at} \quad \gamma \ll 1\,;$$

$$\frac{\beta U}{N} = -\frac{9}{10}\gamma \qquad \text{at} \quad \gamma \gg 1\,.$$

The range of existence of a strongly non-ideal Coulomb plasma (Coulomb fluid) is limited by the maximum value of $\gamma \sim 155 \div 171$ and by the crystallization limit (see Section 5.1.4). Therefore, at $\gamma \le 0.75$, the plasma is described by asymptotic expressions [306] based on the perturbation theory, and at $\gamma \ge 155 \div 171$ it forms a Coulomb crystal. In the intermediate range, as revealed by the results of an analysis of numerical experiments, the plasma may be treated as a disordered ion lattice described by relations (5.8) and (5.9).

Even the most simple correction of approximations, which is applicable at $\gamma < 1$, results in a marked increase in their capabilities in the $\gamma \approx 1$ range [307, 308].

A series of known approximations give the correlation function in the form of

$$g(x) = 1 - \exp(-Ax)/(Bx)\,, \qquad x = r/r_{\rm D}\,. \tag{5.31}$$

Such a linearized form violates the condition (5.26). The transition to the non-linear form

$$g(x) = \exp[-\exp(-Ax)/(Bx)] \tag{5.32}$$

violates condition (5.27). In [293, 309, 310], a method is proposed which consist of the simultaneous inclusion of conditions (5.25) to (5.27), where only the positive part of $g(r)$ is employed while the parameter B becomes a normalizing constant. The latter was selected because it satisfied Eq. (5.27). For example, a linearized Debye–Hückel approximation (LDH) with $A = 1$, $B = \Gamma^{-1}$ assumes the form LDH*:

$$g(x) = 1 - (R/x)\exp(R - x)\,, \quad x \ge R;\ g(x) = 0, \quad x < R\,, \tag{5.33}$$

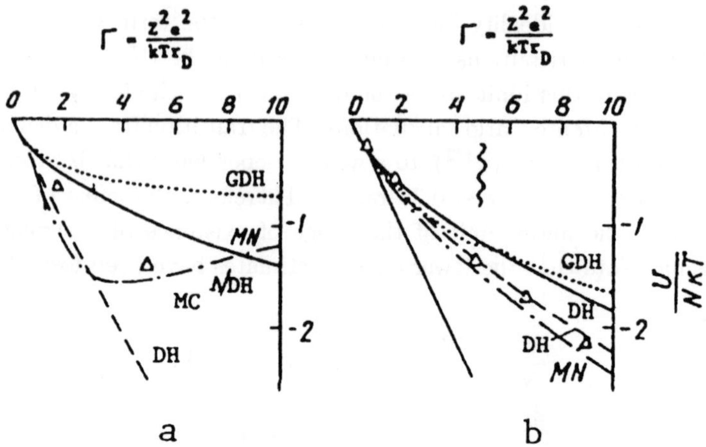

Fig. 5.5. The Internal energy of OCP as given by a series of asymptotic approximations before (a) and after (b) correction, in accordance with the relationships (5.25) to (5.27) [309].

where $R = (1+3\Gamma)^{1/3} - 1$ and $\Gamma = (Ze)^2/Tr_D$. In this approximation,

$$U/(NT) = -1/4[(1+3\Gamma)^{2/3} - 1]\,. \tag{5.34}$$

The results of four approximations before and after the correction, which is analogous to that described above, are presented in Figs. 5.5 and 5.6: linearized and non-linearized Debye approximations (LDH and DH); Debye approximation in a grand canonical ensemble (GDH); and approximations which asymptotically allow for the dependence of the screening distance (MN) [311]. One can see that this correction improves considerably the agreement with computer experiment (MC), even in cases when the oscillations of the correlation functions occur in computer solutions (behind the wavy line in Fig. 5.5).

Gryaznov and Iosilevskii [308] have constructed a model describing the oscillations of $g(x)$ (LDH**):

$$g(x) = 1 - (A/x)\exp(-x)\cos Bx\,, \quad x \geq R;\ g(x) = 0\,, \quad x \geq R\,, \tag{5.35}$$

where A and B are the normalization constants, $R = R(A, B)$. The MSA theory [315], as well as the model (5.35), satisfies the criteria (5.25) to (5.28) and at the same time, as shown by Gillan [314], describes well the results of the Monte Carlo calculations (Fig. 5.6).

Therefore, the satisfaction of the conditions (5.25) to (5.28) is an effective means for checking the validity of the results of thermodynamic calculations

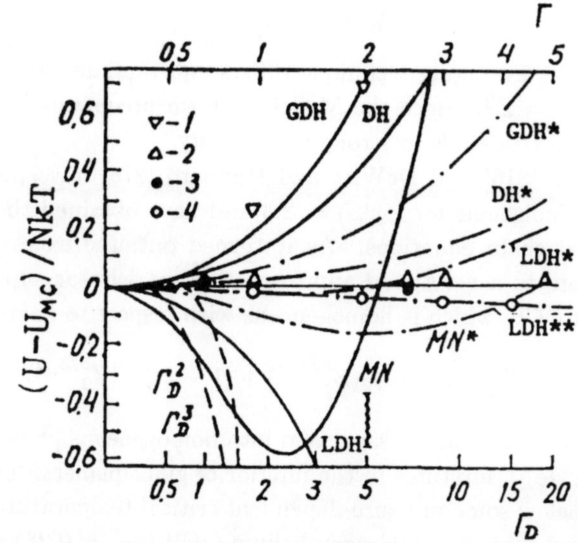

Fig. 5.6. Coulomb interaction energy in OCP [44]. U_{MC} the Monte Carlo method [272, 276]. Expansion in Γ: LDH, first order (Debye–Hückel approximation); Γ_D^2, second order [305]; Γ_D^3, third order [310]. Simplest models: DH, non-linear form of Debye–Hückel approximation; GDH, Debye approximation in a grand canonical ensemble [293]; MN, [311]; LDH*, LDH**, DH*, GDH*, MN*, the same approximations after correction (see text). More complex approximations: *1*, HNC [312]; *2*, HNC [290]; *3*, solution of BBGKY hierarchy equations with due regard for Eq. (5.27) [313]; *4*, MSA [314].

using both numerical and analytical techniques. For instance, the failure of Carley [312] to satisfy Eq. (5.27) affected the results (cf., Fig. 5.6) and pointed to their considerable incorrectness.

5.1.10. *OCP of a Mixture of Ions with Z_1 and Z_2 Charges*

The OCP of a mixture of ions with Z_1 and Z_2 charges is of interest from the viewpoint of astrophysics. In the ion-sphere model, the static energy of the binary mixture is given by the expression

$$u_{\text{ex}}/NT = -0.9[x_1 Z_1^{5/3} + (1-x_1)Z_2^{5/3}]\bar{Z}^{1/3}(e^2/\bar{r}T)$$
$$= -0.9\overline{Z^{5/3}}(\bar{Z})^{1/3}\gamma', \tag{5.36}$$

where $N = N_1 + N_2$, $x_1 = N_1/N$, $\bar{Z} = (Z_1 N_l + Z_2 N_2)/N$, $\bar{r} = (4\pi N/3)^{-1/3}$, $\gamma' = e^2/\bar{r}T$. Using the prevailing view on the mean ion charge, Eq. (5.36) suggests the following law of charge averaging [303]:

$$Z_{\text{eff}}^2 = \overline{Z^{5/3}}(\bar{Z})^{1/3}. \tag{5.37}$$

Note that this averaging technique differs substantially from the Debye-Hückel one: $Z_{\text{eff}}^2 = \overline{Z^2}$. Since the Madelung term predominates in u_{ex}, it is Eq. (5.37) that corresponds to strong non-ideality.

Hansen *et al.* [316] and DeWitt and Hubbard [278] have performed the Monte Carlo calculations for $Z_2/Z_1 = 2.3$ and have obtained the solution to the hypernetted chain equations. As it turned out, the excess free energy could be written, to a very good approximation, as a linear superposition of free energies of OCP, which is homogeneous with respect to charge [274]:

$$F_{\text{ex}}/NT = f = x_1 f_0(Z_1^{5/3}\gamma') + (1 - x_1) f_0(Z_2^{5/3}\gamma'). \tag{5.38}$$

A binary ionic mixture can be separated into homogeneous phases [317], such as the H^+ and He^{++} mixtures in the interior of giant planets. This occurs at temperatures below some pressure-dependent critical temperature T_c. Hansen *et al.* [316] found that, in a hydrogen-helium OCP ($x_{\text{He}} = 0.28$) at a pressure of 60 Mbar, the critical temperature $T_c = 6,300$ K. Under these conditions, $r_s = 0.85$. It is significant that if the linear superposition (5.38) is quite accurate, phase separation is impossible. Of considerable importance in describing the separation effect is the inclusion of the electron background correlation. As shown by Pollock and Alder [318], the separation may result from a more effective local quasi-neutrality in homogeneous phases than in a mixture thereof.

5.2. Multicomponent Plasma. The Results Derived from the Perturbation Theory

The main disadvantage of the one-component model of plasma resides in an extremely simplified inclusion of the opposite sign charge, which represents a structureless compensating background. More perfect models of plasma provide an explicit inclusion of the structure and interaction of charges of all signs, along with a mandatory description of the quantum effects under conditions of Coulomb interaction. The point is that the neglect of the quantum-mechanical features of the problem leads to the principal difficulties of the classical description of the motion of particles over small distances ($\sim \lambda_e$). This is in view of the divergence of the co-ordinate part of Gibbs' probability upon the bringing together of opposite charges [319].

In quantum theory, such a union is naturally eliminated by a growth in the mean momentum and kinetic energy upon the localization of charges. This ensures the stability of the entire system. In doing so, the quantum effects lead to the formation of bound states (such as molecules, atoms and ions) which, in turn, affect the interaction of free charges. Numerous attempts have been made to maintain classical formalism by introducing a cutoff in the Coulomb potential at small distances with a forced elimination of configurations with closely spaced charges. Whilst doing so, the final answer incorporates the cutoff parameter. Meanwhile, the models proper lose the thermodynamic stability at $\gamma \geq 1$ when the cutoff radius becomes comparable with the interparticle distance.

A consistent quantum-mechanical discussion of the problem proceeds from a Hamiltonian description of the full interaction between all charges. This corresponds to the "physical" model of a multi-component plasma, where the discrete spectrum contribution proves finite and occurs simultaneously with the contribution by continuous free spectrum of charges [320]. The "physical" model is the most general and consistent one for real plasma. However, the practical calculations within this model are very laborious and have not yet been used widely. The point is that the application of the physical model to a partly ionized plasma assumes that, as an element, the quantum-mechanical calculation of the internal structure of bound states is analogous to that of atoms and ions using the Hartree-Fock method.

In a rarefied plasma, configurations with closely spaced particles are quite unlikely, which permits considerable simplifications. The main simplification consists in a separate description of the state of discrete and continuous spectra. The first corresponds to the internal structure of atoms and ions, and the second to an electrically charged component. Such an approximation forms the basis of the model. It is currently the most popular model in plasma physics and is referred to as "chemical" [293]. This is because the number of particles of a different type, $\{N_i\}$, are governed by the conditions of chemical ionization equilibrium, $\sum \mu_i = 0$:

$$[\partial F(V,T,N_i)/\partial N_k]_{V,T} = 0\,, \tag{5.39}$$

where all hypotheses on the structure of particles and their interactions are contained in the expression for free energy $F(V,T,N_i)$:

$$F(V,T,N) = F^k + F^c + F^b + F^l\,. \tag{5.40}$$

The discrete spectrum contribution F^b in this model is separate. It is calculated independently on the continuous spectrum contribution presented by the kinetic part F^k and various corrections for interparticle interaction, F^c.

If radiation is in local thermodynamic equilibrium with material, the free energy of photon gas and its contribution become appreciable upon extreme high heating of the material or strong rarefaction ($F^l = -(4\sigma/c)V^4$). For tungsten of solid-state density, the contribution of thermal radiation becomes appreciable at $T \geq 10^7$ K ($p \geq 10^4$ Mbar).

The free energy of ideal gas can be written as

$$F^k = kT \sum N_k \left\{ \mu_k - \frac{F_{3/2}(\mu_k)}{F_{1/2}(\mu_k)} \right\}, \quad F_p = \frac{1}{\Gamma(p+1)} \int_0^\infty \frac{t^p dt}{e^{t-2}+1}, \tag{5.41}$$

where the reduced (in kT units) chemical potential μ_k is a measure of the degree of plasma degeneracy and is determined by the relationship

$$F_{1/2}(\mu_k) = (N_k/g_k V)\lambda_k^3 \sim n_k \lambda_k^3 .$$

For heavy particles, $n_k\lambda_k^3 \ll 1$ and $\mu_k \ll 1$. Hence the transition to the classical limit of Boltzmann statistics:

$$F^k = kT \ln[(N_k/g_k V)\lambda_k^3 + 1] . \tag{5.42}$$

In accordance with Eq. (5.42), the inclusion of electron component degeneracy reduces the degree of plasma ionization while the kinetic part of electron pressure is increased.

The chemical model usually involves the use of relations derived for a fully ionized plasma using the methods of modern perturbation theory [320]. It is developed until an arbitrary order is reached with respect to the expansion parameter, and is equipped with diagram techniques that facilitate the classification and identification of terms in the respective series.

The main correction for plasma is calculated by the summation of the so-called "ring" diagrams [320]. For plasma degeneracy, it corresponds to the Gell–Mann-Bruekner model [321]. However, in the Boltzmann limit, it leads to the Debye-Hückel model:

$$\beta F_2^c = -\sum_i N_i \frac{1}{3} \Gamma P(\gamma) . \tag{5.43}$$

Here, the parameter of non-ideality of a multi-component plasma is

$$\Gamma = 2\pi^{1/2} e^3 \beta^{3/2} (N/V)^{1/2} \left(\sum_i z_i^2 N_i \theta_i \Big/ \sum_i N_i \right)^{3/2}, \tag{5.44}$$

where the parameter $\theta_i = F_{-1/2}(\mu_i)/F_{1/2}(\mu_i)$ allows for the degeneracy of the i-th component of plasma. The additional quantum-mechanical correction in Eq. (5.43) allows, in a high-temperature plasma ($\beta e^2\langle Z^2\rangle < \lambda_i$), for the effects caused by the uncertainty principle (Fig. 5.7 [322, 324]) and corresponds to the effective repulsion of charges at small distances:

$$P(\gamma) = 1 - (3/16)\pi^{1/2}[(Z_e^4 N_e^2 \gamma_{ee} + 2Z_e^2 Z_i^2 N_e N_i \gamma_{ei} + Z_i^2 Z_j^2 N_i N_j \gamma_{ij})/N^2\langle Z^2\rangle^2] + (1/4)\gamma_e^2\{[Z_e^4 N_e^2 + Z_e^2 Z_i^2 N_e N_i(1 + m_e/m_i) + Z_i^2 N_i^2(m_e/m_i)]/N^2\langle Z^2\rangle^2\}, \quad (5.45)$$

where $\gamma_i = \lambda_i/\rho_D$; $\gamma_{ei} = \lambda_{ei}/\rho_D = \hbar^2(2\mu_{ei}kT)^{-1/2}/\rho_D$; $\mu_{ei} = m_e m_i/(m_e + m_i)$; and $\rho_D = [4\pi\beta^2 e^4 N \sum_j Z_j^2 N_j(\theta_i/V)]^{-1/2}$ is the Debye screening distance. The expansion term following Eq. (5.43) is referred to as the ladder term and it describes the binary interactions of charges in terms of dynamic screened Coulomb potential. The general expression for ladder correction is given by Ebeling *et al.* [320] and it includes the multi-dimensional integrals of complex structure. In a high-temperature approximation $\gamma_{ii} < \Gamma < \gamma_{ei}, \gamma_{ee} < 1$ (corresponding, for a hydrogen plasma, to $50 \leq T \leq 2,000$ eV), the ladder correction takes the form [320] of

$$\beta F_3^c = -\sum N_i \frac{1}{12}\frac{\Gamma^2}{N^2\langle Z^2\rangle^2}\{Z_e^6 N_e^2(\ln\gamma_{ee} + 0.887) + 2Z_e^3 N_e \sum Z_i^3 N_i \times (\ln\gamma_{ei} + 0.887) + \sum\sum Z_i^3 Z_j^3 Z_i N_j [\ln Z_i Z_j \beta e^2/\rho_D + 0.887]\}. \quad (5.46)$$

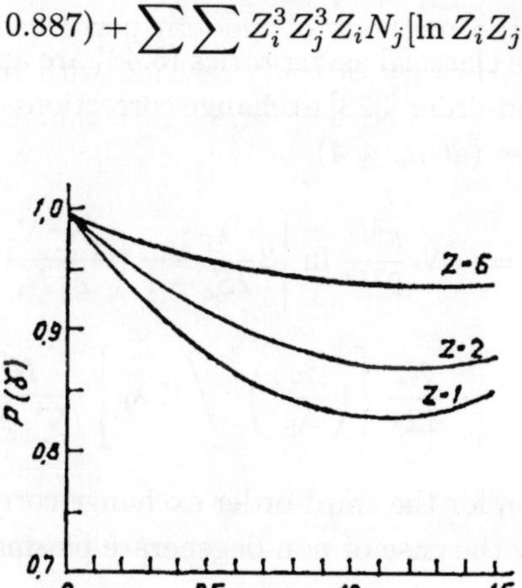

Fig. 5.7. The effect of the uncertainty principle on the equations of state (5.43) and (5.45) [322].

The first term in this expression describes the electron-electron interactions and the second, the electron-ion interactions. The third term describes the ion-ion interactions. In order to describe quantum and diffraction effects in this and subsequent terms of the perturbation theory, a number of approximation expressions of a much more complex structure have been proposed [46].

Along with the Coulomb interaction, the methods of the perturbation theory enable one to calculate the corrections for the exchange interaction between the free charges of identical spins. The first-order correction for arbitrary statistics takes the form of

$$\begin{aligned}\beta F^{c}_{\mathrm{ex}_1} &= -N_e\frac{1}{2}\gamma_e^2\frac{N_e}{\sum_i Z_i^2 N_i}\int_{-\infty}^{\mu_e}\frac{d\mu'_e F^2_{-1/2}(\mu'_e)}{F^2_{1/2}(\mu'_e)}\\ &= -\frac{N_e}{2}\gamma_e^2\frac{N_e}{\sum_i Z_i^2 N_i}\psi(\mu_e)\,. \end{aligned}\tag{5.47}$$

The exchange integrals for the Boltzmann and degenerate plasmas are calculated analytically, which give rise to the following asymptotics (Fig. 91):

$$\lim_{\mu_e\ll -1}\beta F = N_e\frac{1}{4}\gamma_e^2\frac{N_e}{\sum_i Z_i^2 N_i^2}\,;\tag{5.48}$$

$$\lim_{\mu_e\gg 1}\beta F = -N_e\frac{1}{2}\gamma_e^2\frac{N_e}{\sum_i Z_i^2 N_i^2}\frac{1}{2}\frac{4}{\pi}\mu_e^2 - \frac{2}{3}\pi\ln\mu_e + c_3\left[\left(\frac{4}{3}\sqrt{\pi}\right)\mu_e^{3/2}\right]^2\tag{5.49}$$

It can be seen that the classical asymptotics (5.48) are applicable in the $-\infty \le \mu_e \le 1.5$ range. Second-order [323] exchange corrections can only be calculated for the Boltzmann case (at $\mu_e \le 4$):

$$\begin{aligned}\beta F^{c}_{\mathrm{ex}_2} &= -N_e\frac{\pi^{3/2}}{2^{7/2}}\ln\left[2\frac{\lambda_{\mathrm{F}}\gamma_e}{2S_e+1}\left(\frac{Ze^2N_e}{Z_i^2N_i}\right)^{1/2}\right]\\ &\quad -\frac{N_e}{12\pi}\left[\left(\frac{\beta e^2}{\lambda_{\mathrm{F}}}\right)^3\Big/ 12\lambda_{\mathrm{F}}\right]\frac{N_e^2}{(\sum_i N_i)^2}\,.\end{aligned}$$

The general expression for the third-order exchange correction, as obtained by Ebeling *et al.* [320] in the case of non-degenerate plasma, reduces to

$$\beta F^{c}_{\mathrm{ex}_3} = -N_e\left[\left(\frac{\beta e^2}{\lambda_{\mathrm{F}}}\right)^3\Big/ 12\lambda_{\mathrm{F}}\right]\frac{N_e^2}{(\sum_i N_i)^2}\frac{1}{12\pi}\,.$$

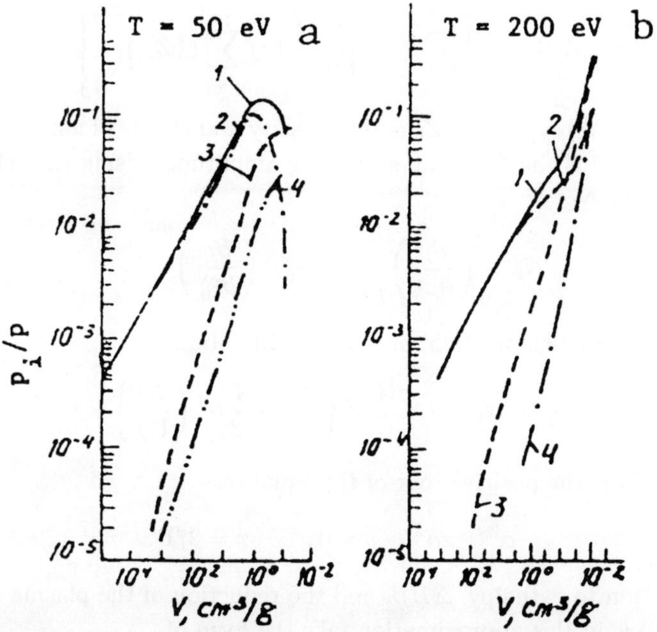

Fig. 5.8. Coulomb and exchange corrections in hydrogen plasma [322] for $T = 50$ eV (a) and $T = 200$ eV (b): $4 - p_{3L}$, ladder correction (5.46); $2 - p_{\rm ring}$, ring correction (5.43); $4 - p_{1ex}$, exchange correction (5.44); $1 - p_4$, total value of Coulomb corrections.

The relative contribution by corrections to the pressure of hydrogen plasma is illustrated in Fig. 5.8.

The models obtained within the perturbation theory are asymptotic and applicable, strictly speaking, only at low values of the expansion parameters $\Gamma \ll 1$, $\gamma \ll 1$. According to Graboske *et al.* [322], the range of validity of the ring model (5.43) is estimated at $\Gamma \leq 0.5$ ($\mu \leq 2$) whereas Ebeling *et al.* [320] give a more modest estimate of $\Gamma \leq 0.1$. For hydrogen, according to Graboske *et al.* [322], the interference effects are adequately described by Eq. (5.45) at $\gamma < 1.1$. In the region of increased non-ideality ($\Gamma \geq 1, \gamma \geq 1$), the perturbation theory results are inapplicable, and here one must either employ non-parametric methods or use extrapolations. For the latter, the use of a ring approximation in a grand canonical ensemble of statistical mechanics [308] is convenient.

The expression for the thermodynamic potential of plasma in a grand canonical ensemble to a ring approximation takes the form [308] of

$$-\beta\Omega = pV\beta = V\left[\sum_k \xi_0^k + \frac{1}{12\pi}\left(4\pi f \sum_k \xi_0^k Z_k^2\right)^{3/2}\right], \tag{5.50}$$

where $\xi_0^k = (1/\lambda_k^3)\exp(\mu_k\beta)$, Z_k is the fugacity and charge of ion of the species "k" and $f = e^2\beta$ is the Coulomb scattering amplitude. Using the relationship for the density N_k,

$$N_k = \left(\frac{\partial p}{\partial \mu_k}\right)_{T,V} = N_0^k\beta\left(\frac{\partial p}{\partial \xi_0^k}\right)_{T,V}. \tag{5.51}$$

In the case of single ionization, one can readily obtain

$$\beta p = N_0 + 2N_e\left\{\alpha^2\left(\frac{2}{\Gamma}\right) + \frac{\Gamma}{3}\alpha^3\left(\frac{2}{\Gamma}\right)\right\}, \tag{5.52}$$

where $\alpha(2/\Gamma)$ is the positive root of the equation

$$\alpha^3 + x\alpha^2 - x = 0\,; \qquad x = 2/\Gamma\,. \tag{5.53}$$

The correction to enthalpy, $\Delta H\beta$, and the reduction of the plasma ionization potential, ΔI, in this approximation take the form of

$$\Delta H\beta = -8N_e\left(1 - \alpha^2 - \frac{\Gamma}{3}\alpha^3\right), \tag{5.54}$$

$$\beta\Delta I = \chi = \Delta\mu_a - \Delta\mu_e - \Delta\mu_i = 2\ln[1 + (1/2)\Gamma\alpha]\,. \tag{5.55}$$

In the $\Gamma \to 0$ limit, this approximation has the correct asymptotics (5.43) and in the limit of high Γ, possesses acceptable extrapolation properties which permit a description of shock wave compression experiments of up to $\Gamma \sim 2.5$. For this reason, the model (5.50)–(5.55) was used in the engineering calculations of the thermophysical properties of working media in a gas-phase nuclear reactor [308].

5.3. Pseudopotential Models. Monte Carlo Calculations

The perturbation theory methods, as discussed in Section 5.2, are based on the expansion in small parameters and because of this are strictly speaking, valid in the limit of weakly non-ideal plasma. From the viewpoint of describing the thermodynamics of plasma under conditions of marked non-idealities, the Monte Carlo techniques [286] are effective in doing so. They do not involve any expansions in small parameters and are therefore, especially attractive in the case of dense gases and liquids, as well as one-component plasma (Section 5.1).

Meanwhile, the form of interparticle interaction potential does not cost any doubts at all. Based on the principles of statistical physics, this method enables one to perform direct computer calculations of the mean thermodynamic values (5.1). It is important that, within this method, the interparticle interaction potential is assumed to be pre-assigned and, in most concrete calculations, two-particle. Therefore, given such an approach, all physical hypotheses refer to this form of interaction potential. This is with due regard for the fact that, following the choice of this form, the Monte Carlo technique enables one to carry out all thermodynamic calculations through to the end.

However, the application of general methods to multi-component plasma runs into specific difficulties. This is because it allows, within classical Monte Carlo formalism, for the quantum effects in electron-ion interactions at close distances. In real plasma, the latter play a decisive role because they ensure stability and lead to the emergence of bound states [286, 325].

The inclusion of the specific quantum-mechanical features of the problem within classical formalism is accomplished using a pseudopotential model. In the model, the description of electron-ion interaction is effected through the introduction of effective potential $\Phi\,(r, T)$, which differs from the initial potential $\Phi(r)$ at only small distances $r \leq \lambda_e$. This difference is governed by the quantum uncertainty of the electron's position within the thermal de Broglie wavelength, λ_e, the exchange effects (for like particles), and the possibility of bound state formation at low ($kT <$ Ry) temperatures. In the case of the pseudopotential approach, the quantum statistical sums reduced to an expression that is classical in form, thus enabling one to employ the classical Monte Carlo calculation technique. In this case, it is possible to allow for the so-called pair quantum effects governing the difference of the two-particle pseudopotential from the initial potential.

5.3.1. *The Choice of Pseudopotential*

We now consider two particles with the interaction potential $\Phi(r)$. The probability density of finding these particles at a distance r in classical statistics is proportional to $\exp[-\beta\Phi(r)]$. In quantum statistics, it is expressed by the Slater sum

$$S_2(r,T) = 2\lambda^3 \sum_{\alpha} |\psi_\alpha(r)|^2 \exp(-\beta E_\alpha)\,, \tag{5.56}$$

where ψ_α and E_α are the wave functions and respective eigenvalues of the energy of two particles, and the summation in Eq. (5.56) is made over all states of discrete and continuous spectra.

When the pseudopotential $\tilde{\Phi}(r,T)$ is defined as a potential giving, in the classical case, the same particle distribution in space as the potential $\Phi(r)$ gives in the quantum case, we obtain

$$\tilde{\Phi}(r,T) = -kT \ln S_2(r,T) . \tag{5.57}$$

In the $T \to \infty$ limit, $\tilde{\Phi}(r,T)$ coincides with $\Phi(r)$, while at finite temperatures they only coincide at great distances. In doing so, it is important that the $\Phi(r,T) - \Phi(r)$ difference is found within a short range. This enables one to construct a thermodynamic perturbation theory [54] for such deviations.

An important part in the development of the pseudopotential model of plasma belongs to the classical studies performed by representatives of the school of W. Ebeling [320]. Ebeling *et al.* [320] and Zelener *et al.* [325] presented the results of calculations of pseudopotentials of electron-proton (Φ_{ep}) and electron-electron interactions. These were obtained by direct summation (5.56) using the ψ_α and E_α of an isolated hydrogen atom (Fig. 5.9). The analytical results also available for $\Phi(r,T)$ in the high-temperature limit are [320, 325, 326].

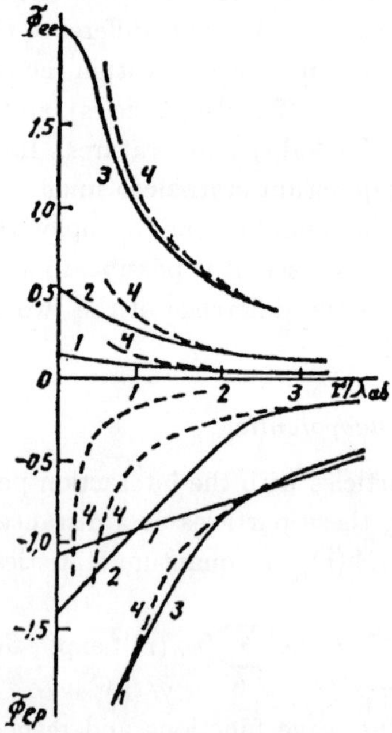

Fig. 5.9. Pseudopotentials Φ_{ep}, Φ_{ee} [325]: 1, $T = 10^3$ K; 2, 10^4 K; 3, 10^5 K; 4, Coulomb's law.

As distinct from $\Phi(r)$, the pseudopotential at $r \to 0$ has a finite value. This is dependent on the concrete electronic structure of atom whose structure is governed by the self-consistent solution of quantum-mechanical many-body problem and, therefore, cannot be described by the pair approximation (5.57). A disregard for this fact led Barker [327] to commit serious qualitative errors in the pseudopotential model, which manifested themselves in the emergence of non-physical complexes due to the great depth of the pseudopotential (5.57). Indeed, the assumption of pair additivity of interaction is, in principle, unfit the interaction between the dissociation in describing and ionization products, which is saturated in character. Therefore, in order to describe chemically reacting systems, one must modify the pseudopotential model by including, explicitly, bound states.

To this end, the relationship [325, 328]

$$S_2(r) = S_2^b(r) + S_2^f(r) \tag{5.58}$$

is introduced, where $S_2^f(r)$ and $S_2^b(r)$ correspond to the states of continuous and discrete spectra, respectively. The latter was selected to obtain a convergent reflection for the statistical sum having the form of Eq. (5.73):

$$\sum_b = \lambda_e^{-3} e^{\beta E_0} \int S_2^b(r) r^2 dr = e^{\beta I} \sum_{n=0}^{\infty} (e^{-\beta E_n} - 1 + \beta E_n) \,. \tag{5.59}$$

In this case, the pseudopotential takes the form of

$$\Phi^*(r,T) = -\beta^{-1} \ln S_2^f = \lambda_e^3 \left\{ \sum_{E_\alpha = E_0}^{0} |\psi_\alpha(r)|^2 (1 - \beta E_\alpha) \right.$$

$$\left. + \sum_{E_\alpha = 0}^{\infty} |\psi_\alpha(r)|^2 \exp[-\beta E_0] \right\} , \tag{5.60}$$

where in the first summand, the summation is made over the states of discrete spectrum and in the second, over states of continuous spectrum. In the case of hydrogen, the following expression is valid [286] for the depth of pseudopotential $\tilde{\Phi}_{\text{ep}}^*(0)$ at $r = 0$:

$$\beta \Phi_{\text{ep}}^*(0,T) = -\ln\{\pi^{1/2} \xi^3 [\xi(3) + \beta \text{Ry} \xi(5)] + 2\pi^{1/2} \xi\} \,,$$

$$\xi = 2(\beta \text{Ry})^{1/2}, \xi(k) = \sum_{n=1}^{\infty} n^{-k}, \xi(3) = 1.202, \zeta(5) = 1.0369 \,.$$

The constructed pseudopotentials Φ^*_{ei} were calculated for a number of temperatures and chemical elements, and the results are presented by Zamalin *et al.* [286].

In view of a weak dependence of Φ^*_{ei} on the temperature and the type of chemical element, the range of discussed parameters for Φ^*_{ei} contains a simple approximation, which forms the basis of the pseudopotential model of plasma to a "zero" approximation [286]:

$$\beta\Phi^0_{ei}(x,\beta) = \begin{cases} -\varepsilon & r \le \sigma \\ -x^{-1} & r \ge \sigma \end{cases} \quad \sigma = e^2\beta\varepsilon^{-1}, x = r/(\beta e^2)$$
$$\beta\Phi^0_{ee}(x) = \beta\Phi^0_{ii}(x) = x^{-1}\,, \tag{5.61}$$

where the numerical parameter of the model, $\varepsilon = 2$, is selected using the experimental data (a dependence of pseudopotential on density was not investigated in [286]). In view of the simple form of Φ^0 in Eq. (5.61), the model in question satisfies the similarity relations. This allows the results to be presented in a compact form (Figs. 5.10, 5.11) and greatly reduces the amount of calculations required. This fact makes the practical use of the model in concrete thermodynamic calculations move simple, whereby it is convenient to make use of the approximation expression [286] for $\varepsilon = 4$:

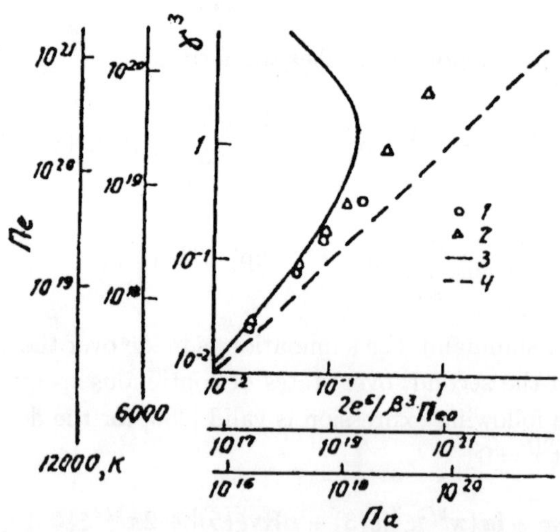

Fig. 5.10. Monte Carlo calculations of the equation of state for plasma to the pseudopotential approximation (5.61) [286], $\gamma^3 = Z^2(Z+1)e^6N_e\beta^3$: *1*, $\varepsilon = 2$ in Eq. (5.61); *2*, $\varepsilon = 4$; *3*, Debye dependence; *4*, ideal plasma.

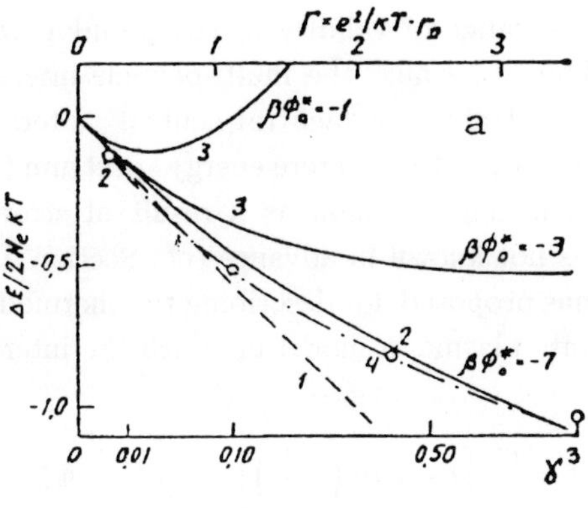

(a)

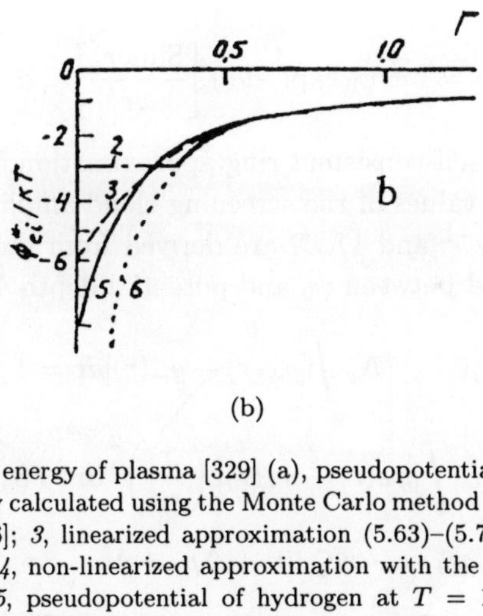

(b)

Fig. 5.11. Dimensionless energy of plasma [329] (a), pseudopotential of plasma (b). *1*, Debye approximation; *2*, energy calculated using the Monte Carlo method for pseudopotential (5.61) with $\beta\Phi^*_{ei}(0) = -3$ [286]; *3*, linearized approximation (5.63)–(5.71) for different depths of pseudopotential (5.63); *4*, non-linearized approximation with the depth of pseudopotential (5.63) $\beta\Phi^*_{ei}(0) = -6$; *5*, pseudopotential of hydrogen at $T = 10^5$ K [286]; *6*, Coulomb potential; *7*, pseudopotential (5.61); *8*, pseudopotential (5.63).

$$\left.\begin{aligned} \frac{\beta p}{\rho} &= 1 - 0.66\gamma^{3/2} + 0.59\gamma^3 - 0.2\gamma^{3/2}\,; \\ \frac{\beta F}{N} &= -0.89\gamma - 0.45\gamma^2 + 0.54\gamma^3\,, \quad \gamma^3 = Z^2(Z+1)e^6 N_e \beta^3 \end{aligned}\right\}. \tag{5.62}$$

The differences between the real pseudopotential (5.60) and zero approximation model (5.61) can then be allowed for with the aid of the methods of the thermodynamic perturbation theory developed for arbitrary local potential [286].

On the whole, the range of validity of the pseudopotential Monte Carlo model is limited. This is because the multi-particle interactions, degeneracy effects and "charge-neutral" and "neutral-neutral" interactions are ignored. Furthermore, one must know the discrete energy spectrum (5.51), (5.57) which, may be distorted in a dense plasma as a result of strong interaction and, generally speaking, is not known in advance (cf., Section 5.4).

Iosilevskii [329] has proposed, for describing the thermodynamics of a partly ionized non-degenerate plasma, a model in which the interparticle interaction takes the form of

$$\Phi_{ei}^*(r) = \frac{e^2}{r}\left[1 - \exp\left(-\frac{r}{\sigma}\right)\right]\ ;\quad \Phi_{ee}^* = \Phi_{ii}^* = \frac{e^2}{r}\,. \tag{5.63}$$

Following Gryaznov and Iosilevskii [309], we shall write the binary correlation functions as

$$g_\pm(r) = g_{\alpha\beta}(r) = 1 \pm \psi_0 \exp(-\nu r)\left[\frac{\mathrm{Sh}(\omega r)}{\omega r}\right]\,,\quad \alpha,\beta = e, i, \tag{5.64}$$

as derived within the self-consistent ring approximation for potential (5.63) in the $\Gamma \to 0$ limit. The values of the screening cloud amplitudes, ψ_0, and of the screening distances, ν^{-1} and $1/\omega^*$, are derived from the screening condition and approximate bond between ψ_0 and potential depth $\Phi_{ei}^*(0)$ [330]:

$$N_e \int [g_+(r) - g_-(r)]d\mathbf{r} = 1\,, \tag{5.65}$$

$$N_e \int [g_+(r) - g_-(r)]\left(\frac{r}{r_{\mathrm{D}}}\right)^2 d\mathbf{r} = 3\,, \tag{5.66}$$

$$\psi_0 \approx -\beta\Phi_{ei}^*(0) + \Delta\mu_e + \delta\mu_i\,. \tag{5.67}$$

Corrections to the potential energy $\Delta\Pi^*$ and internal energy ΔE^*, as well as to the pressure and chemical potential, take the form [329] of

$$\Delta\Pi^* = N_e V^{-1}\int \frac{e^2}{r}[g_+(r) - g_-(r)]d\mathbf{r}\,, \tag{5.68}$$

$$\Delta E^* = N_e^2 V^{-1}\int [\Phi_{ei}^* g_+(r) - \Phi_{ii}^*(r) g_-(r)]d\mathbf{r}\,, \tag{5.69}$$

$$\Delta p = (3V)^{-1}(2\Delta E^* - \Delta\Pi^*)\,, \tag{5.70}$$

$$\Delta\mu_e = \Delta\mu_i \approx (2N_e)^{-1}\Delta E^*\,,\quad N_e = N_i = n_e V\,. \tag{5.71}$$

The relationship (5.70) follows from the virial theorem while Eq. (5.71) expresses the condition that the correlation, which occurs upon the inclusion of an additional charge, is proportional to the magnitude of the latter. Concrete calculations using this model boil down to the solution of algebraic equations. In the weak non-ideality limit, the result obtained changes into a Debye result and differs from the latter, with a density increase, by smaller correction values which become positive at $\sigma/r_{\mathrm{D}} \geq 1$. As can be seen in Fig. 5.6, which shows the dimensionless energy of a free charge system [329], for like potentials the results of simple calculations using the model (5.63)–(5.71) are in adequate agreement with the results of the Monte Carlo calculation. Apparently, this is due to a satisfaction of the general conditions of local electroneutrality in the model (5.63)–(5.71) which proved to be very important (as in the case of the one-component plasma model).

In selecting the sole numerical parameter of the model (5.63)–(5.71), that is, the potential depth $\Phi^*_{ei}(0)$, one must make use of experimental data, which is done in the case of the Monte Carlo model (5.61). It turned out that the optimum description could be ensured by choosing $\Phi^*_{ei}(0) \approx \varepsilon = -kT$. This energy separates the electronic states into free and bound ones. Note that, with such a choice of $\Phi^*_{ei}(0)$, one manages to describe, in a uniform manner, both the experiments in cesium (which served as a basis for such a choice) and those in argon and xenon, the results of which were not previously used in selecting Φ^*_{ei}.

5.4. The Bound State Contribution. The Confined Atom Model

An adequate inclusion of bound states presents one of the most complicated problems in describing a non-ideal low-temperature plasma. In the chemical model, the bound states appear separated (5.40) from the continuous spectrum and are described by statistical sums:

$$F_b = -kT \sum_k N_k \ln \Sigma_k\,; \quad \Sigma_k = \sum_{n=1}^{\infty} g_n^k \exp\{-\beta E_n^k\}\,, \tag{5.72}$$

where g_n^k and E_n^k denote the statistical weight and excitation energy of the n-th energy state, respectively. In order to find these quantities, use is usually made either of spectroscopic measurement data obtained for rarefied plasma, or of the results of quantum-mechanical calculations of isolated atoms and ions. The statistical sum is divergent and calls for forced constraint, which is reflective of the presence of plasma environment. Many recipes for such a constraint have been proposed recently on the basis of qualitative physical considerations, such as over the Debye screening distance, on the mean interparticle distance, and on the last realized (in spite of the microfield effect) quantum number. These models were reviewed by Armstrong *et al.* [331]. For most substances, the value of the energy of first excited states is comparable to that of the ionization potential. Hence their contribution to Σ_k only appears appreciable at high temperatures when the plasma is already ionized to a considerable extent and contains few neutrals. Because of this, the concrete mechanism of constraint for a rarefied plasma is of less importance than the very fact of the presence of such a constraint. This circumstance explains, in part, the extremely small number of available studies into the thermodynamics of discrete spectrum of low-temperature plasma, as well as the particular degree of approximation of the models used. As the pressure rises, the degree of plasma ionization drops. This increases the sensitivity of thermodynamic functions to specific methods of calculation Σ_k [332] and calls for a more thorough analysis of the effect which non-ideality has on the bound state contribution.

The inclusion of mechanical interaction under the perturbation theory leads [333] to the convergent expression (hydrogen plasma)

$$\Sigma_k = \sum_n g_n(e^{-\beta E_n} - 1 + \beta E_n) \tag{5.73}$$

used in the pseudopotential model discussed in Section 5.3.

In addition, in a strongly compressed plasma, the interparticle interaction causes considerable shift, deformation and splitting of energy levels, that is, the phenomena which cannot be calculated under the perturbation theory and call for the solution of a full-scale quantum-mechanical problem with due regard for all forms of interaction.

The most simple method is to calculate the bound states of a single electron in potentials with different structures, which simulate the plasma environment. The results of such calculations [334] are given in Fig. 5.12. It shows a sharp variation of the energy structure in a relatively narrow range of atomic compressions. The thermodynamic consequences of such a deformation of the energy spectrum for the model (5.74) have been analyzed by Graboske *et al.* [324] and illustrated in Fig. 5.13, indicating the relative contributions by various thermodynamic corrections. It is seen that in the range of increased plasma densities, even for hydrogen one should expect an appreciable variation of plasma compressibility due to a deformation of the bound states. Unfortunately, this simple model of hydrogen plasma does not permit a comparison with experimental data, which are only available for many-electron atoms.

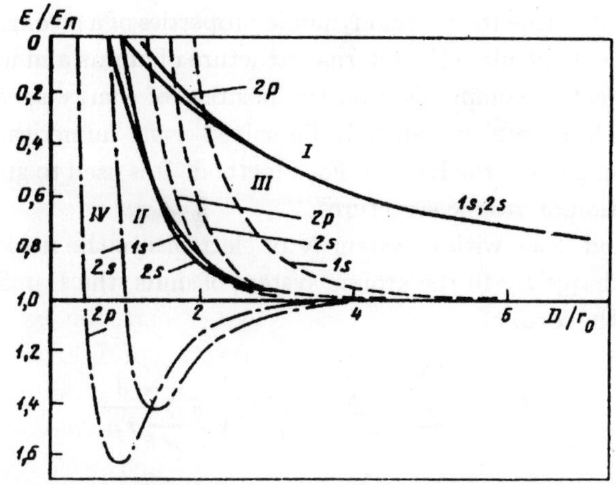

Fig. 5.12. Deformation of energy spectrum for hydrogen: *I*, Debye–Hückel potential; *II*, cutoff Coulomb potential; *III*, boundary condition for wave function $f(r_0) = 0$; *IV*, boundary condition $(\partial f/\partial r)|_{r=r_0} = 0$ (condition of periodicity), D is the Debye screening distance, r_0 is the boundary atomic radius.

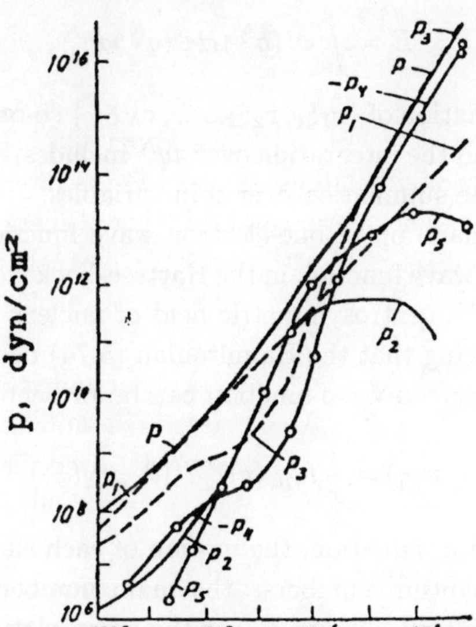

Fig. 5.13. Equation of state for hydrogen plasma, $T = 2$ eV [324]: p_1, the ideal gas ion pressure; p_2, the pressure allowing for the deformation of discrete spectrum (Section 5.4); p_3, the electron pressure (5.41); p_4, Coulomb correction; p_5, inclusion of the proper size of atoms; p_Σ, total pressure.

In order to describe the thermodynamic properties of a strongly compressed plasma, one must calculate the internal structure of atoms and ions, as well as include the effect of compression on the position of energy spectra of bound electrons. Such a problem can only be solved using numerical methods, of which the leading one is the Hartree-Fock method. It is used to great advantage in the calculation of atomic structures.

This method deals with a system N of electrons in the nuclear field of an atom with a charge Z. In the atomic system of units, the Hamiltonian of the system takes the form of

$$\hat{H} = -\sum_{i=1}^{N}\left[\frac{1}{2}\Delta_i + U(r_i)\right] + \sum_{j<k}^{N}\frac{1}{r_{jk}}, \tag{5.74}$$

where $U(r_i)$ is the effective potential which, in the case of isolated atom, is equal to Z/r_i.

Proceeding from the variational principle of quantum mechanics, we shall determine the many-electron wave function from the condition of minimum of the functional

$$E = \int \psi^*(q^N)\hat{H}\psi(q^N)dq^N, \tag{5.75}$$

where q^N is a combination of $\{\mathbf{r}_1\xi_1, \mathbf{r}_2\xi_2, \ldots, \mathbf{r}_N\xi_N\}$ co-ordinates and spin variables of electrons, and the integration over dq^N includes, along with integration over co-ordinates, the summation over spin variables.

A determinant made up of one-electron wave functions is selected as the approximate atomic wave function in the Hartree-Fock method. Assuming that each electron is in the centrosymmetric field of nucleus and of the remaining electrons, and observing that the Hamiltonian (5.74) does not depend on spin variables, the one-electron wave function can be presented as

$$\psi_{(q)} = \frac{1}{r} f_{n_i l_i}(r) Y_{l_i}^{m_i}(\theta, \varphi)\chi(S_i). \tag{5.76}$$

Therefore, in this approximation, the motion of each electron is characterized by the following quantum numbers: the main number n, the orbital l the magnetic m, and the spin number s. For the given n and l, $2(2l+1)$ electron states exist, which correspond to the various m and S. It is assumed that all electrons with pre-assigned n and l have identical radial parts $f_{nl}(r)$. In the absence of spin-orbit interaction, the atomic state is characterized by a complete orbital moment L and spin moment S. They are equal to zero if, for every n and l, all the $2(2l+1)$ states are occupied, that is all the nl shells are filled. When an atom has unfilled shells, then for the atomic wave function to be an eigenfunction for the operators of the complete orbital and spin moments, that is, for the atom to have pre-assigned L and S, it is necessary to seek that function in the form of a linear combination of determinants, the coefficients of which are found from the condition of equality of the complete orbital and spin moments to the pre-assigned values.

The application of the variational principle leads to a set of integro-differential equations (Hartree-Fock equations):

$$\left[\frac{d^2}{dr^2} + U_{n_i l_i}(r) - \frac{l_i(l_i+1)}{r^2} - \varepsilon_{n_i l_i}\right] f_{n_i l_i}(r) + \sum_{n_j = n_i} \varepsilon_{n_i l_i n_j l_j} f_{n_j l_j}(r) = G_{n_i l_i}(r)\,, \tag{5.77}$$

where $U_{n_i l_i}(r)$ is the Coulomb potential of interactions of electrons with the nucleus and with each other, $G_{n_i l_i}(r)$ is the non-local part of the potential or exchange term, and the quantum numbers n_i and l_i vary according to the atomic configuration. The eigenvalues of $\varepsilon_{n_i l_i}$ and non-diagonal Lagrangian multipliers $\varepsilon_{n_i l_i n_j l_j}$, introduced to satisfy the condition of orthogonality of the radial functions $f_{n_i l_i}(r)$ and $f_{n_j l_j}(r)$ at $n_i \neq n_j$, are derived from the condition of orthonormality,

$$\int_0^\infty f_{n_i l_i}(r) f_{n_j l_j}(r) d\mathbf{r} = \delta_{n_i n_j}\,, \tag{5.78}$$

and boundary conditions, of which one is the result of finiteness of the wave function at the origin of co-ordinates $f_{n_i l_i}(0) = 0$. The other condition is selected based on the concrete statement of the problem. For instance, in an isolated atom approximation, the electron wave function must decrease exponentially at great distances. In a "bound" (or "confined") atom approximation [146, 335], which is one of the ways for an effective inclusion of the external effect on the atom, the condition

$$f_{n_i l_i}(r)|_{r=r_c} = 0\,, \tag{5.79}$$

which corresponds to the interaction potential

$$U(r) = \begin{cases} -\dfrac{Ze^2}{r}\,, & r < r_c\,, \\ \infty, & r > r_c \end{cases} \tag{5.80}$$

is imposed upon the radial wave function. We now consider the case of a ternary plasma consisting of atoms, single-charged ions and electrons. Assuming that the atoms are spheres with a variable radius r_c and for the sake of simplicity, ignoring the size of electrons and ions, we shall regard a subsystem of finite-sized atoms as a system of hard spheres which do not interact when the distance between them exceeds $2r_c$. The free energy of such a model can be written as

$$F(N_i, V, T) = F_{\text{id}} + F_{\text{hs}} + \Delta F_{\text{coul}}\,.$$

The first summand is the free energy of ideal plasma. It is distinguished by the fact that the statistical sum for atom depends, in accordance with the boundary condition (6.8), on the radius. The second summand is the contribution by the hard sphere repulsion, which also depends on r_c, in terms of the dimensionless parameter $\nu = N_a(4\pi r_c^3/3)$. In order to include this contribution, use is made of the results of calculations of the thermodynamics of the system of hard spheres. The calculations were performed using the methods of molecular dynamics whose results are described by the Pade representation [336]:

$$F_{\text{hs}} = N_a kT \nu \frac{3\nu - 4}{(\nu - 1)^2}\,. \tag{5.81}$$

The corrections to pressure and chemical potential

$$\frac{\Delta p_{\text{hs}}}{N_a kT} = 2\nu \frac{\nu - 2}{(\nu - 1)^3}\,, \qquad \frac{\Delta \mu_{\text{hs}}}{kT} = \nu \frac{3\nu^2 - 9\nu + 8}{(1 - \nu)^3}\,. \tag{5.82}$$

Therefore, the free energy in this model depends on r_c. This is due to the compression of atoms on the one hand, and to their interaction as hard spheres on the other. An equilibrium value of atomic radius can be determined from the condition of minimum of free energy

$$\partial F / \partial r_c = 0\,. \tag{5.83}$$

The r_c dependence of the statistical sum for atom is derived from the solution of the set of Hartree-Fock equations for the ground and excited atomic states with boundary condition (5.79) at different r_c (Fig. 5.14). The solution of Eq. (5.83) gives an equilibrium value of $r_c(V, T)$, which renders the model thermodynamically closed.

As distinct from solid-state cell models [160], this approximation is constructed within the quasi-chemical method of description with an explicit inclusion of the translational degrees of freedom of individual particles. In doing so, the electrons are separated into two species located inside and outside the cells whose volume, $(4/3)\pi r_c^3$, is but a part of the average volume per particle.

The thermodynamic calculations performed for the model (5.77)–(5.83) demonstrate that the effective repulsion and deformation of the discrete atomic

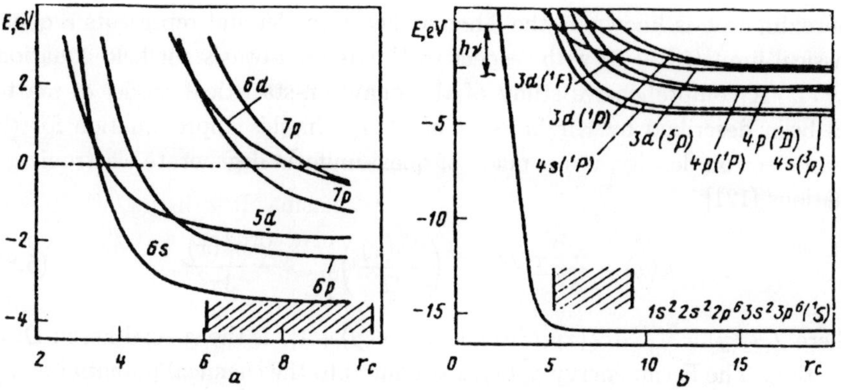

Fig. 5.14. Quantum-mechanical calculation of the energy spectrum of compressed cesium (a) and argon (b) based on the bound atom model (5.77)–(5.83). The shaded region is the experimental range [106, 128, 137]; r_c is the atomic cell radius in a_0 $h\nu = 2.14$ eV.

spectrum in the selected potential have an appreciable effect on both the thermal and caloric equations of state for plasma. Given the pressure and temperature, the density calculated under this model has a smaller value than that obtained in a ring Debye approximation, which is the same as the enthalpy at fixed pressure and volume. The latter fact is important because it is one of the principal qualitative results heretofore experimentally obtained in a non-ideal plasma (cf., Section 5.6).

The effect of the energy level deformation shows up largely in the optical properties of plasma since the atomic photoionization by visible radiation occurs from highly excited energy levels, the distortion of which occurs at even relatively low compressions. We shall discuss this effect at length in Chapter 8.

5.5. Quasi-Classical Approximation

As pressure increases, molecular and ion-molecular formations in the plasma disintegrate. This results in the separation of the outer shell electrons and the electron shells of atoms. The ions reorganize themselves and, at extremely high pressures, disintegrate to bring about a quasi-uniform distribution of charge in the atomic cell [121]. In this case, it is possible to change from a quantum-mechanical description of bound electrons using the language of wave functions and discrete energy spectrum to a quasi-classical description in terms of the distribution of mean electron density $n(x)$. The pertinent approximation is known as the Thomas-Fermi model and represents a quasi-classical limit ($\hbar^2 \to 0$) with respect to Hartree self-consistent field equations (5.77). A comprehensive study of the quantum-statistical model of matter has been described by Kirzhnits *et al.* [121]. In this approximation for the mean electron density, use is made of quasi-uniform degenerate electronic gas relations [121]:

$$n(x) = \frac{\sqrt{2}}{\pi^2} T^{3/2} J_{1/2}\left(\frac{p_{\rm F}^2(x)}{2T}\right)_{T\to 0} \to \frac{p_{\rm F}^2(x)}{3\pi^2}\,, \tag{5.84}$$

where $J_{1/2}(y) = \int_0^\infty d\tau\{(\tau y^{1/2} d\tau)/[\exp(\tau - y) + 1]\}$ is the special Fermi-Dirac function. The Fermi energy $p_{\rm F}^2(x)/2$ is related to the chemical potential by the relationship

$$\mu = \frac{p_{\rm F}^2}{2} + U(r)\,. \tag{5.85}$$

In the thermodynamic Thomas-Fermi model, the matter is broken down into electroneutral spherical Wigner-Seitz cells containing a nucleus and z electrons surrounding the latter. These electrons are in a self-consistent potential developed by the electrons and nucleus and they satisfy the Poisson's equation

$$-\nabla^2 U = 4\pi n \tag{5.86}$$

with the boundary conditions

$$\underset{r\to 0}{U(r) \to} \frac{1}{r}\,; \quad \underset{r\to R}{U(r) \to} (R - r)^2\,, \tag{5.87}$$

where the cell radius R is governed by the condition of electroneutrality $\int n(r) d\mathbf{r} = Z$. Numerical integration (5.84)–(5.87) enables one to determine the intra-atomic density $n(x)$, from which all thermodynamic functions of the electronic gas of the atomic cell are recovered in accordance with the expression for free energy

$$F = \frac{\sqrt{2}}{\pi^2} T^{5/2} \int d\mathbf{r} \left[\frac{p_{\rm F}^2}{2T} J_{1/2}\left(\frac{p_{\rm F}^2}{2T}\right) - \frac{2}{3} J_{3/2}\left(\frac{p_{\rm F}^2}{2T}\right)\right] + \int U(r) n(r) d\mathbf{r}\,.$$

In order to obtain the summary thermodynamic characteristics of the model, the contribution by the motion of nuclei described either to ideal-gas (5.42) or quasi-harmonic approximations [337], or using the one-component plasma model [337] (cf., Section 5.1), is added to the electron terms at $T \geq 1$ keV, one must further allow for the pressure and energy of equilibrium radiation. Equations (5.84) to (5.87) are characterized by self-similarity with respect to the nuclear charge Z:

$$ZV; \quad Z^{-4/3}T, \quad Z^{-10/3}p, \quad Z^{-4/3}E, \tag{5.88}$$

which simplifies the practical use of numerical calculations under the Thomas-Fermi model [338–340] considerably while enabling one to allow for the specific features of a particular element by a simple variation of the scale. Modifications of this model are associated with a more detailed inclusion of correlation and quantum effects. The correlation corrections are caused by the difference of the Hartree self-consistent field from the true field within the atomic cell. These corrections are the result of the anti-symmetry of the electron wave functions, and are interpreted as exchange and correlation effects. In addition, force correlation effects occur because of the inaccuracy of the independent particle pattern assumed in the model. The quantum-mechanical corrections are due to quasi-classical formalism. They are divided into a regular, with respect to $\hbar^2$, part (referred to as quantum) reflecting the presence of a non-local correlation between $n(x)$ and the potential $U(x)$ due to the uncertainty principle, and an irregular correction reflecting the non-monotonicities of physical quantities due to discrete energy spectrum. It is significant that the most modern modifications of the Thomas–Fermi model [121] are associated with the introduction of oscillation correction while the inclusion of exchange, correlation and quantum corrections [339] present a traditional approach employed in the physics of high energy densities.

The relative magnitude of the correlation and exchange effects is controlled by the dimensionless parameters [121] $\delta_{\text{cor}} \sim (n^{1/3}/p_{\text{F}}^2)^\nu$ and $\delta_{\text{quant}} \sim \delta_{\text{exch}} \sim n/p_{\text{F}}^4$. In the degeneracy region (where $n^{2/3} \gg T$, $p_{\text{F}} \sim n^{1/3}$, $\delta_0 \sim n^{-1/3}$, $\nu = 2$), they amount to $\delta_{\text{cor}} \sim n^{-1/3}$ and $\delta_{\text{exch}} \sim n^{-2/3}$, and in the classical region ($n^{2/3} \ll T, p_{\text{F}} \sim T^{1/2}$, $\delta_0 \sim n^{-1/3}/T$, $\nu = 3/2$), to $\delta_{\text{cor}} \sim n^{1/2}/T^{3/2}$, $\delta_{\text{exch}} \sim n/T^2$. The range of validity of the quasi-classical method in describing the atomic cell electrons, estimated [121] on the basis of these criteria, is defined at $T = 0$ by the condition $n \gg 1$ and in hot matter by the condition $n \ll T^3$, which corresponds to the conditions of smallness of the binary interaction energy of electrons as compared to their kinetic energy.

The physical conditions of validity of the quasi-classical model correspond to extremely high pressures $p \gg p_{\text{at}}$ ($p_{\text{at}} = e^2/a_0^4 = 300$ Mbar is the atomic unit of pressure) and temperatures $T \gg 10^5$ K, which are realized in the interior of superdense stars and in other outer space objects but are still inaccessible to experimental techniques under terrestrial conditions. This is because they call for extremely high local concentrations of energy [106]. At present, record-breaking pressures and temperatures are attained by dynamic methods using the powerful shock wave technique. The data of these experiments pertain to pressures of tens of megabars, that is, they are very much below the range of formal validity of quasi-classical models that is set by the smallness of the appropriate literal criteria. Therefore, the existing estimates of the lower range of validity of the Thomas-Fermi model [341] are based on the extrapolation of the results of those experiments and are purely qualitative. Al'tshuler *et al.* [341] note that the inclusion of quantum, exchange and correlation corrections (oscillation corrections were not included) improves appreciably the extrapolation which they believe, in this case, to be possible for a cold ($T = 0$) plasma up to pressures of $p \geq 300$ Mbar and, at $T \sim 10$ keV, up to about 50 Mbar.

Recently, however, such estimates were questioned. The point is that the commonly used version of the quasi-classical model [388–342] ignores the so-called shell effects caused by the individual peculiarities of the population of atomic energy levels and bands and describes "averaged" characteristics of matter. The elegant studies performed by Kirzhnits *et al.* [121] have shown that the shell effects can be qualitatively described within quasi-classical approximation by including an irregular (with respect to $\hbar^2$) correction corresponding to the oscillating part of electron density, which was erroneously discarded previously.

The inclusion of shell effects greatly changes the equation of state for superdense plasma and brings, in the high-pressure region (where this approximation is justified), about the emergence of sharp non-monotonicities in thermodynamic functions [121]. In addition, discontinuities occur in the equations of state. These are caused by electron phase transitions of the first kind when energy levels are "forced out" from the discrete spectrum into the continuous spectrum. This in turn, causes peculiarities in other physical quantities. Note that the sizable contribution by shell effects at hypermegabaric pressures follows from the results of quantum-mechanical calculations using the more accurate method of attached plane waves [160]. The foregoing effects are predicted in a wide range of parameters and must disappear only at $n \gg Z^4$ in the uniformity region when all atomic energy levels pass over to the continuous spectrum [160].

Therefore, the question of the lower range of validity of the quantum-static model for superdense plasma is open. Meanwhile, the behavior of matter in the region of $p \geq 300$ Mbar appears more diversified than previously assumed on the basis of simplified representations [338–342]. At the same time, the simplicity and universality of the Thomas-Fermi model (5.88) render it indispensable in the case of an "averaged" description of plasma properties at ultrahigh pressures when the use of more complex approaches involves a large amount of laborious quantum-mechanical calculations. Based on various modifications of the quasi-classical model, extensive tables of thermodynamic functions have been published [338-342] which cover the range of parameters far beyond the range of validity of this approximation. In particular, Kalitkin and Kuz'mina [342] propose to use their data for a low-density "gas" plasma, for which the "chemical" model of plasma (Section 5.2) is traditionally used. A detailed analysis of the validity of the Thomas-Fermi model for this range of parameters is given by Iosilevskii and Gryaznov [343].

Of special importance in using the quasi-classical method to describe a plasma of reduced (as compared to the density of a solid) density are specific errors introduced by the cell model proper. In this model, all correlations are automatically restricted by the atomic cell size and, therefore, cannot increase the mean distance between the nuclei. This fact, in particular, sets limits to the validity of the Thomas-Fermi model in describing the plasma under conditions which are typical for this state of matter when the Debye screening sphere contains a large number of charges $\Gamma \ll 1$. Therefore, the limit of weakly non-ideal ("Debye") plasma is absent from the results of the Thomas-Fermi theory.

Figures 5.15 and 5.16 show the comparison between thermal and caloric equations of state for lithium calculated from the "chemical" model of plasma with due regard for ionization under Saha's equation (Section 5.2) and in a

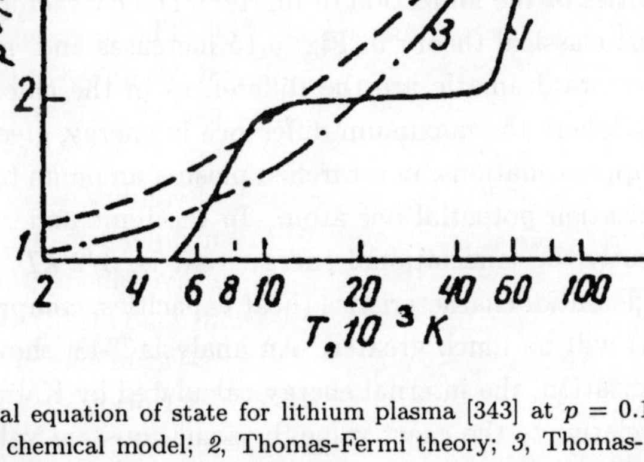

Fig. 5.15. Thermal equation of state for lithium plasma [343] at $p = 0.1$ Mpa: *1*, calculation from Saha's chemical model; *2*, Thomas-Fermi theory; *3*, Thomas-Fermi theory with corrections [339, 342].

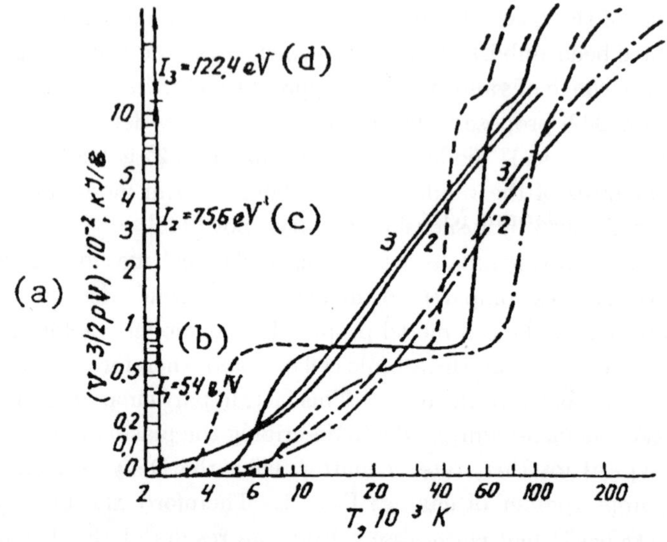

Fig. 5.16. Caloric equation of state for lithium [342]. Broken curves, pressure of 100 Pa; solid curves, 0.1 MPa; point-dotted curves, 100 MPa. I_1, I_2, I_3 are the potential successive ionization. The rest of the designations are the same as in Fig. 5.15.

Thomas-Fermi approximation with quantum and exchange corrections [339, 342]. The step-wise behavior of the thermodynamic functions in Figs. 5.15 and 5.16 is caused by ionization processes, and the steepness of these steps increase as the density decreases ($\rho \to 0$). The Thomas-Fermi theory, while producing a "smoothed" description of thermodynamic functions, fails to convey the peculiarities of the ionization of matter. The characteristic maximum error of the quasi-classical theory in Fig. 5.15 increases and, at $p \to 0$, tends to unity. Even more dramatic are the differences in the caloric equation of state (Fig. 5.16), where the maximum difference in energy, given by Saha and Thomas-Fermi approximations, in a rarefied plasma amounts to a value of the order of the ionization potential per atom. In the limit under consideration, this exceeds greatly the translational energy, that is, 3/2 kT. Naturally, the differences in differential characteristics (heat capacities, compressibility moduli and so forth) will be much greater. An analysis [343] shows that, in the limit of total ionization, the internal energy calculated by Kalitkin *et al.* [340, 342] is shifted, relative to the exact value, by some constant value and reaches a value of 35–60% for lithium.

As a result of the absence of long-range correlations from the cell model, instead of giving the Debye term in the caloric equation of state ($\sim \rho^{3/2}$), the quasi-classical approximations give a negative term ($\sim \rho^{4/3}$) which corresponds to the nuclear charge interaction with electrons uniformly distributed over the cell volume.

5.6. Density Functional Method

At present, the method of density functional (MDF) for energy (thermodynamic potential) is the most effective method to describe strongly interacting dense systems. According to the Hohenberg-Kohn theorem and its generalizations, the energy (thermodynamic potential) and wave function of ground state with a non-uniform electron system are single-value functionals of the distribution of electron number density $N_e(\mathbf{r})$. The consistent use of MDF to describe the properties of a dense electron-ion system combines the simplicity of a quasi-classical method with an accurate inclusion of the exchange-correlation effects and separation of bound states. This method, however, is free from considerable mathematical complications associated with the consistent use of the Hartree-Fock method. It is not surprising that MDF and its modifications were used to advantage in numerous spheres of physics to describe the electronic structure of atoms and molecules, clusters and other small particles, as well as the surface phenomena and properties of solid and liquid metals and non-ideal plasma. It is convenient to use the MDF to investigate the equations of state for dense systems.

The description of MDF and some of its applications can be found, for example, in [344–346]. Below, we will dwell on the most simple formulation of MDF as a non-uniform electron fluid at $T = 0$, and then briefly discuss the calculation of the equation of state for cellular plasma.

The energy of non-uniform electron system in external field $V(r)$ is the functional of its density:

$$E\{N_e(r)\} = \int V(r)N_e(r)d\mathbf{r} + \frac{1}{2}\int \frac{N_e(r)N_e(r')}{|\mathbf{r}-\mathbf{r}'|}d\mathbf{r}d\mathbf{r}' + G\{N_e(r)\}\,. \quad (5.89)$$

Here, $G\{N_e(r)\}$ is the contribution of kinetic energy and exchange-correlation effects. The functional G has a universal form, which is not known exactly, so that use is made of its fairly accurate and simple approximations.

When the electron number density is allowed to vary slowly, that is, $d\ln N_e(r)/dr \ll k_{\mathrm{F}}(r)$, where $k_{\mathrm{F}}(r)$ is the local Fermi wave vector, one can then use the gradient expansion of G:

$$G = \int d\mathbf{r}[g_0(N) + g_2(N)|\nabla N_e|^2 + \cdots].$$

The kinetic and exchange-correlation energies are treated separately, with G written as the sum $G = E_t + E_{xc}$. For a degenerate electron gas,

$$E_t = \frac{3}{10}(3\pi^2)^{2/3} \int N_e^{5/3}(r)d\mathbf{r} + \frac{1}{72} \int |\nabla N_e(r)|^2 N_e^{-1}(r)d\mathbf{r}, \qquad (5.90)$$

where the second term is the Weizsaecker-Kirzhnitz gradient correction. A large number of studies have been devoted to the investigation of E_{xc}. In a local approximation, one of the most simple expressions for E_{xc} is the interpolation formula of Pines and Nosierez:

$$E_{xc} = -\int N_e(r)\left[\frac{3}{4}\left(\frac{3}{\pi}\right)N_e^{1/3}(r) + 0.0474 + 0.0155\ln(3\pi^2 N_e(r)^{1/3})\right]d\mathbf{r}. \qquad (5.91)$$

The functional (5.89)–(5.91) may be minimized with the aid of a set of trial functions, $N_e(r)$.

It is interesting to note that even the most simple approximations of the type of (5.89)–(5.91) yield very satisfactory results. For example, the functional written above gives a qualitatively adequate description of the surface properties of alkali metals and even of small metal particles as well. For small particles, the role of the external potential $V(r)$ is played by the field of ion core of the particle $N_e(r) = N_{\mathrm{L}}\theta(R-r)$, where N_{L} is the metal density and R is the particle radius.

We write the Euler-Lagrange equation for the functional $E[N_e(r)]$, in view of the conservation condition $\int N_e(r)d\mathbf{r} = N$, to derive the self-consistent system of Kohn-Sham equations:

$$\left[-\frac{1}{2}\nabla^2 + V_{\mathrm{eff}}(r)\right]\psi_i(r) = \varepsilon_i\psi_i(r), \qquad (5.92)$$

where

$$V_{\mathrm{eff}}(r) = \varphi(r) + \frac{\delta E_{xc}[N_e(r)]}{\delta N_e(r)}; \quad \varphi(r) = V(r) + \int \frac{N_e(r')}{|\mathbf{r}-\mathbf{r}'|}d\mathbf{r}'.$$

The electron number density is derived with the population N of lower eigenstates:

$$N_e(r) = \sum_{i=1}^{N} |\psi_i(r)|^2. \qquad (5.93)$$

Therefore, the problem consists in solving Eq. (5.92). Subsequently, it reaches self-consistency with the aid of (5.95). As distinct from the Hartree-Fock equations, the Kohn-Sham equation is exact and the exchange-correlation potential $V_{xc} = \delta \cdot E_{xc}/\delta \cdot N_e$ is local and universal.

The theory is generalized to a two-component system at finite temperature. The free energy F is the functional of electron and ion distribution:

$$F = F[N_e(r), N_i(r)] = \int f[N_e(r), N_i(r)]d\mathbf{r}\,.$$

We restrict ourselves, as above, to the first terms of gradient expansion to derive

$$f[N_e(r), N_i(r)] = g(N_e, N_i) + \frac{1}{2}\varphi(N_e - N_i) + \sum_{a,b} \Phi_{a,b} \nabla n_a \nabla_b\,,$$

where $g(N_e, N_i)$ is the local density of free energy (quasi–one-dimensional part of the functional $f(N_e, N_i)$) and

$$\varphi(r) = \int [N_e(r') - N_i(r')]|\mathbf{r} - \mathbf{r}'|^{-1} d\mathbf{r}'$$

is the electrostatic potential that satisfies Poisson's equation. In the gradient term, the subscripts a and b run through the values of e and i.

Klyuchnikov and Lyubimova [346] and Liberman [347] investigated the equation of state for metals within the cell model. A nucleus, shielded by its Z electrons, is located at the center of a spherical space of radius $\bar{r} = (3/4\pi \cdot N_i)^{-1/3}$. The distribution of electrons inside the cell is defined by the set of solutions of the Kohn-Sham equation:

$$\frac{1}{2r^2}\frac{d}{dr}\left(r^2 \frac{du_l}{dr}\right) + \left[E - 2V - \frac{l(l+1)}{2r^2}\right] \cdot u_i = 0\,, \quad r \le \bar{r}\,. \tag{5.94}$$

For bound states, the boundary condition is $u_l(r) = 0$. For free states $E = k^2/2$, solution (5.94) must be joined at $r = \bar{r}$, with the wave function,

$$u_{kl}(r) \sim \cos\delta_l(k) \cdot j_l(kr) - \sin\delta_l(k) \cdot n_l(kr), \quad r > \bar{r}\,,$$

where j_l and n_l are the Bessel and Neumann functions, respectively.

The potential $V(r)$ has the form

$$V(r) = -\frac{z}{r} + \frac{4\pi}{r}\int_0^r N_e(r')r'^2 d\mathbf{r}' + 4\pi \int_r^R N_e(r')r' d\mathbf{r}' + V_{xc}^0(N_e(r))\,, \quad r < \bar{r}\,.$$

The electron distribution is determined analogously to (5.93), but with due regard for the weighting factors described by the Fermi distribution at the

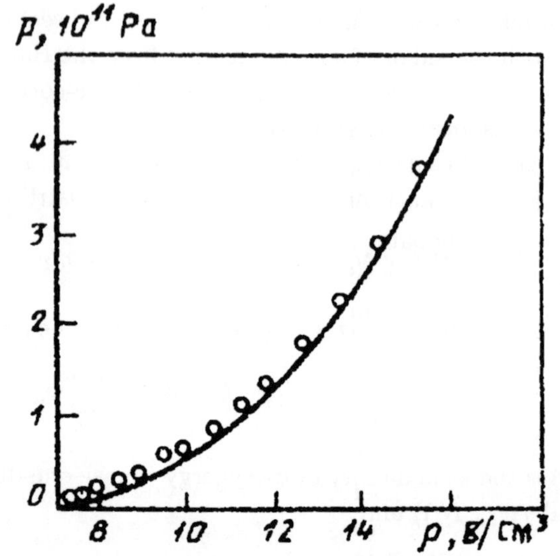

Fig. 5.17. Equation of state for zinc at zero temperature: the continuous curve indicates the experimental data of [107]; the points indicate the calculation results of [347].

pre-set temperature. The solution explicitly divides the electrons into free and bound ones. The pressure of the electron component of the system is equal to that of Fermi gas of density $N_e(\mathbf{r})$ on the spherical boundary of the cell. The ion's contribution to pressure must be calculated separately, which is an independent problem.

Liberman [347] calculated the equation of state for cold copper, zinc and nickel. For copper and zinc, this equation agrees well with the results of dynamic experiments [107] (Fig. 5.17), but less with the measurement results for nickel.

Note the investigations of the thermodynamics of cellular plasma in the Hartree–Fock-Slater approximations [348–350]. These methods follow from MDF. The papers cited give the results of wide-range calculations of the equation of state for a number of metals, which are compared to the available experimental data.

5.7. A Comparison of Thermodynamic Models

By now, a fairly large body of experimental data have been accumulated on the thermodynamics of non-ideal plasma of various elements under conditions of developed ionization. The great majority of these data were obtained using dynamic methods (cf., Chapter 3), such as shock and adiabatic compression of cesium, as well as compression of argon, xenon and copper in powerful shock waves.

The data obtained for cesium plasma [123, 128, 217, 343] relate to various regions of the phase diagram (Fig. 3.2), which partly overlap at the interfaces and agree with each other within the claimed experimental accuracy. Experiments in adiabatic compression [123] enabled one to move forward, as compared to static measurements [217, 342], to higher temperatures ($T \leq 4,000$ K) which, however, proved insufficient for marked thermal ionization of the plasma. Leading under these conditions is the charge-neutral interaction whose contribution to the equation of state stays within the measuring error. The principal conclusion [123] which follows from these experiments consists of the absence of phase separation (cf., Section 5.7) caused by the metal-dielectric transition. Experiments in cesium compression by direct and reflected shock waves allowed a further expansion of the temperature range, $T \sim (2.6 - 20) \cdot 10^3$ K (Fig. 3.18), where the Coulomb interaction is strong, $\Gamma \sim 0.2$–2.2, and defines the physical properties of plasma with developed ionization.

Considerably higher plasma parameters were attained as a result of explosive compression of heavy inert gases [106, 137, 145, 146], namely, pressures of up to 60 kbar and temperatures of up to $6 \cdot 10^4$ K. The plasma densities of $p \sim 0.4$ g/cm^3, $n_e \approx 3 \cdot 10^{21}$ cm^{-3} thus obtained approach the density of condensed xenon. In the experiment by Mintsev *et al.* [143], they exceed ($\rho \sim 4.5$ g/cm^3) that density. In doing so, the characteristic interparticle distances in plasma [146] amounted to 6–7 Å. This is comparable to the ion and atom size of 3–4 Å, with the maximum non-ideality parameters $\Gamma \sim 5$ being close to the maximum possible values (Fig. 1.1) for non-degenerate plasma. The experimental data for argon and xenon [106, 137, 145, 146] relate to the region of developed single ($x_{\text{Ar}} \leq 0.7$) and double ($x_{\text{Xe}} \leq 1.8$) ionizations. It also allows condensed materials to advance from the gas side to the region of densities (cf., Fig. 1.5) while approaching the extremely high pressures ($p \leq 1.3$ Mbar) and compressions ($\rho \leq 9.6$ g/cm^3) of a plasma obtained by shock compression of liquid xenon and heated to $T \leq 30 \cdot 10^3$ K [344].

All thermodynamic measurements (Figs. 5.18–5.22, Tables 5.2–5.5) exhibit a clearly defined tendency towards understating the measured enthalpy or internal energy values (index I in the abovementioned drawings), as compared to the traditional plasma calculation represented by a ring Debye approximation in a grand canonical ensemble of statistical mechanics (5.50)–(5.52) (index II). In Figs. 5.18–5.22 and Tables 5.2–5.5, the experimental data are compared to a number of other plasma approximations, such as the ideal plasma model ($F_b = 0$, $\Sigma_k = 2$, III; $F_b = 0$, in Eq. (5.40) Σ_k is in accordance with Eq. (5.73), IV), Debye theory in a small canonical ensemble (V, Eq. (5.43)), pseudopotential Monte Carlo model (Eqs. (5.58)–(5.61), VI), and the "bound" atom model (Eqs. (5.57)–(5.83), VII). The relative contribution by other models to the equation of state was analyzed.

The analysis performed shows that although the majority of the theoretical models used are not at variance with the experimental isotherms of cesium plasma within the measurement accuracy, they are incapable of providing a consistent but simultaneous description of thermal and caloric information. Inert gas plasmas correspond to appreciably higher parameters of non-ideality. In

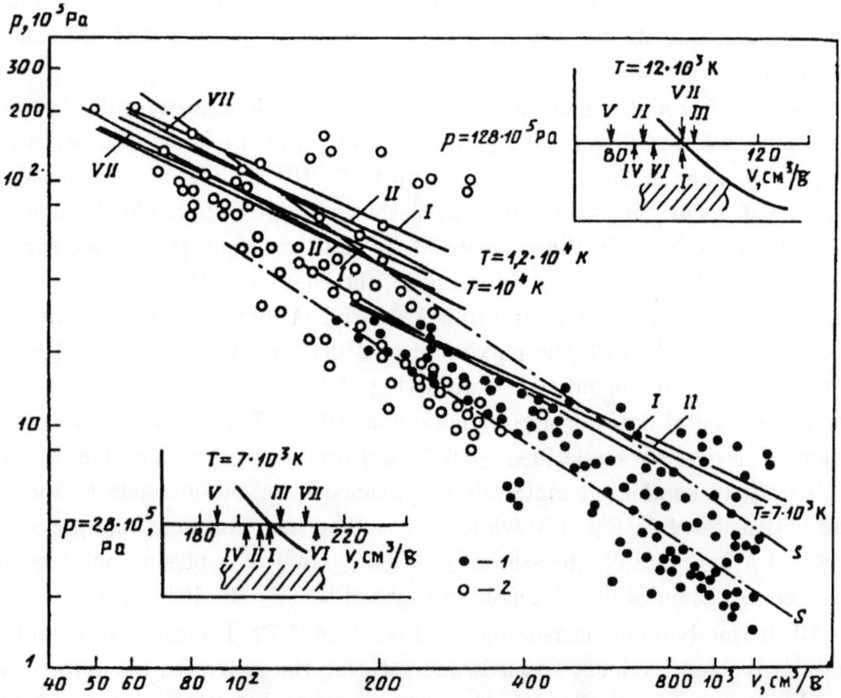

Fig. 5.18. The results of shock compression of cesium plasma: T, isotherms; S, isentropes; *1*, incident shock wave; *2*, reflected shock wave.

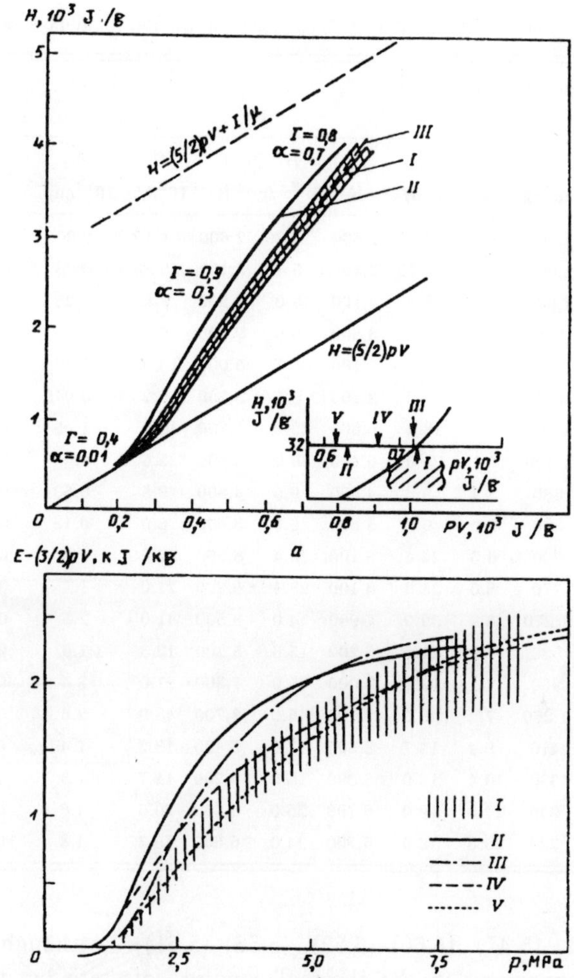

Fig. 5.19. Isochore of cesium plasma. The error band is shaded. $V = 10^3$ cm^3/g (a); $V = 2 \cdot 10^2$ cm^3/g (b). I, experiment; II, ring Debye approximation (5.50)–(5.52); III, "bound" atom model (5.77)–(5.83); IV, pseudopotential model (5.63)–(5.71); V, models (5.77)–(5.83) and (5.63)–(5.71).

this case, the traditional plasma models ((5.41)–(5.47), (5.50)–(5.52), (5.58)–(5.61)) are at variance with both the caloric and thermal measurements.

An analysis of thermodynamic data points to the presence of interparticle repulsion in a strongly compressed plasma, which is not described by plasma

Table 5.2. Thermodynamic parameters of cesium behind incident shock wave.

Experiment I							Theory						
							II		III				
p_0, atm	T_0, K	u, km/s	V_1, cm^3/g	p_1, atm	H_1, $10^2 J/g$	T_1, K	H, $10^2 J/g$	T, K	H, $10^2 J/g$	N_e, $10^{13} cm^{-3}$	α	Γ	T, K
0.130	780	0.83	1,200	1.40	4.20	2,600	4.20	2,600	4.10	0.006	0.002	0.20	2,600
0.147	780	0.935	1,000	2.0	5.10	3,100	5.40	3,100	5.10	0.03	0.007	0.34	3,100
0.107	760	1.23	1,030	2.80	8.3	4,100	10.0	4,200	8.6	0.25	0.056	0.62	4,200
0.160	780	1.05	800	2.9	6.3	3,500	6.6	3,500	6.0	0.09	0.015	0.48	3,500
0.074	750	1.53	1,160	3.1	12.4	5,100	15.7	5,000	13.0	0.56	0.14	0.73	5,000
0.24	830	0.95	630	3.3	5.3	3,200	5.5	3,200	5.2	0.046	0.006	0.41	3,200
0.063	740	1.87	1,060	4.2	18.0	5,600	22.0	5,700	18.0	1.05	0.25	0.83	5,700
0.051	700	2.23	1,130	5.1	25.0	6,400	30.0	6,600	25.6	1.6	0.41	0.85	6,400
0.18	790	1.33	580	5.4	9.6	4,600	9.5	4,600	9.5	0.53	0.07	0.84	4,600
0.38	860	1.03	390	6.0	6.2	3,600	6.5	3,600	6.0	0.13	0.01	0.59	3,600
0.078	750	2.1	800	6.5	22.4	6,100	26.4	6,300	22.0	1.8	0.32	0.93	6,200
0.08	740	2.05	770	6.6	21.6	6,100	25.4	6,200	21.0	1.7	0.29	0.94	6,100
0.044	690	2.77	1,270	6.9	39.0	7,900	44.0	8,500	41.0	2.5	0.7	0.76	8,000
0.18	790	1.49	530	7.0	11.8	5,200	15.6	5,200	12.3	1.0	0.12	0.93	5,200
0.056	710	2.65	950	7.1	32.0	7,000	35.0	7,300	31.0	2.4	0.5	0.89	7,000
0.044	690	2.83	1,260	7.2	40.0	8,100	45.0	8,700	43.0	2.6	0.73	0.75	8,200
0.24	820	1.55	410	9.4	12.7	5,500	17.0	5,400	13.2	1.45	0.13	1.04	5,400
0.29	840	1.49	350	10.4	12.0	5,300	15.0	5,200	11.7	1.3	0.1	1.05	5,200
0.36	860	1.92	240	22.0	19.0	6,700	25.0	6,800	20.0	4.6	0.24	1.35	6,700
0.44	870	1.87	200	25.0	18.0	6,700	24.0	6,600	18.3	4.8	0.21	1.4	6,600

theories ((5.41)–(5.47), (5.50)–(5.52), (5.58)–(5.61)). Although one can obtain a good agreement in the thermal equation of state by modifying the value of corrections F_b given by these theories, such a procedure leads to an even greater discrepancy in caloric information. It was found [128, 145, 146] that a consistent description of thermal and caloric data could be attained by way of modifying the contribution made by bound states to the thermodynamic functions of dense plasma, an effect ignored by most traditional plasma approximations.

Irrespective of this conclusion, the experimental results indicate that the Debye theory and the theories close to it, (5.43)–(5.47), over-estimate the corrections to thermodynamic functions for interaction in continuous spectrum. The best properties of extrapolation to the $\Gamma \geq 1$ region are exhibited by

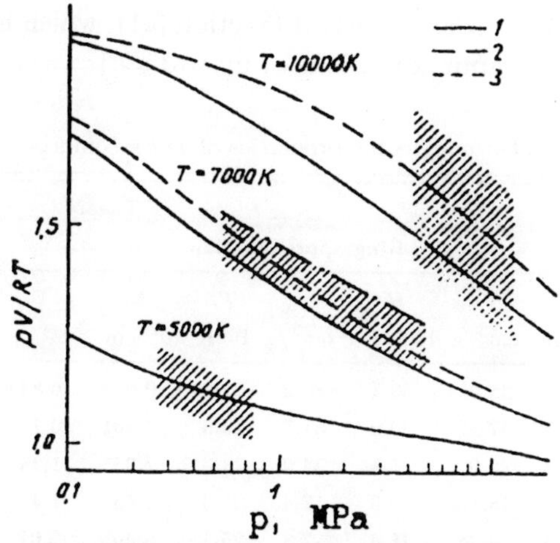

Fig. 5.20. Thermal equation of state for cesium plasma. Solid, the ring Debye approximation; dashed, the pseudopotential model (5.63); dash-and-dotted, the bond atom model; the region investigated experimentally are marked.

Table 5.3. Thermodynamic parameters of cesium behind incident shock wave.

Experiment I								Theory						
								II					III	
p_0, atm	T_0, K	U, km/s	V_1, cm^3/g	V_2, cm^3/g	p_2, atm	H_2, 10^2J/g	T_2, K	H, 10^2J/g	N_e, 10^{18}cm^{-3}	α,	Γ,	T K	H 10^2J/g	T, K
0.230	620	0.90	680	250	12.0	9.0	4,400	8.5	0.75	0.04	1.0	4,500	10.4	4,500
0.155	790	1.10	820	260	15.0	13	5,600	13	2.1	0.12	1.2	5,600	17.0	5,600
0.094	760	1.94	700	170	60	42	9,600	42	16	0.60	1.4	9,900	45	10,700
0.047	690	2.52	1,150	300	64	73	15,600	74	14	0.95	0.66	15,700	73	17,300
0.166	790	1.70	450	110	70	32	8,500	30	16	0.38	1.7	8,600	35.0	9,000
0.320	840	1.58	280	75	95	27	8,700	26	19	0.32	1.9	8,500	33.0	8,800
0.180	800	1.98	420	100	100	42	10,000	41	25	0.56	1.6	10,300	44	11,000
0.064	710	2.75	860	240	104	90	19,700	90	19	0.97	0.54	19,500	85	21,400
0.130	790	2.30	490	144	130	64	15,200	71	28	0.90	0.95	15,500	70	17,400
0.440	870	1.87	190	50	200	38	11,000	37	43	0.45	2.1	10,500	41	11,200

the theories giving non-ideality corrections that do not exceed the corrections of ring approximation in a grand canonical ensemble (5.50)–(5.52). The conclusion is then in qualitative agreement with the results of one-component model studies using the Monte Carlo method (Section 5.1), which is indicative of the validity of asymptotic approximations of up to $\Gamma \sim 1$.

Table 5.4. Thermodynamic properties of shock-compressed argon.

Experiment					Theory							
					Ring approximation				Bound atom			
p_0, atm	D, km/s	u, km/s	p, atm	V, cm^3/g	H, 10^{-3}J/g	V, cm^3/g	T, 10^3K	N_e, 10^{20}cm^{-3}	Γ,	$N\sigma^3$,	V, cm^3/g	T 10^3K
0.78	7.78	6.94	0.692	86.0	30.1	88.2	22.0	0.642	0.843	0.00422	90.66	21.6
1.0	6.5	5.59	0.597	87	20.9	81.7	19.5	0.401	0.798	0.0057	85.5	19.1
3.0	7.93	6.97	2.73	25.0	31.1	24.0	24.3	2.23	1.35	0.016	25.6	23.0
5.0	6.51	5.52	2.96	18.8	20.9	17.45	21.3	1.63	1.4	0.0277	19.9	19.9
10.0	7.6	6.52	8.19	8.78	28.4	7.8	25.1	5.405	2.01	0.055	9.06	22.3
20.0	7.35	6.28	15.4	4.47	26.6	4.05	25.3	8.68	2.51	0.113	5.19	20.5
20.0	7.65	6.4	16.3	5.01	28.6	4.00	26.0	9.77	2.54	0.11	5.0	21.3

Table 5.5. Thermodynamic properties of shock-compressed argon.

Experiment					Theory							
					Ring approximation				(5.63)–(5.71)			
p_0, atm	D, km/s	u, km/s	p, atm	V, cm^3/g	H, 10^3J/g	V, cm^3/g	T, 10^3K	N_e, 10^{20}cm^{-3}	Γ,	$N\sigma^3$,	V, cm^3/g	T 10^3K
1.0	4.58	4.11	1.02	19.03	10.4	18.8	21.0	1.33	1.31	0.0163	21.0	21.6
3.0	7.81	7.00	8.99	6.43	30.2	6.00	42.0	9.71	1.38	0.051	7.07	41.7
5.0	6.35	5.39	9.49	5.56	19.7	3.77	33.1	11.4	1.99	0.0812	4.62	33.7
10.0	6.23	5.4	19.2	2.38	19.1	1.83	34.0	22.0	2.64	0.168	2.34	35.1
10.0	8.93	7.69	39.2	2.48	39.1	1.94	55.8	34.9	1.78	0.158	2.40	55.4
10.0	4.11	3.52	8.27	2.56	8.32	2.04	21.6	7.2	2.91	0.150	2.4	23.4

In the region of high plasma densities under consideration, the mean interparticle distances are comparable to the characteristic sizes of atoms and ions. This fact, along with a strong Coulomb interaction of free charges, may bring about a marked deformation of energy levels.

The description of this effect calls for the use of the quantum-mechanical model of "bound" atom (5.78)–(5.83), which is unconventional from the standpoint of plasma physics. It allows and allowing for the effect of the plasma environment on the discrete spectrum of atoms and ions in a strongly compressed

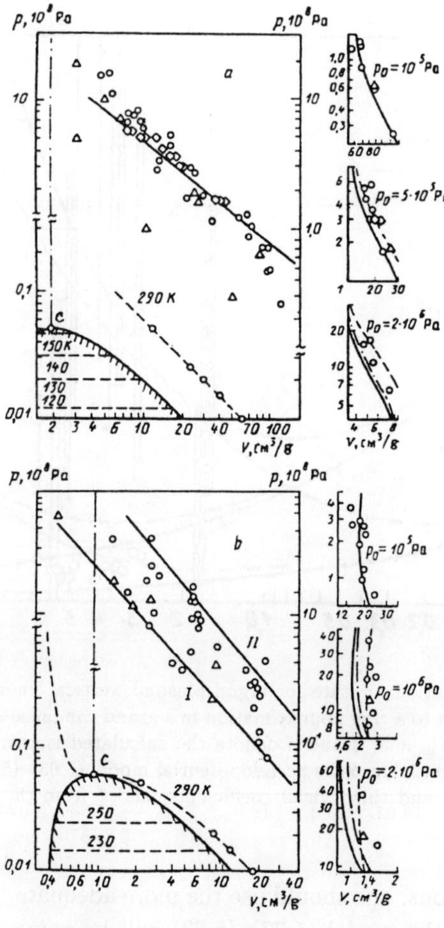

Fig. 5.21. Phase diagram of argon (a) and xenon (b). The two-phase region boundaries are marked with K_C as the critical point. The dots represent the experimental results: ∘ [146]; Δ [145]; ⋄ [137]. The broken line shows the isotherms of the initial states. The wave lines mark the boundaries of single (I) and double (II) ionizations. On the right is a comparison between the experimental shock adiabats (dots) and theoretical data. The solid curves represent the calculation to a ring Debye approximation in grand canonical ensemble (5.50)–(5.52); the dash-and-dot lines represent the additional inclusion of atomic interaction within the second virial coefficient; the broken line in (a) represents the "bound" atom model (5.77)–(5.83), and in (b) the pseudopotential model (5.63)–(5.71).

plasma. This is seen from a comparison of the model with the experiment: the model reflects correctly the tendency revealed in the course of experiment while somewhat over-estimating the repulsion effect at near-critical plasma densities.

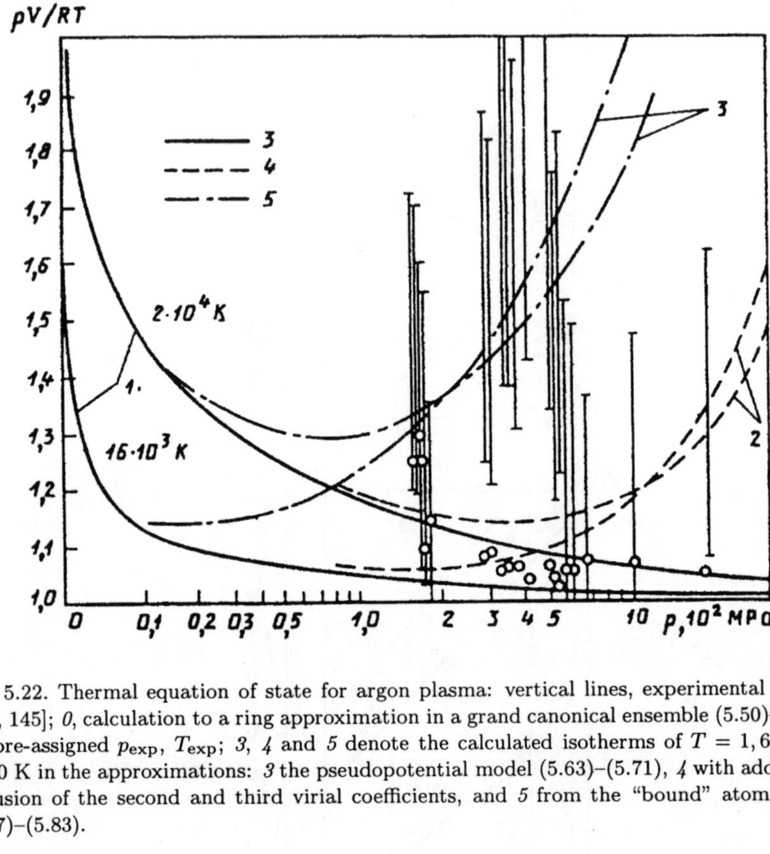

Fig. 5.22. Thermal equation of state for argon plasma: vertical lines, experimental results [137, 145]; *0*, calculation to a ring approximation in a grand canonical ensemble (5.50)–(5.52) for pre-assigned $p_{\rm exp}$, $T_{\rm exp}$; *3*, *4* and *5* denote the calculated isotherms of $T = 1,600$ and 2,000 K in the approximations: *3* the pseudopotential model (5.63)–(5.71), *4* with additional inclusion of the second and third virial coefficients, and *5* from the "bound" atom model (5.77)–(5.83).

Under these conditions, one should use the more adequate "soft" sphere model [160]. Hereinafter, the model (5.77)–(5.83) will be employed to describe the optical characteristics of non-ideal plasma.

The need to include the variation of discrete electron spectrum in the thermodynamics of dense plasma is clearly demonstrated in experiments in shock compression of liquid argon and xenon at a pressure range of up to 1.3 Mbar [344, 348]. The interpretation of these measurements has shown (Fig. 5.23) that a failure to include the discrete spectrum (excitation energy) leads to a sharp over-estimation of the pressures being calculated (curve C). Curves A and B allow for the thermodynamic effects of thermal excitation of electrons from the valence band 5s to the conduction band 5d, and for the variation of the forbidden band width from 7 to 4 eV with an increase in density [344, 347]. A peculiar feature of the quasi-chemical description of plasma under

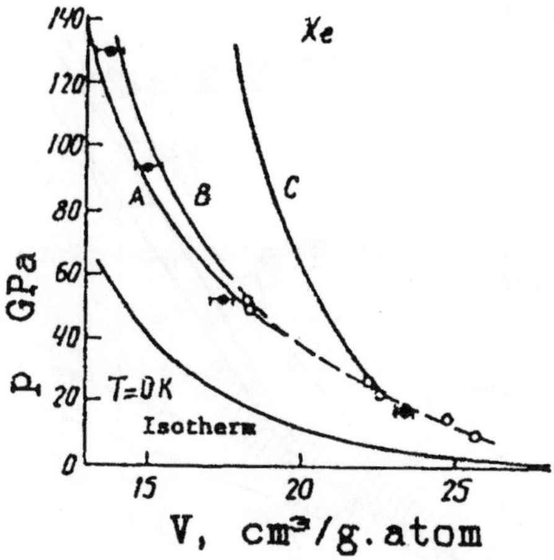

Fig. 5.23. Comparison of the shock compressibility of xenon: •, [344]; ∘, [346]; C, calculation disregarding electronic excitation [347]; A and B, two variants of the inclusion of discrete spectrum deformation.

conditions of strong non-ideality is the conventionality of the particle separation into free and bound ones. Therefore, the effect regarded as a distortion of the contribution by excited states may, in the case of a different separation into species, be interpreted as a manifestation of the quantum nature of electron-ion interaction at close distances. In order to describe this interaction, use was made [330, 335] of the pseudopotential (5.63) on the basis of which, using the conditions of local electroneutrality (5.65), (5.66), the semi-empirical model of non-ideal plasma was constructed. The choice of the sole parameter of the model, that is, the pseudopotential depth $\Phi^*_{ei}(0)$, which is equal to the energy separating the particles into free and bound ones, was instrumental in producing a consistent description of the totality of currently available results of thermodynamic experiments in "gas" plasma [128, 137, 143, 145, 343].

We shall now dwell on the comparison between the quasi-classical model in Section 5.5 and the plasma experiment. This is done in Fig. 5.24 for the gas plasma of cesium based on the results of the shock wave [128] and electric explosion [343] experiments. In accordance with the conclusions of Section 5.5, the inaccuracy of the Thomas-Fermi approximation is at its maximum in the region of temperatures corresponding to the transition of ionization from the filled electron shell (the beginning of the secondary ionization of cesium). Note that the plasma investigated by Bushman *et al.* [128] and Dikhter and Zeigarnik [343] featured a marked degree of non-ideality, $\gamma \leq 1$.

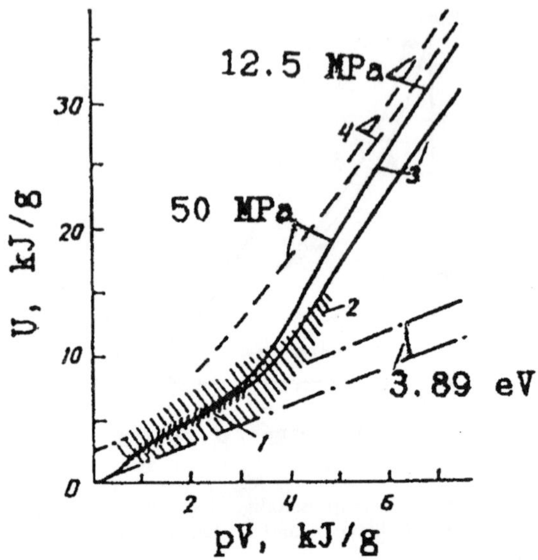

Fig. 5.24. Caloric equation of state for cesium plasma [343], $p = 12.5$–50 Mpa: *1*, experiment [128]; *2*, experiment [343]; *3*, calculation to a ring approximation in a grand canonical ensemble (5.50)–(5.52); *4*, Thomas-Fermi approximation with quantum and exchange corrections.

The ultrahigh pressure region was investigated in [118, 119] using the technique of powerful shock waves. A comparison of the results with those obtained with the aid of plasma techniques is given in Fig. 5.25 [349]. The plasma parameters are specified for some points.

Curves *1*, *2*, *4* and *5* in Fig. 5.25 indicate the data of [349]; *1*, the modified Hartree–Fock-Slater method [349, 350]; *3*, the self-consistent field method [348]; and *4*, the Thomas-Fermi theory with quantum and exchange corrections [340, 342]. Curve *2* indicates the plasma model of bound (confined) atom (see Section 5.4) in view of electron degeneracy, the Coulomb non-ideality, the excitation of bound states, and the short-range repulsion [349]; *5* — ditto, in view of the contribution made by equilibrium radiation.

One can see that the plasma model of bound atom reproduces the states of ultrahigh energy density reasonably well. Unfortunately, the precision of measurements in [119] was insufficient for a thorough analysis of the advantages and singularities of models for a description of superdense compression and, in particular, of the role of shell effects in thermodynamics. However, these experiments enable one to follow the asymptotic properties of the theories identified above. In the experiments of [119], the parameter of plasma

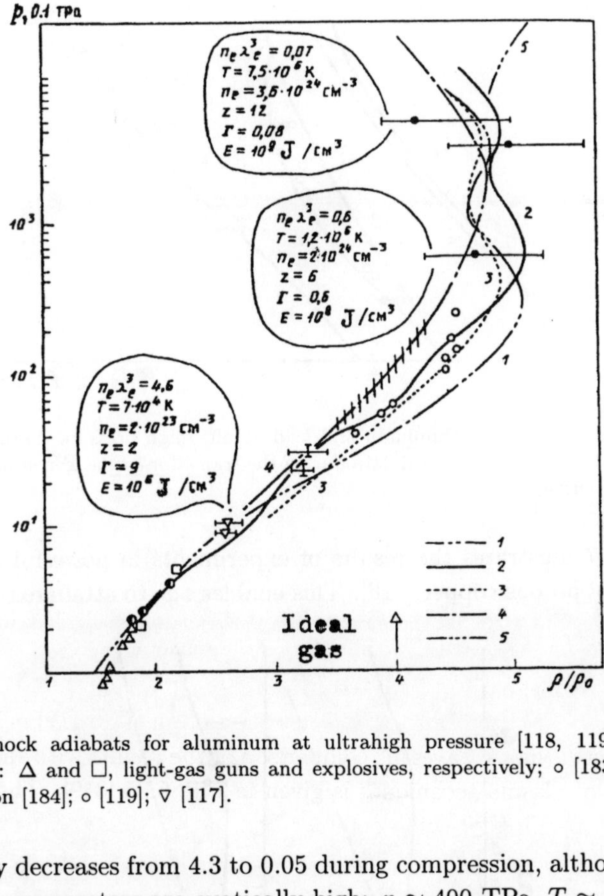

Fig. 5.25. Shock adiabats for aluminum at ultrahigh pressure [118, 119] (see text for designations): Δ and □, light-gas guns and explosives, respectively; ○ [183, 187]; • [119]; hatched region [184]; ○ [119]; ▽ [117].

non-ideality decreases from 4.3 to 0.05 during compression, although the thermodynamic parameters are exotically high: $p \approx 400$ TPa, $T \approx 7 \cdot 10^6$ K and $N_e \approx 3.6 \cdot 10^{24}$ cm^{-3}. The specific energy density is also close to 1 GJ/cm^3. One can see in Fig. 5.25 that the pressure range of a thousand terapascals is of great interest in terms of revealing the role of shell effects, where the contribution of bound states according to different models is the least defined. Figure 5.26 gives the results of measurements [118] of the relative (iron as standard) compressibility of aluminum and copper, as well as a comparison of these results with the quasi-classical model [340, 342] (curve *2*) and with the cell model of self-consistent field [348] (curve *1*). The experimental results clearly point to the substantial contribution by the bound states which cause non-monotonicity in the thermodynamic functions of dense plasma. Naturally, this is not described by the Thomas-Fermi model with quantum and exchange corrections [340, 342].

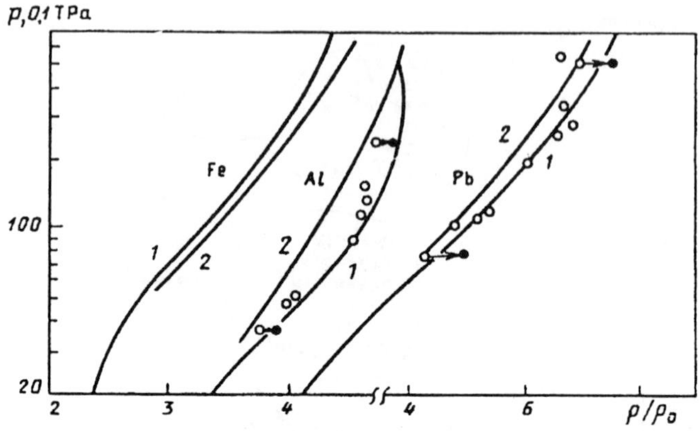

Fig. 5.26. Shock adiabats for aluminum and lead at ultrahigh pressure [118]. The arrows indicate the shift of the experimental data during the transition of the iron standard adiabat from curve *1* to curve *2*.

Figure 5.27 illustrates the results of experiments in powerful shock wave compression of porous copper [116]. This enables one to attain extremely high

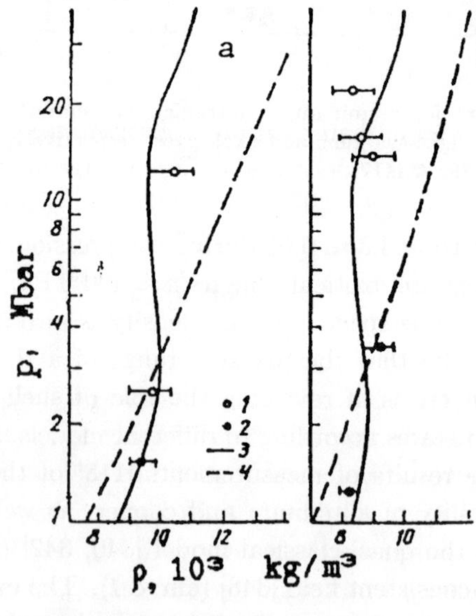

Fig. 5.27. Thermodynamics of non-ideal copper plasma in the region of extremely high pressures [360]. Initial porosity: (a) $m = \rho_0/\rho_{00} = 3$; (b) $m = 4$. Solid curves, shock adiabats calculated under a variant of the "bound" atom model (5.77)–(5.83); dashed, Thomas-Fermi theory with corrections [340, 342]. Points, experiment [116, 362].

concentrations of thermal energy. A superdense, quintuply ionized weakly degenerate ($N_e\lambda_e^3 \sim 0.7$) plasma with extremely high parameters of $p \sim$ 20.9 Mbar, $N_e \sim 2 \cdot 10^{23}$ cm^{-3} and $\Gamma \sim 2$ is realized at the maximum temperatures of the order of $\sim 2.5 \cdot 10^5$ K. Figure 5.27 demonstrates a reasonable description of the properties of such a medium with the aid of the "bound" atom model (5.77)–(5.83). We see that as the Coulomb interaction decreases with a rise in temperature and pressure, the transition to a quasi-ideal, multiply ionized plasma is realized. In doing so, the difference between the experimental results (as well as Saha's model (5.77)–(5.83)) and the Thomas-Fermi approximation increases progressively to reach, under extreme conditions, almost an order of magnitude in pressure. This is supported by the conclusion of Iosilevskii and Gryaznov [343] on the absence of ideal plasma asymptotics from the said approximation.

5.8. Phase Transitions in Non-ideal Plasma

In view of the serious difficulties in producing a consistent theoretical description of plasma with strong interparticle interaction in the entire range of parameters, a number of researchers (cf., [160, 240, 295, 319, 320, 325] and the references cited therein by the authors) make use of model considerations based on physical simplifications and extrapolations of views (developed for weakly non-ideal plasma) on the quantum and collective effects under conditions of Coulomb interaction. While being purely approximate, such a heuristic approach is convenient and was widely used for qualitative analysis of the possible behavior of plasma in situations characterized by the absence of measurements of its physical properties. Characteristically, a number of proposed models in the region of increased non-ideality lose the thermodynamic stability, which is attributed to the possibility of phase transition and the separation of the system into phases of different densities.

A model analysis (cf., Section 5.1.4.) of strongly compressed degenerate systems of electrons leads to the possibility of phase transition with the formation of an ordered electronic structure [295] of the "Wigner crystal" (Fig. 5.28). The properties of this crystal, as well as its phase diagram, have been studied in detail by numerous researchers (cf., reviews in [240, 320, 363–366] and their respective references) who considered the cases with zero and finite temperatures. In the latter case, the separation curve of the electron liquid and the Wigner crystal has a critical point while the stability limit of such a crystal at $T = 0$ is quite indeterminate [363–365]. Ivanov *et al.* [367] link the phase transitions in a one-component plasma with the melting of metals. The

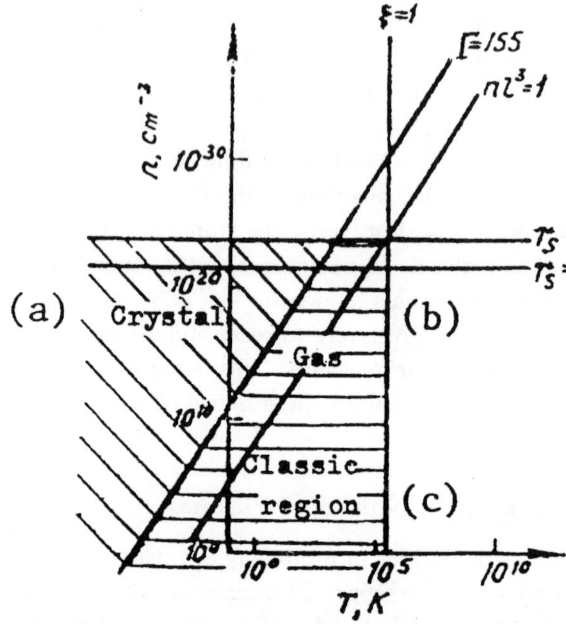

Fig. 5.28. The range of existence of electronic Wigner crystal [295, 320]: $\gamma = (e^2/kT)[3/(4\pi n_e)]^{-1/3}$; $\xi = e^2 m^{1/2}/[(kT)^{1/2}\hbar]$; $l = e^2/kT$; $r_S = [3/(4\pi n_e)]^{1/3}(me^2/\hbar^2)$.

dielectric permeability of strongly compressed Coulomb systems and the stability of non-ideal plasma were analyzed in detailed reviews by Maksimov and Dolgov [364] and Kirzhnits [368].

A more complex and diversified phenomenon is the case of a real multicomponent plasma. With due regard for the fact that polarization is the principal effect in a dense plasma, Mott [369] has considered a model according to which the interatomic electrons are in the screened potential $V(r) = -(e^2/r)\exp(-r/r_0)$. The numerical solution of Schrödinger's equation for this potential enables the critical value [369] $r_c = 0.84a_0 = 0.84(\hbar^2/me^2)$ to be derived, which point the ground state disappears and corresponds to the so-called "Mott's" metal-dielectric transition.

The possibility of anomalies in thermodynamic functions upon metallization of dense vapors was discussed by Landau and Zel'dovich [232], who had assumed that metallic systems might contain two interfaces: one for the liquid-gas transition and the other for the metal-dielectric transition (Fig. 5.29). The correlation between these transitions and anomalies was analyzed by Norman [363].

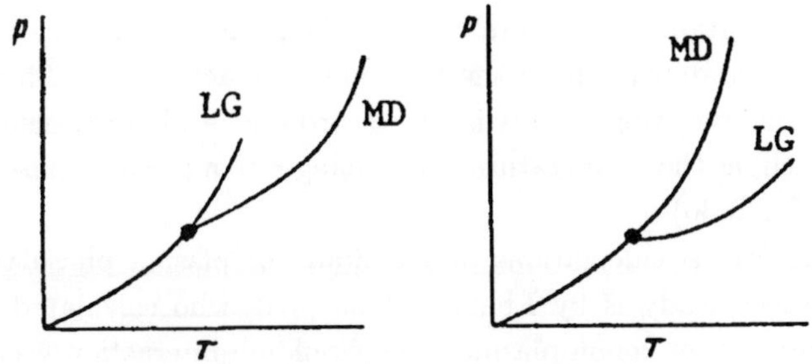

Fig. 5.29. Possible phase diagrams for metals [232].

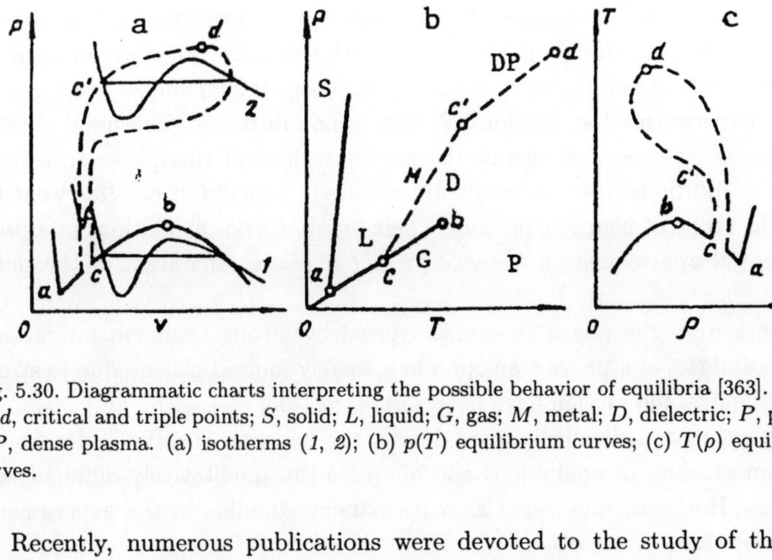

Fig. 5.30. Diagrammatic charts interpreting the possible behavior of equilibria [363]. *a*, *b*, *c*, *c'*, *d*, critical and triple points; *S*, solid; *L*, liquid; *G*, gas; *M*, metal; *D*, dielectric; *P*, plasma; *DP*, dense plasma. (a) isotherms (*1*, *2*); (b) $p(T)$ equilibrium curves; (c) $T(\rho)$ equilibrium curves.

Recently, numerous publications were devoted to the study of the LG, liquid-gas phase transition; MD, metal-dielectric phase transition analysis of phase transformations in a non-degenerate, strongly non-ideal plasma. Their authors used simple models [236, 240, 320, 352] as a basis to analyze the numerous emerging possibilities. The principal effect responsible for the condensation of a multi-component plasma is the polarization attraction of opposite charges. Meanwhile, the stabilization of the emerging phase is supported by quantum effects (interference) and at high densities, by the degeneracy and overlap of electronic shells of ions.

Two phase transitions are predicted for the model (5.89). At low densities and temperatures, the determining factor is the interatomic interaction leading to the gas-liquid transition (curve *1* in Fig. 5.30). With an increase in density, "ionization by pressure" occurs alongside a secondary violation of thermodynamic stability due to the Coulomb interaction F_{ei}. Therefore, the metal-dielectric transition in this model is proof of a phase transition of the first kind which, as the temperature rises, changes to a plasma phase transition (curve *2* in Fig. 5.30)

Of the numerous calculations of the limits of plasma phase transitions, the most perfect study is by Ebeling *et al.* [370] who calculated the phase boundaries of a dense xenon plasma. The Coulomb interaction was described by a wide-range Pade approximation (reproducing the asymptotics of Debye-Hüeckel and Hell–Mann-Brueckner), and the contribution of atoms and ions was described by the hard-sphere and Van der Waals models. It is demonstrated that with this approach, a first-order phase transition is realized between weakly and strongly ionized states with the phase boundary of negative slope and critical point with $T_c \approx 12,600$ K, $p_c \approx 9.9$ GPa and $\rho_c \approx 3.7$ g/cm^3. Recent experiments (see Section 5.7) with xenon in the neighborhood of those boundaries yield no evidence of the manifestations of this phase transition. Possibly, additional measurements are required. Zamalin *et al.* [286] went beyond the scope of heuristic models. Their Monte Carlo calculations in a pseudopotential approximation likewise predict a phase separation of the dense plasma.

Apart from the phase anomalies caused by strong Coulomb interaction, the possibilities of a phase transition in a weakly ionized plasma due to strong charge-neutral interaction have been discussed [236].

Since they are obviously approximate, the heuristic theories suffer from a great uncertainty of predictions and often lead to qualitatively different conclusions. However, they serve as a permanent stimulus in the experimental search for these exotic effects.

The possible hydrodynamic effects which result from the dynamic stimulation of plasma are analyzed in Chapter 10. They indicate the types of phase transitions which may be recorded upon shock or isentropic variation of pressure [371]. In the latter case, the phase transition leads to an abrupt change in isentropic compressibility depending on the properties of the competing phases (cf., Chapter 10). If the value of this change is sufficiently great, expansion shock waves occur which are similar to those recorded in iron and ionic crystals [107]. In general, phase anomalies would occur, upon compression or expansion of metals, at the kinks of experimental and other thermodynamic curves discussed in Section 5.7. The experiments in adiabatic expansion of shock-compressed metals, where one can follow, in a single series of experiments,

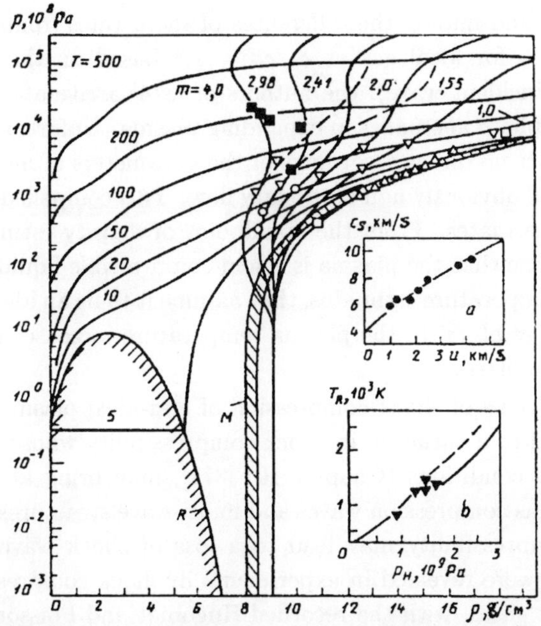

Fig. 5.31. Phase diagram for copper (using the semi-empirical equation of state from [179]): m, shock adiabats for samples of various porosity; T, isotherms (in 10^3 K); M, R, melting and boiling phase boundaries, two-phase regions are shaded; a, velocity of sound on the shock adiabats; b, temperature in release wave.

the behavior of material over an extremely wide range of parameters and can expect to come across most of the phase anomalies predicted by theory (Figs. 5.31), are especially revealing. The data from experiments in isentropic compression and expansion of metals (Chapter 3), as well as from experiments in rapid [372] and slow [373] electric explosion of metals using electric current, produce no direct indications to some anomalies which could be interpreted as phase transitions in non-ideal plasma. Note, in this connection, that in the case of dynamic experiments one manages to record both the evaporation and an effect which produces but little influence on the hydrodynamics of flow, such as the melting of material (Figs. 3.17 and 3.18). In view of this, we will specifically note the study [170], where the optical measurements of the plasma temperature in a release wave likewise failed to reveal plasma anomalies.

Rusakov *et al.* [374] presented the results of studies an the dynamics of separation of shock-compressed materials in long (60–200 mm) cylindrical and cone-shaped channels. Upon measuring the rates of travel of luminescent plasma clusters and finding the differences of those rates from self-similar solutions (obtained for ideal gas at $c_p/c_v = 1.4$ based on the assumption of stationarity of incident wave), the authors of [374] arrive at a conclusion on the presence of phase anomalies in expanding plasma. Unfortunately, Rusakov *et al.* [374] report no direct recordings of the parameters of moving plasma in this complex and obviously non-stationary flow. This compelled them to resort to qualitative estimates. From the standpoint of density estimates, Rusakov *et al.* [374] assume that the plasma is a non-compressible liquid and, from the standpoint of temperature estimates, they assume it to be an ideal gas. According to Rusakov *et al.* [374], the plasma temperature increases upon adiabatic expansion of the plasma.

Under conditions of shock compression of non-ideal plasma, phase transitions would lead to a variation in shock compressibility whose increase, given some additional conditions (Chapter 10) [371], may bring about the formation of continuous compression waves and multi-wave structures, while a sharp decrease in compressibility may lead to a loss of shock wave stability. No such anomalies were revealed in experiments in shock compression of cesium and heavy inert gases, with the recorded Hugoniot and Poisson adiabats (cf., Section 5.7) having the form of smooth curves.

In addition, the (Section 5.7) behavior of the variation of the thermodynamic functions of non-ideal plasma, which is observed in experiment, corresponds to a one-phase case. The results of electrophysical (Chapter 7) and optical (Chapter 8) measurements, which are interpreted within one-phase plasma models, are also indicative of the absence of marked phase transformations in the range of parameters being investigated.

Note that, based on existing knowledge [106, 160], phase transitions in dynamic experiments may occur over very short (of the order of 10^{-10}-10^{-6} s) time periods, which are considerably less than that found in the static case. This makes (although it does not rule out altogether) phase "quenching" unlikely under conditions of non-stationarity.

Numerous studies are devoted to the search for an electronic Wigner crystal in a solid-state plasma. A thorough review of these studies is found in [296]. In common metals, the high electron number density ($2 \leq r_s \leq 7$) precludes the formation of the Wigner crystal. Hence the search for the latter was concentrated on highly compensated semi conductors and magnetic dielectrics.

By alloying n-type semiconductors with acceptors, one can greatly reduce the number density of free electrons and the ion background approaches close to uniform. In semiconductors of the type of ZnSb and $Hg_{0.8}Cd_{0.2}Te$ with

$n = N_D - N_A \approx 10^{14}$ cm^3 and $N_H/N_D \approx 0.99$, approximately 300 impurity ions are found inside the electron orbit, which provides the condition of validity of the OCP model. However, no clear proof of Wigner condensation is produced in this case. Apparently, this is due to the earlier Mott localization of electrons on donors or the destructive effect of fluctuating microfields.

In the case of magnetic isolators, the Wigner's transition is apparently suppressed by the splitting of energy bands according to Hubbard's model.

The presence of threshold magnetic field in ZnSb, below which no activation energy was revealed during measurements of the Hall effect and electrical conductivity, was treated as proof of electron condensation. The results of these measurements are analyzed in detail by Tsidil'kovskii [296], who produces an argument against Wigner condensation, which is based on the fact that one deals with impurity electrons that are highly susceptible to the effect of a random potential. Tsidil'kovskii [296] further treats the inferences about electron condensation in crystals of n-$Hg_{0.8}Cd_{0.2}Te$ [59, 366] as being poorly grounded. Apparently, the electrons are localized on impurities due to the magnetic field.

Speaking of plasma condensation, some authors [240, 363] have appealed to the series of studies [375–377], which revealed "anomalous" effects during the arrival of a detonation wave from condensed explosive to the free surface of explosive charge. Coock [375–377] gives an interpretation of the observed effects. According to him, a "Wigner crystal" is formed as a result of "chemical ionization" in the front of detonation wave and, after this wave leaves the explosive charge, the crystal is in the metastable state. These plasma concepts formed the basis [376] of the model of detonation of condensed explosives. According to the model, the detonation-guiding process is the electron thermal conductivity and not the shock wave.

A close analysis of the results of the studies by Coock *et al.* [375–377] has shown [378, 379] that the interpretation offered is not unambiguous and gives rise to objections. Meanwhile, the conclusions on plasma phase transition lack substantiation. This is because Coock *et al.* [375–377] do not offer any results of the direct measurements of density, which must be the first to be affected by such transition.

The exposure of plasma to pulse X-ray radiation with a wavelength of 10^{-2} and 0.9 Å revealed [378] the absence of an anomalously high density of plasma. Meanwhile, its measured values proved to be eight to ten times higher than the free air density. This is in good agreement with the calculation of states behind the air shock wave front [380] performed for a quasi-ideal gas approximation.

The recording of the optical radiation emerging from plasma has shown [378, 141] that, as distinct from the conclusions by Coock *et al.* [375–377], it is not anomalous and corresponds to a brightness temperature of (8–10) $\cdot\ 10^3$ K (which is close to standard calculation [380]). Likewise, the optical absorption coefficient of plasma does not contain any qualitative anomalies and is described by the rarefied plasma model (see Chapter 8). Experiments involving the detonation wave arrival into vacuum [378] and helium [379] revealed the absence of plasma glow at the time when, according to concept of Coock [375–377], this glow must have intensified. In order to explain the observed rate of motion of plasma cluster, Coock [375–377] employed artificial reasoning about reactive thrust during recombination, while these rates are easily calculated on the basis of the theory of "break-up of discontinuity" at the explosive air interface.

Therefore, the experiments and analysis of available data demonstrate that the thermodynamic, optical, electrophysical, hydrodynamic and mechanical properties of plasma cluster are fully described by the ideal plasma model. This observation is based on the assumption that this cluster is formed by way of compression and an irreversible heating of air in the front of a wave emerging upon an expansion of the products of detonation of condensed explosive. The erroneous conclusion [375–377] on plasma condensation appears to be mainly due to a low rate of side expansion of the glass channels of tubes because of their considerable mass. This was erroneously interpreted as the "stability" of plasma.

Hall *et al.* [298] obtained a plasma with a density which are several times that of a solid-state plasma, and a temperature of about 1 eV as a result of collision of two shock waves. The latter were excited by high-power [(4–5) 10^{12} W/cm^2] laser radiation with a wavelength of 0.53 μm and pulse duration of about 100 p_s. The singularities of the X-ray spectra of radiation from an external source, transmitted through plasma, were interpreted by Hall *et al.* [298] as the presence of short-range order aluminum associated with Wigner concentration in a highly compressed plasma. In interpreting the results of these interesting experiments, one must perform a numerical simulation of the complex process of plasma generation and assess the non-equilibrium effects during the melting of aluminum.

Gilbert *et al.* [297] used Penning's cell to study the structure of an ion cloud containing 1,500 Be^+ ions with $N \sim 10^8$ cm^{-3}, $T \sim 10^{-2}$ K and $20 \leq \gamma \leq 200$. The laser fluorescence technique was used to register the periodic structure. The authors had associated it with structure anomalies in a non-ideal plasma which is far from being electrically neutral.

The occurrence of electronic phase transformations, due to the transition of the inner electronic shells of an atom or ion from a discrete spectrum to a continuous spectrum, is possible in a strongly compressed plasma. The phase transitions associated with the shell redistribution of electrons in the course of compression [121, 160] were theoretically analyzed using the band theory methods by Royce [381] and Arkhipov [382]. They were experimentally recorded by Bushman and Fortov [160] and Al'tshuler and Bakanova [383]. Brush *et al.* [384] and Kirzhnits *et al.* [21] have predicted electronic transformations in the ultramegabar range of pressures, which correspond to phase transition of the first kind, with the given electronic shell being in a discrete spectrum in one phase and in a continuous spectrum in the other. Naturally, such electronic transitions correspond to a series of phase transitions caused by the "ionization by pressure" of a sequence of electronic shells. The evaluation of the parameters of these transformations using the techniques of the quasi-classical theory [121] leads to ultrahigh pressures (over 300 kbar). The prospects of attaining such pressures were discussed in Chapter 3.

Six

Electrical Conductivity of Partly Ionized Plasma

The electrophysical properties of matter are defined primarily by the state of the electron component. During the transition of metal from the liquid to weakly conducting gas phase, the state of charged particles varies from a state of almost free electrons of liquid metal to that of electrons bound strongly in atoms. In doing so, the value of the coefficient of electrical conductivity varies by many orders of magnitude. Thanks to the studies carried out in the last 10 to 15 years, one can construct a pattern of variation of the coefficient of electrical conductivity in the entire range of parameters.

6.1. Electrical Conductivity of Ideal Partly Ionized Plasma

6.1.1. *Electrical Conductivity of Weakly Ionized Plasma*

The electrical conductivity σ is defined by the concentration of electrons and their mobility μ:

$$\sigma = e\mu N_e\,. \tag{6.1}$$

In an ideal plasma, these quantities are defined by well-known expressions. The concentrations of electrons, N_e, and ions, N_i, are related to each other by Saha's formula (cf., Chapter 4), which can be transcribed into

$$N_i N_e = K_1 N_a\,, \qquad K_1 = (2\Sigma^+/\Sigma)[mT/(2\pi\hbar^2)]^{3/2}\exp(-I/T)\,,$$

where N is the atomic concentration, Σ and Σ^+ are the inner statistical sums of atom and ion, and I is the atomic ionization potential. There are no complex ions in ideal plasma. Therefore, $N_i = N_e$. At low temperatures, the degree of ionization is low: $N_e \ll N_a$. Under these conditions,

$$N_e = N_e^0 = \sqrt{K_1 N_a}\,. \tag{6.2}$$

In a weakly ionized plasma, one can ignore the electron-ion and inter-electron interactions while allowing only for the interactions of electrons with the atoms (molecules) of gas. If one can regard this gas as ideal and the interaction of electrons with its atoms as a sequence of independent binary collisions, such a system of electrons is then well-described by the Lorentz gas model. The electron mobility in an atmosphere of massive, chaotically distributed scatterers is given by Lorentz's formula (cf., [387]). We now reproduce, schematically, the derivation of this formula.

We proceed from the Boltzmann's equation for the velocity distribution function of electrons, $f(\mathbf{v})$. Under steady-state and spatially homogeneous conditions, Boltzmann's equation for electrons in electric field $\mathbf{F}$ takes the form of

$$-(e\mathbf{F}/m)\partial f(\mathbf{v})/\partial \mathbf{v} = I_{\text{coll}}(f)\,.$$

The left-hand side describes the field effect and the right-hand side, the variation of the number of electrons in an element of phase volume due to collisions. I_{coll} is the collision integral. Since the electron mass is much less than the atomic mass ($m \ll M$), the form of $f(\mathbf{v})$ is close to spherically symmetric. This enables one to represent it as

$$f(\mathbf{v}) = f_0(\mathbf{v}) + \cos\theta f_1(\mathbf{v})\,,$$

where θ is the angle between the directions of velocity and field. Under the conditions of thermodynamic equilibrium the symmetric part of the distribution function, $f_0(\nu)$, is Maxwellian. The non-symmetric part, $f_1(\mathbf{v})$, is essential for calculating the electron mobility.

On substituting the expansion of $f(\mathbf{v})$ in the initial kinetic equation, we obtain the equation

$$-(eF/m)\partial f_0/dv = I_{\text{coll}}(f_1)$$

in $f_0(v)$. The collision integral, $I_{\text{coll}}(f_1)$, can be written away readily if one allows for the fact that the electron velocity varies greatly with respect to the direction upon each collision. After the collision, this direction can be regarded as being independent of its velocity prior to collision. Then,

$$I_{\text{coll}}(f_1) = -\nu(\mathbf{v})f_1(\mathbf{v})\,, \qquad \nu(\mathbf{v}) = N\mathbf{v}q(\mathbf{v})\,,$$

where $q(\mathbf{v})$ is the transport cross-section of electron-atom scattering and $\nu(\mathbf{v})$ is the electron-atom collision frequency.

In the electric field $\mathbf{F}$, the electrons, which are in chaotic heat motion, drift along this field. The mean electron velocity over the time which exceeds greatly the time between collisions is referred to as the drift rate w. For w, one can apparently write

$$w = \int \mathbf{v} \cos\theta f(\mathbf{v}) d\mathbf{v} = \int \mathbf{v} \cos^2\theta f_1(\mathbf{v}) d\mathbf{v} ,$$

since f_0 does not contribute to w. By definition, the electron mobility $\mu = w/F$. Therefore, by substituting the expression for $f_1(\mathbf{v})$, which follows from the kinetic equation in the formula for w, we obtain the expression for mobility to a Lorentz approximation:

$$\mu = -(e/m) \int (\partial f_0/\partial \mathbf{v}) \cos^2\theta \mathbf{v} d\mathbf{v}/\nu(\mathbf{v}) .$$

This expression is valid if a whole series of conditions are met. Note some of the conditions which are most vital for further presentation. The gas of atoms must be sufficiently rarefied for the binary collision approximation, $Nq^{3/2} \ll 1$, to be valid. The temperature must be sufficiently high and the thermal wavelength of electron, sufficiently small. This will enable one to ignore the interference of electron on neighboring atoms ($Nq\lambda \ll 1$). The potential energy of the Coulomb interaction of electrons must be very much less than the kinetic energy, $\gamma = e^2 N_e^{1/3}/T \ll 1$. The plasma must be non-degenerate: $\hbar^2 N_e^{2/3}/mT \ll 1$. The inter-atomic correlation can be neglected, $N_a b \ll 1$, $N_a a/T \ll 1$, where a and b are the coefficients of the van der Vaals' equation for gas.

It is more convenient to turn to an integration for the value of energy, $E = mv^2/2$, and substitute the Maxwellian expression for f_0 immediately. Hence one finally has

$$\sigma = e\mu N_e , \qquad \mu = \frac{4}{3\sqrt{\pi}} \frac{e}{mT^{5/2}} \int_0^\infty \frac{E^{3/2}}{\nu(E)} \mathrm{e}^{-E/T} dE . \tag{6.3}$$

When the mean collision frequency and cross-section are introduced, one can write

$$\mu = e/m\bar{\nu} , \qquad \bar{\nu} = (3\sqrt{\pi}/4) N_a \bar{q}(T) \sqrt{2T/m} , \tag{6.4}$$

where

$$\frac{1}{q(T)} = \frac{1}{T^2} \int_0^\infty \mathrm{e}^{-E/T} \frac{E dE}{q(E)} .$$

If $q(E)$ are known, the mean cross-sections $\bar{q}(T)$ can be readily calculated. A sizable set of reference data on electron-atom and electron-molecule scattering

Table 6.1. Averaged transport cross-sections $\bar{q}(T)$, $10^2\ a_0^2$, of scattering of electrons from atoms of alkali metals.

Alkali metal atom	T, 10^3 K							
	1.0	1.2	1.4	1.6	1.8	2.0	2.2	2.4
Li	16.5	14.4	12.6	11.1	9.91	8.93	8.11	7.46
Na	15.0	14.0	12.9	11.7	10.6	8.73	8.73	7.97
K	15.3	13.6	12.1	10.9	9.84	8.96	8.20	7.56
Cs	14.1	12.8	11.7	10.8	10.1	9.42	8.89	8.41
	2.6	2.8	3.0	3.2	3.4	3.6	3.8	4.0
Li	6.88	6.41	5.99	5.63	5.32	5.05	4.80	4.59
Na	7.31	6.73	6.23	5.79	5.41	5.07	4.77	4.51
K	7.00	6.52	6.10	5.73	5.41	5.12	4.87	4.64
Cs	8.98	7.63	7.32	7.04	6.79	6.57	6.37	6.20

Source: [49].

cross-sections is provided by Kieffer [338]. Table 6.1 gives the values of cross-sections for the atoms of alkali metals.

One can easily perform a numerical estimation of the coefficient of electrical conductivity using the following formula:

$$\sigma = 3.8 \cdot 10^2 \frac{N_e}{N_a} \frac{1}{\bar{q}\sqrt{T}} \text{ Ohm}^{-1}\ \text{cm}^{-1}, \tag{6.5}$$

where q is the average cross-section, 10^{-16} cm^2, and T is the temperature K.

6.1.2. *Electrical Conductivity of Partly Ionized Plasma*

Equation (6.4) describes very well the electron mobility before the electron-atom collision frequency ν becomes comparable with the electron-ion collision frequency ν_i. Let us now dwell on the peculiarities associated with the determination of ν_i (Spitzer [389]).

The cross-section of thermal electron scattering by an ion with a charge Z is defined by the square of Coulomb scattering amplitude $f = Ze^2/T$. It is referred to as the Landau length. The quantity $4\pi f^2 = 4\pi(Ze^2/T)^2$ only allows for collisions leading to large-angle scattering. The specific character of the long-range Coulomb interaction resides in the prevalence of large-angle scattering, which is taken into account by means of the "Coulomb logarithm", $\ln\Lambda$. In a classical plasma, the quantity Λ is defined as the ratio between the Debye length and the Landau length. Therefore, the frequency of collisions of an electron having energy E with ions is given by the expression

$$\nu_i(E) = 2\pi e^4 Z^2 E^{-3/2}(2m)^{-1/2} N_i \ln\Lambda \,.$$

The expression for the electrical conductivity of fully ionized plasma (ions with a charge Z and electrons) can be readily written based on the assumptions that the ions are immobile, uncorrelated scatterers and the electrons do not interact with each other. Such a model corresponds to the Lorentz gas model. The latter was used to describe the electron motion in the gas of neutral atoms:

$$\nu_i = \frac{\pi^{3/2}}{4\sqrt{2}} N_i \sqrt{\frac{T}{m}} \left(\frac{Ze^2}{T}\right)^2 \ln\Lambda \,. \tag{6.6}$$

Using Eqs. (6.1) and (6.4) and the condition of electroneutrality of plasma, $N_e = ZN_i$, one can readily obtain

$$\sigma = [2(2T)^{3/2}/(\pi^{3/2} Ze^2 m^{1/2})](1/\ln\Lambda)\,, \qquad \ln\Lambda = \ln(3/\Gamma)\,, \tag{6.7}$$

where $\Gamma = Ze^2/(r_{\rm D}T)$; the electron Debye length is $r_{\rm D}^{-1} = (4\pi e^2 N_e/T)^{1/2}$.

The specific nature of the Coulomb long-range interaction shows up in the strong effect produced by inter-electron correlations on the electrical conductivity. This is found even in the range of very small values of Γ. In order to include these correlations into the right-hand part of the kinetic equation, the integral of the electron-electron collisions must be added to that of the electron-ion collision. This equation was solved numerically by Spitzer and Härm [389], and corrections were calculated to the Lorentz formula (6.7). These corrections are given by the factor $\gamma_{\rm E}$ dependent on the ion charge Z:

Ion charge Z	1	2	4	16	∞
Factor $\gamma_{\rm E}$	0.582	0.683	0.785	0.923	1.000

The inter-electron interactions cause a decrease in the coefficient of electrical conductivity. When the electric field is switched on, the distribution function, which was spherically symmetric initially, stretches along the field. In opposing this, the electron-electron interactions somewhat symmetrize the electron distribution in the velocity space. This leads to a decrease in the transport coefficients. In the range of high values of Z, the factor $\gamma_{\rm E}$ tends to unity because the importance of inter-electron collisions decreases.

Spitzer and Härm have derived a formula for the electrical conductivity of fully ionized plasma, usually referred to as the Spitzer's formula:

$$\sigma_{\rm Sp} = \gamma_{\rm E}(Z) \frac{2(2T)^{3/2}}{\pi^{3/2} Z e^2 m^{1/2} \ln \Lambda} . \tag{6.8}$$

For singly charged ions ($Z = 1$), the factor $\gamma_{\rm E}(1) = 0.582$. Then, numerically,

$$\sigma_{\rm Sp} = 1.53 \cdot 10^{-4} T^{3/2} (\ln \Lambda)^{-1} \, {\rm Ohm}^{-1} \cdot {\rm cm}^{-1} , \tag{6.9}$$

where the temperature is expressed in kelvins.

For a classical ideal plasma, Eq. (6.8) is an exact expression. For this, the Coulomb logarithm $\ln \Lambda$ must be great, $\ln \Lambda \gg 1$.

At high temperature, the amplitude of the Coulomb scattering Ze^2/T is comparable to the proper radius of complex ion. Thus, the ion scattering becomes non-Coulomb. If an ion is fairly heavy (for example, the ion of xenon), its radius is great and for $T \approx {\rm Ry}$, the non-Coulomb corrections are significant.

A plasma may be regarded as strongly ionized if the electrons collide more frequently with the ions than with the atoms. After equating the frequencies of these collisions, we have

$$q N_a = (\pi e^4 / T^2) N_e \ln \Lambda . \tag{6.10}$$

In Fig. 6.1, the $T - \rho$ diagram for cesium is divided into a number of regions. Each region has a different charge transfer mechanism. Equation (6.10) gives the equation for the curve separating regions VII and VI, that is, the regions of strongly and weakly ionized plasma. Since the cross-section of the Coulomb scattering usually exceeds that of the electron-atom scattering considerably, a plasma can be regarded as being weakly ionized at only very low degrees of ionization.

Until recently, the electrical conductivity of plasma in the region of intermediate degrees of ionization was calculated using the Ghapman-Enskog method. In the first approach, it is not the reciprocal of the frequency $\nu^{-1}(E)$ that is averaged in a first approximation in the expression for mobility, but the frequency $\nu(E)$ proper. Therefore, the expression

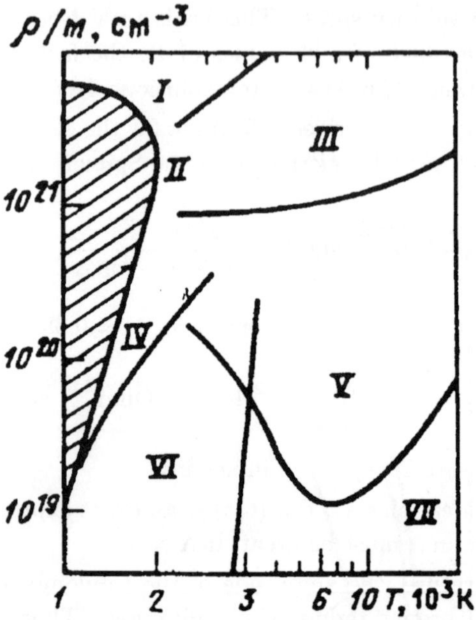

Fig. 6.1. $\rho - T$ Diagram for cesium [29]: I, liquid metal; II, "metal–non-ideal plasma" transition; III, non-ideal fully ionized plasma; IV, non-ideal plasma of metal vapors (weakly ionized plasma); V, weakly non-ideal strongly ionized plasma; VI, weakly non-ideal plasma of metal vapors; VII, ideal strongly ionized plasma.

$$\tilde{\nu}_i = \frac{2\pi}{3} N_i \left(\frac{8T}{\pi m}\right)^{1/2} \left(\frac{Ze^2}{T}\right)^2 \ln \Lambda \tag{6.11}$$

appears in the case of electron-ion collisions instead of the Lorentz frequency ν_i (6.6). The Chapman-Enskog method is characterized by poor convergence if the $q(E)$ cross-section depends non-monotonically on E.

In view of this, a number of interpolation formulas were proposed along the line of the Frost formula:

$$\sigma = \frac{4}{3\sqrt{\pi}} \frac{e^2 N_e}{m T^{5/2}} \int_0^\infty \exp\left(-\frac{E}{T}\right) E^{3/2} [\nu(E) + \nu_i(E)/\gamma_{\mathrm{E}}]^{-1} dE\,. \tag{6.12}$$

We now turn to an explanation of the structure of Eq. (6.12). Frost had proceeded from the Lorentz approximation (6.3) rather than from the Chapman-Enskog series. If one could ignore the interparticle correlations of electrons and ions, then, in order to determine the electron transit time, it would be sufficient to sum the frequencies of collisions with atoms and ions, $\nu(E) + \nu_i(E)$, in a ternary plasma. In doing so, one would not consider the correlations between charges. Its inclusion had led to the emergence of the factor γ_E in Eq. (6.12). In this case, their corrective inclusion is laborious. The inclusion of γ_E in Eq. (6.11) renders the latter exact in the weakly ionized and fully ionized limits, and gives adequate interpolation to the intermediate region.

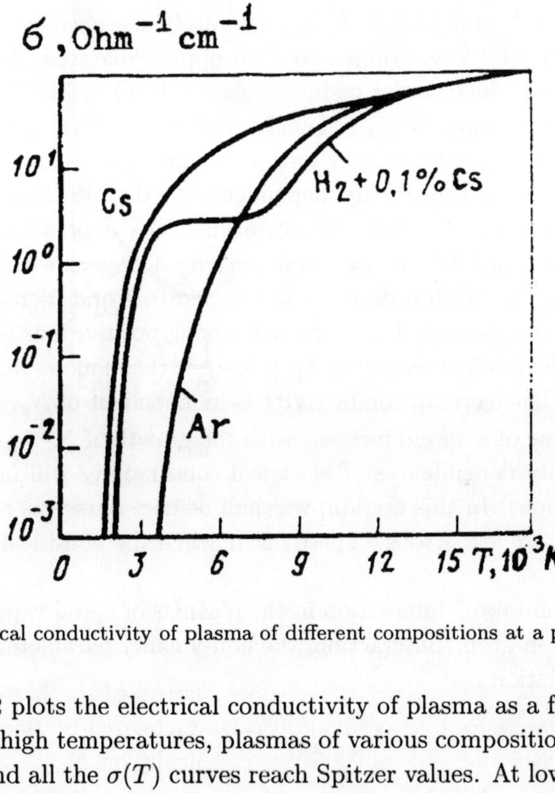

Fig. 6.2. Electrical conductivity of plasma of different compositions at a pressure of 1 atm.

Figure 6.2 plots the electrical conductivity of plasma as a function of temperature. At high temperatures, plasmas of various compositions lose their individuality and all the $\sigma(T)$ curves reach Spitzer values. At low temperatures, the parameter I/T in a weakly ionized plasma assumes principal importance. Hence the high values of σ for cesium plasma and a sharp increase in the value of σ for hydrogen plasma. This takes place when the latter is seeded with cesium.

6.2. The Coefficient of Electrical Conductivity of Weakly Non-ideal Plasma

With an increase in density (or a decrease in temperature), the effects of non-ideality show up in plasma. In a weakly ionized plasma, non-ideality is caused by charged particle-atom interactions. Non-ideality may affect both the concentration of free electrons and their mobility. The effect of non-ideality on N_e was discussed in Section 4.3. In a weakly non-ideal plasma, this effect is due to a reduction of the ionization potential and the emergence of cluster ions. In a plasma of metal vapors, both the reduction of the ionization potential and the emergence of cluster ions lead to an increase in N_e and σ. Whilst doing so, the former effect is still small ($\Delta I \ll T$) in moderately dense vapors. However, the inclusion of molecular ions brings about an appreciable variation of N_e. The region of the $\rho - T$ diagram for cesium is shown at VI in Fig. 6.1.

When the expression (4.43) is substituted in Eq. (6.1) for N_e, which is derived with due regard for cluster ions, one can observe how the emergence of cluster ions affects the density dependence of the electrical conductivity. At high temperatures and low concentrations, the expression $N_e \sim N_a^{1/2}$ (Eq. (6.2)) is valid and the electrical conductivity decreases with an increase in density, $\sigma \sim N_a^{-1/2}$. With a decrease in temperature and increase in concentration, cluster ions emerge. If A_3^+ prevails among positive ions and A^- among negative ions (the binding energy of A_2^- is low), it then follows from Eqs. (4.32) and (6.5) that the electrical conductivity is independent of N_a. If the A_4^+ ion prevails, the value of σ would increase with the growth of N_a ($\sigma \sim N_a^{1/2}$). The increasing density dependences of electrical conductivity will be discussed in subsequent sections. In this section, we shall restrict ourselves to a discussion of effects which are less strong. These occur when the non-ideality parameter $\gamma_{ai} < 1$.

Since the "ion-atom" interaction in the plasmas of metal vapors is stronger than the "electron-atom" interaction, the non-ideality parameter can be introduced as the relation

$$\gamma_{ai} = N_3^+/N_2^+ = N_a K_4/K_3 K_5\,.$$

The condition $\gamma_{ai} = 1$ defines the boundary separating the regions V and VI in Fig. 6.1. If $\gamma_{ai} > 1$, then $N_3^+ > N_2^+$. This means that one should expect the emergence of A_4^+, and may be A_m^+, $m \gg 1$.

The first experimental study, the results of which point to the effect of ionic complexes on electrical conductivity, was carried out by Morrow and Crags [228]. The measurements in sodium vapors were performed at $p \approx 0.5p_s$, $T = 1{,}200$–$1{,}500$ K. The results of the calculation, performed by Khrapak [254] using the procedure described in Section 4.3, are in good agreement with the experiment (Fig. 6.3). In Fig. 2.11, the results of analogous calculations [49] are compared against the values of the electrical conductivity measured

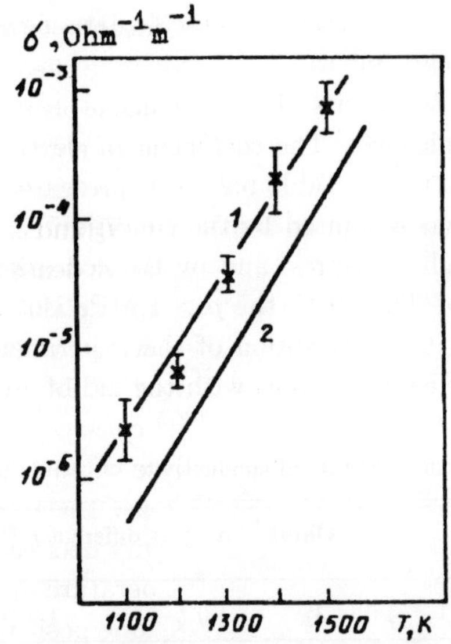

Fig. 6.3. Coefficient of electrical conductivity of sodium plasma at a pressure $p = p_s$: the points, experiment [228]; calculation curves — *1*, eight-component composition of plasma; *2*, three-component composition.

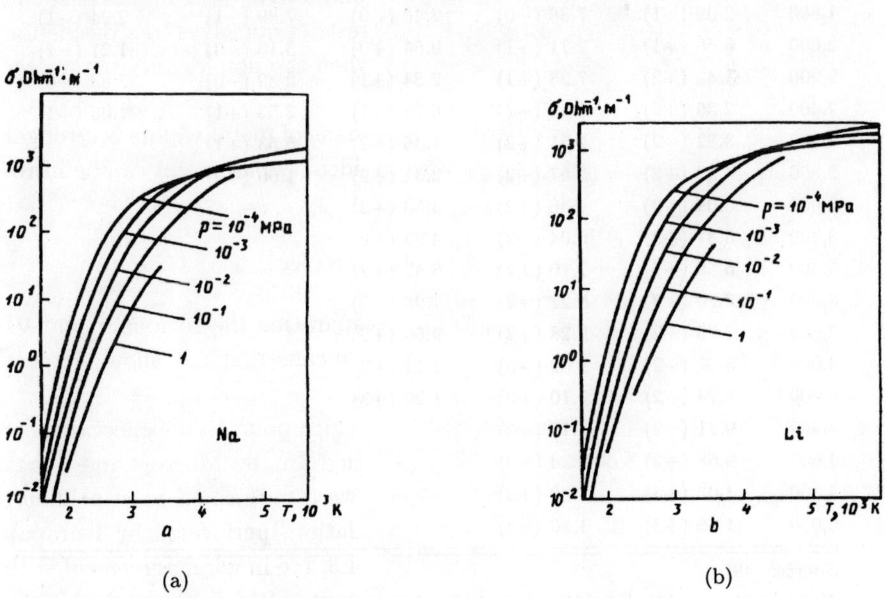

Fig. 6.4. Isobars of electrical conductivity of sodium (*a*) and lithium (*b*) plasmas [254].

for cesium vapors. A comparison of the calculation and measurement results points to their good agreement.

Figure 6.4 gives the isobars of the coefficient of electrical conductivity of sodium and lithium plasmas. The coefficient of electrical conductivity for cesium plasma is tabulated in Table 6.2. The pressure and temperature range covered by calculations is limited by the condition $\gamma_{ai} \leq 1$ in the case of low temperatures and high pressures, and by the condition $\ln \Lambda \geq 3$ in the case of high temperatures. Therefore, the $p > 1$ MPa isobars are absent. Indeed, even at $p = 2$ MPa, the description of the experimental isobar of the electrical conductivity of cesium vapor, with the aid of analogous calculations, is

Table 6.2. Electrical conductivity of cesium plasma.

T, K	σ $\text{Ohm}^{-1} \cdot \text{m}^{-1}$ at different p (MPa)				
	1.0 (−4)	1.0 (−3)	1.0 (−2)	1.0 (−1)	1.0
1,000	4.97 (−4)	1.57 (−4)	4.95 (−5)	–	–
1,200	2.71 (−2)	8.56 (−3)	2.70 (−3)	8.42 (−4)	–
1,400	4.84 (−1)	1.54 (−1)	4.87 (−2)	1.54 (−2)	–
1,600	4.19 (+0)	1.37 (−0)	4.37 (−1)	1.40 (−1)	5.11 (−2)
1,800	2.09 (+1)	7.38 (−0)	2.43 (−0)	7.89 (−1)	2.94 (−1)
2,000	6.56 (+1)	2.71 (+1)	9.54 (+0)	3.18 (−0)	1.21 (−1)
2,200	1.42 (+2)	7.23 (+1)	2.84 (+1)	9.89 (−0)	3.83 (−0)
2,400	2.36 (+2)	1.49 (+2)	6.78 (+1)	2.53 (+1)	1.00 (+1)
2,600	3.32 (+2)	2.51 (+2)	1.35 (+2)	5.53 (+1)	–
2,800	4.23 (+2)	3.67 (+2)	2.31 (+2)	1.06 (+2)	–
3,000	5.09 (+2)	4.86 (+2)	3.53 (+2)	–	–
3,200	5.87 (+2)	6.04 (+2)	4.92 (+2)	–	–
3,400	6.58 (+2)	7.10 (+2)	6.42 (+2)	–	–
3,600	7.20 (+2)	8.27 (+2)	7.98 (+2)	–	–
3,800	7.75 (+2)	9.28 (+2)	9.56 (+2)	–	–
4,000	8.27 (+2)	1.02 (+3)	1.11 (+3)	–	–
4,200	8.74 (+2)	1.10 (+3)	1.26 (+3)	–	–
4,400	9.21 (+2)	1.17 (+3)	–	–	–
4,600	9.68 (+2)	1.24 (+3)	–	–	–
4,800	1.02 (+3)	1.30 (+3)	–	–	–
5,000	1.06 (+3)	1.36 (+3)	–	–	–

Source: [49].

Note: Figures in brackets designate decimal powers. For example, $4.97(-4) = 4.97 \cdot 10^{-4}$.

markedly poorer. We speak of the portion of the isobar of electrical conductivity, shown in Fig. 2.10, which corresponds to high temperatures and in which the plasma of cesium vapors can still be regarded as weakly non-ideal. An adequate description is only attained (curve *2* in Fig. 2.10) upon inclusion of density corrections to the electron mobility. We now turn to a discussion of these corrections.

Amidst heavy uncorrelated scatterers, the electron mobility is given by the Lorentz formula. The Lorentz formula describes the effect of binary collisions, that is, the interactions which take place when the radius of the acting forces is very much less than the electron free path. It is well-known that in order to apply this approximation, no two particles should be found in the sphere of interaction, that is, the following inequality should be satisfied:

$$(4\pi/3)Nq^{3/2} \ll 1\,. \tag{6.13}$$

At densities $N = 10^{20}$ cm^{-3}, the quantity $(4/3)Nq^{3/2}$ in a cesium plasma reaches a value close to unity ($q = 300 - 400\pi a_0^2$; Table 6.1). When the inequality (6.13) is violated, one must allow for the collisions between three, four and more particles. Ternary collisions, whose frequency is proportional to N^2, lead to a reduction of the free path length.

However, this does not mean that in a plasma of metal vapors the electron mobility μ becomes, with a rise in density, less than the mobility μ_0 calculated for the same density from the Lorentz formula. It is not always the case that one can describe the electron-atom interaction when the atom is represented by a hard sphere with a radius of the order of $q^{1/2}$. If the atomic polarizability is sufficiently high, the "electron-atom" interaction is one of polarization. The sign of the scattering length may serve as an indication of this (it must be negative). The dense vapors of metals fall into this category of strongly polarizable media. Therefore, their compaction may cause an increase in mobility rather than its decrease. This effect is well-known in the physics of dense fluids. The latter is referred to as "high mobility" fluids. These include, for instance, dense inert gases (argon, krypton and xenon, the polarizability of their atoms being high) and their liquids. These effects were reviewed by Atrazhev and Iakubov [391]. As a result, the summary field of scatterers is smoothed. In describing this effect, the following considerations are important. The potential of the "electron-isolated atom" interaction can be expressed as

$$V(r) = \frac{2\pi\hbar^2}{m} L_c \delta(r) - \frac{\alpha e^2}{2(r^2 + r_a^2)^2}\,. \tag{6.14}$$

Here, the short-range component is written in the form of the Fermi pseudopotential (1.11). The quantity L_c is the length of electron scattering by the potential $V(r)$. The polarization component (6.14) features correct asymptotic behavior and the cutoff parameter r_a corresponds to the size of the outer electron shell, that is, the atomic radius. The amplitude of the angle θ scattering, $f(\theta)$, and the length of scattering by potential (6.14), L, take the form of

$$f(\theta) = -L - \frac{\pi\alpha}{2a_0\lambda}\sin\frac{\theta}{2}, \qquad L = L_c - \pi\alpha/(4a_0r_0), \tag{6.15}$$

where $\lambda = \hbar(2mE)^{-1/2}$ is the wavelength of electron with energy E. Therefore, in accordance with the principal ideas of effective radius approximation developed by O'Malley, the scattering amplitude is only governed by two quantities: the scattering length L and the atomic polarizability α. Equation (6.15) is valid for slow electrons when the following inequalities are satisfied:

$$\lambda \gg L_c, \qquad \lambda \gg r_a, \qquad \lambda^2 \gg \alpha/a_0 .$$

An electron moving in a dense medium polarizes the latter. Internal fields occur which reduce the electron-atom interaction. The polarization effect can be readily included by introducing the dielectric permeability $\varepsilon = 1+(8\pi/3)\alpha N$ and replacing the potential (6.14) by that of the electron-atom interaction in a medium,

$$V(r) = \frac{2\pi\hbar^2}{m}L_c\delta(r) - \frac{\alpha e^2}{2(r^2+r_a^2)^2\varepsilon} .$$

If $(8\pi/3)\alpha N \ll 1$, the length of scattering in the medium [391] is:

$$L_m = L + 4\pi^2\alpha^2 N/(3a_0r_a) . \tag{6.16}$$

This suggests that if $L < 0$ (that is, attraction prevails in the electron-atom interaction), $|L_m| < |L|$. In other words, the interaction weakens with an increase in density. Atrazhev and Iakubov [391] have shown that the polarization effect is responsible for the marked increase of reduced mobility ($\mu/\mu_0 \gg 1$) in a number of dense gases and liquids.

The atoms of alkali metals feature very high values of polarizability ($\alpha \sim 100a_0^3$). However, the effective radius theory is not directly applicable in describing the electron scattering by these atoms. This is because such a description is far more complex than it appears to be. Nevertheless, one can produce a qualitatively correct answer by writing the approximate formula for the $q_m(E)$ cross-section of scattering by an atom in a medium:

$$q_m(E) = q(E)(1-\xi)^2, \qquad \xi = 4\pi^{5/2}\alpha^2 N(3a_0r_a\sqrt{q(E)})^{-1}, \tag{6.19}$$

where the parameter ξ allows for a dielectric screening of the polarization interaction. Under experimental conditions [36], the parameter ξ is of the order of 0.2. As a result, the correction to the electron mobility $\mu/\mu_0 = (1-\xi)^{-2}$ is appreciable.

Besides the polarization effect, a number of other factors also affect the electron mobility. The interference effect [392] occurs when the de Broglie electron wavelength λ_e becomes comparable with the electron track length $(qN)^{-1}$. An overlap of two successive scattering events then take place. The correction that emerges can be interpreted as a correction to increase the collision frequency $\nu_{\text{eff}} = \nu(1 + 2\lambda qN)$. As a result, the mobility varies:

$$\mu = \mu_0(1 - \pi^{1/2} N q \lambda)^2$$

(the mobility decreases when compared to the ideal-gas case). However, when the temperature increases, the interference effect disappears. This is because the thermal wavelength $\lambda = \hbar(2mT)^{-1/2}$ decreases.

Therefore, as the density increases, the relative mobility may both decrease and increase. Figure 6.5 gives the results of measurements performed in different gases (see list of references in [4]). In polarizable gases, $\mu/\mu_0 > 1$ and in poorly polarizable gases, $\mu/\mu_0 < 1$. It follows from Fig. 6.5 that the density effects in mobility for atomic and molecular gases, for which $q \approx 10^{-15}$ cm^2,

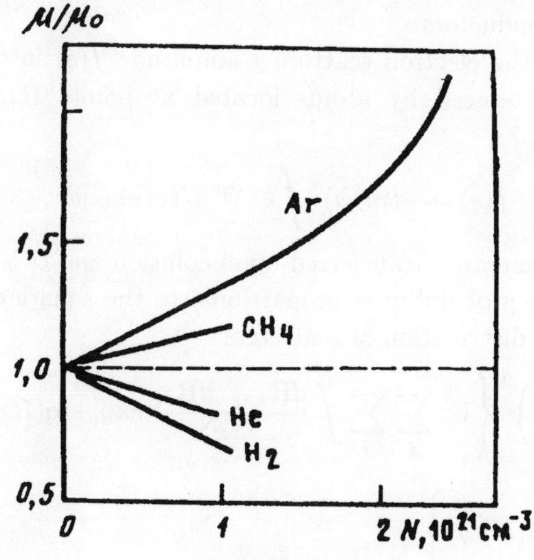

Fig. 6.5. The measured values of μ/μ_0 at $T = 300$ K as a function of density [4].

occur at $N \approx 10^{21}$ cm^{-3}. In the plasma of alkali metals, considerable deviations of μ from μ_0 must occur at $N \geq 10^{19}$ cm^{-3}.

As mentioned above, the values of the coefficient of electrical conductivity, which exceed the Spitzer values σ_{Sp}, cannot be attained under standard conditions. This is due to the additivity of the frequencies ν and νi in the formula for electron mobility $\mu = em^{-1}(\nu+\nu_i)^{-1}$. It should be borne in mind, however, that the additivity of ν and ν_i in a very dense plasma may be disturbed. For this to happen, an electron must manage to interact repeatedly, when passing over the Landau length e^2/T, with atoms, that is, the inequality

$$e^2 qN/T > 1$$

must be valid.

Under these conditions, in a first approximation, the electron-ion collisions have no effect whatsoever on the mobility and they may be ignored. Therefore, the values of σ in a dense medium are not restricted to Spitzer values.

We will not treat the mobility in a medium of correlated scatterers. In a weakly ionized gas, the mobility may be strongly affected by interatomic correlations. The square of the amplitude of electron scattering in the medium must be averaged with due regard for the correlation between chaotically arranged atoms rather than over their distribution. As a result, the expression for the averaged collision frequency varies. This frequency contains the structure factor of the medium (see, for example, [383]). This situation is typical of liquid metals and semiconductors.

Consider now the electron scattering amplitude $f(q)$ in the field $U(\mathbf{r}) = \sum_j \mathbf{V}(\mathbf{r} - \mathbf{R}_j)$ produced by atoms located at points $\mathbf{R}_j$. In the Born approximation,

$$f(q) = -(m/h)^2 \int e^{-i\mathbf{qr}}[U(\mathbf{r}) - \bar{U}]d\mathbf{r}\,,$$

where q is the momentum transferred upon collision and U is the mean field.

The scattering probability is proportional to the square of amplitude f^2, averaged over the distribution of scatterers:

$$\overline{f(q)^2} = \left(\frac{2\pi m}{\hbar^2}\right)^2 \left\{ V_q^2 \sum_i \sum_j \int \frac{d\mathbf{R}_1 \ldots d\mathbf{R}_N}{\Sigma \Omega^N} \exp[-i\mathbf{q}(\mathbf{R}_j + \mathbf{R}_i) + \beta u_a] \right.$$
$$\left. - \delta^2(q)(N\bar{V})^2 \right\}. \tag{6.18}$$

Here, V_q is the Fourier component of the $V(r)$ potential:

$$\bar{V} = \int V(\mathbf{r})d\mathbf{r}\,, \qquad u_a = \frac{1}{2}\sum_{m,n} v(\mathbf{R}_m - \mathbf{R}_n)\,,$$

where v is the potential of interatomic interaction, N is the total number of gas atoms, $N = N\Omega$; Ω is the system volume, and $\Sigma = \Omega^{-N} \int \mathbf{R}_1 \ldots d\mathbf{R_N}$ $\exp(-\beta u_a)$ is the partition function of gas.

The term in curly brackets in (6.18), which corresponds to $i = j$, is equal to NV_q^2. This is a common result of the gas-kinetic theory. With due regard for correlation, the terms corresponding to $i \neq j$ are reduced to the emergence of a term proportional to the next degree of density. Upon introducing the pair function of interatomic correlation $g(r)$ and making transformations in (6.20), the mean square of scattering amplitude per atom may be represented as

$$(N\Omega)^{-1}\overline{f^2} = \left(\frac{m}{2\pi\hbar^2}\right)^2 V_q^2 \left\{1 + N\int \exp(-i\mathbf{q}\mathbf{r})[g(\mathbf{r}) - 1]d\mathbf{r}\right\}$$

$$= \left(\frac{m}{2\pi\hbar^2}\right)^2 V_q^2 S(q)\,.$$

Here, $S(q)$ is the structure factor of the medium and q is the momentum transferred upon collision.

Thus, the expression for collision frequency has the form

$$\nu(v) = 2\pi N v \int q(v,\theta)\cdot S\cdot(2v\cdot\sin(\theta/2))\cdot(1-\cos\theta)\cdot\sin\theta\cdot d\theta\,, \qquad (6.19)$$

where $q(v,\theta)$ is the differential cross-section of electron-atom scattering. Figure 2.22 shows the structure factor of rubidium. One can see that it is a non-monotonic function of q and thermodynamic parameters of the medium. Therefore, the effect of correlation on mobility may vary considerably. Depending on the conditions, the mobility may decrease or increase, as compared to the mobility calculated in an ideal-gas approximation.

Podlubnyi *et al.* [394] treated the coefficient of electrical conductivity of a multi-component plasma. The electron scattering probability was averaged over the multi-component system of scattering centers. The resultant collision frequency has the form of

$$\nu(\nu) = \sum_{a,b}\sqrt{N_a N_b}Q_{ab}v\,, \qquad Q_{ab} = \int d\Omega(1-\cos\theta)f_a f_b^* S_{ab}\,, \qquad (6.20)$$

where $S_{ab}(q) = \delta_{ab} + \sqrt{N_a N_b}\int dr e^{-i\mathbf{q}\mathbf{r}}\cdot(g_{ab}-1)$, the subscripts a and b number the types of particles, N_a and N_b denote the concentration, and f_a and f_b are the scattering amplitudes.

In the Debye-Hueckel approximation, the ion-ion structure factor is given by expression (5.6). Formula (6.20) may be used to calculate the coefficient of electrical conductivity in a plasma, in which ions of different multiplicity are present. In doing so, corrections for the Spitzer factors $\nu_{\rm E}$ must be introduced. Podlubnyi *et al.* [394] have derived

$$\sigma = \frac{2(2T)^{3/2}\gamma_{\rm E}(\bar{Z})}{\pi^{3/2}\bar{Z}e^2m^{1/2}\ln\Lambda}, \qquad \frac{1}{\ln\Lambda} = \frac{1}{3}\int_0^\infty \frac{dx x^3}{\ln(1+ax)}\cdot e^{-x},$$

where $\bar{Z} = \sum_a Z_a N_a / \sum_a N_a$ is the average charge,

$$\gamma_{\rm E}(\bar{Z})^{-1} = \sum_a N_a \gamma_{\rm E}(Z_a)^{-1} Z_a^2/\bar{Z}^2 \sum_a N_a$$

is the average Spitzer factor and $a = 16\bar{Z}^2{\rm Ry}/T\cdot\Gamma^2$. Note that the introduction of the structure factor into the collision frequency eliminates the Coulomb convergence at $q \to 0$.

In Fig. 6.6, the results of the calculations are compared with the results of measurements [143] of the coefficient of electrical conductivity of

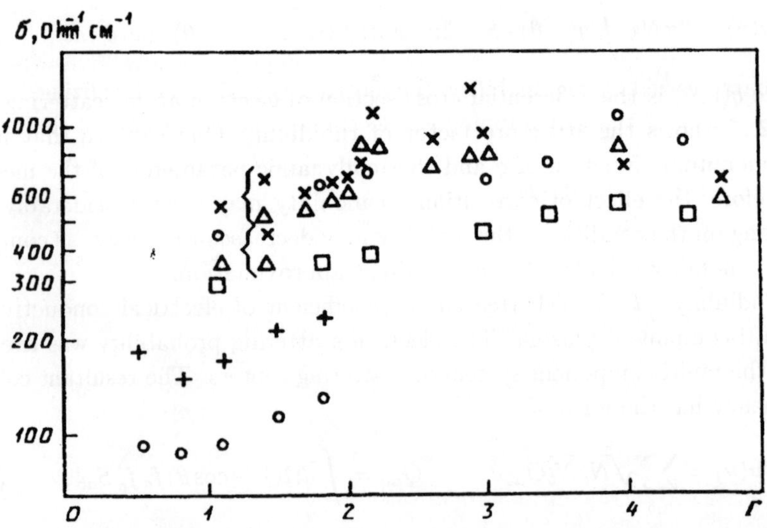

Fig. 6.6. Coefficient of electrical conductivity of non-ideal plasma of argon and xenon [394]: experiment: +, argon [138]; •, xenon [138]; x, xenon [143]; theory [394]; o, argon; △, xenon for the conditions of [138]; •, xenon for the conditions of [143].

high-temperature ($T \approx 7 \cdot 10^4$ K) xenon plasma. Under these conditions, which are discussed in detail in Section 7.5, the average ion charge $\bar{Z} \approx 2$. The experimental crosses in Fig. 6.6 and calculated squares are in good agreement with each other. The situation is much more worse in a comparison of the results of the calculation with the results of measurements in argon and xenon at $T \approx 2.5 \cdot 10^4$ K [138]. Podlubnyi *et al.* [394] have included the non-Coulomb corrections to the frequency of electron-ion scattering in calculations.

Conditions are possible in a weakly ionized plasma, under which the contribution to the collision frequency by the scattering of electron from correlated ion-atom pairs becomes noticeable.

6.3. The Electrical Conductivity of Non-ideal Weakly Ionized Plasma

6.3.1. *The Density of Electron States*

In Chapter 4, we treat the conditions in dense vapors of metals when ions occur in clusters. This affects the ionization equilibrium considerably. As for the electrons, they remain free, that is, they are conduction electrons. However, a different situation is possible. If the density is high and the energy of the electron's attraction to atom is sufficiently great, the fluctuations in the medium density may cause deep potential wells. An electron is captured in such wells and hence there is a transition to the negative energy region. This stabilizes the fluctuation in density. An electron cluster emerges and because it is characterized by low mobility, the electron is said to be localizing.

The effect of the self-trapping of electrons is characteristic of dense disordered systems (see, for example, [395–397]). It is also possible that the electrons are localized in dense vapors of mercury [4]. The distribution of electrons in positive and negative energies is defined by the density of the electron states in a dense medium. The main question that arises is: what is the energy of the electrons which have sufficient mobility to be regarded as conduction electrons? The theory may hardly provide an answer to this question since it only enables the development of rather crude models. The density of electron states $\rho(E)$ defines $N_e(E)dE$, the electron concentration in the given energy range,

$$N_e(E)dE = \rho(E)\exp(\mu_e\beta - E\beta)\,, \tag{6.21}$$

where μ_e is the chemical potential of electronic gas. In an ideal gas,

$$\rho^0(E) = 4\pi(2m)^{3/2}(2\pi\hbar)^{-3}\sqrt{E}\,, \qquad E \geq 0\,; \rho^0(E) = 0\,, \quad E < 0\,. \tag{6.22}$$

In the theory of disordered systems, it is well-known that in a dense medium, $\rho(E)$ may qualitatively differ from $\rho^0(E)$. First, the continuous spectrum boundary shifts through the value of mean field. The latter is interpreted as a decrease in the ionization potential in the plasma. In addition, a "tail" of the density of states $\rho(E)$ emerges, which extends to the region of forbidden states with negative energy (Fig. 6.7). These states occur when an electron falls into the potential wells formed by the fluctuations in the medium density. The states of remote tail $\rho(E)$ correspond to electrons trapped in heavy clusters. Under certain conditions, this is the greater part of electrons which have lost their mobility. Electrons which correspond to negative energies of low absolute values are fewer. These electrons, though not free, may have considerable mobility. It is in their description that they present the most difficulties.

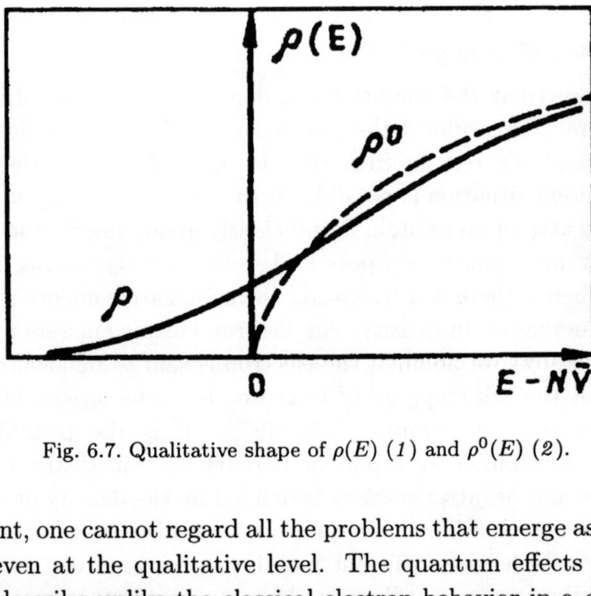

Fig. 6.7. Qualitative shape of $\rho(E)$ (*1*) and $\rho^0(E)$ (*2*).

At present, one cannot regard all the problems that emerge as being solved. This is so even at the qualitative level. The quantum effects are especially difficult to describe unlike the classical electron behavior in a dense medium which is much more simple. One can assume that such a formulation models the situation in a weakly ionized plasma in a qualitative manner.

The density of states of a classical electron placed in a potential field $U(r)$, which is generated by scatterers whose number is equal to $^-$, is defined by the expression

$$\rho(E) = \int \frac{d\mathbf{r}}{\Omega} \int \frac{d\mathbf{p}}{(2\pi\hbar)^3} \langle \delta[E - E(\mathbf{r}, \mathbf{p})] \rangle \,, \tag{6.23}$$

$$E(\mathbf{r},\mathbf{p}) = \mathbf{p}^2/2m + U(\mathbf{r})\,; \qquad U(\mathbf{r}) = \sum_{i=1} V(\mathbf{r}-\mathbf{R}_i)\,.$$

Here, $\mathbf{r}$ and $\mathbf{p}$ denote the electron co-ordinate and momentum, Ω is the system volume, and the averaging is performed over all possible scatterer configurations. If the atomic gas can be regarded as ideal, the mean value of the arbitrary function $Y(\mathbf{R}_1, \mathbf{R}_1, \ldots, \mathbf{R}_\mathsf{N})$ is

$$\langle Y\rangle = \int Y \exp\left[-\beta\sum_{i,k} v_{ik}(\mathbf{R}_i - \mathbf{R}_k)\right] \Omega^{-\mathsf{N}} d\mathbf{R}_1 d\mathbf{R}_2 \ldots d\mathbf{R}_\mathsf{N}\,, \tag{6.24}$$

where N is the total number of atoms in the system and $\nu_{ik}(\mathbf{R}_i - \mathbf{R}_k)$ is the energy of interaction between the i-th and k-th atoms.

In a rarefied gas, the potential energy of electron in the atomic field $U(r)$ can be ignored. It then follows from Eq. (6.23) that $\rho(E) = \rho^0(E)$. In a dense gas, the calculation of $\rho(E)$ is related to the concrete form of the potential $V(r)$. The total and simple results can be obtained by including only the greatest fluctuations in the field U described by the Gaussian approximation. In Eq. (6.23), one can perform an integration over momenta:

$$\rho(E) = \frac{(2m)^{3/2}}{2\pi^2\hbar^3}\int \frac{d\mathbf{r}}{\Omega}\langle\sqrt{E-U(r)}\rangle\,. \tag{6.25}$$

The averaging over all the atomic configurations, Eq. (6.24), can be replaced by an averaging over the random field distribution $P(U)$. Thus,

$$\rho(E) = (2m)^{3/2}(2\pi^2\hbar^3)^{-1}\int_{-\infty}^{\infty} dU P(U)\sqrt{E-U}\,. \tag{6.26}$$

Here, $P(U)$ is the potential energy distribution function, $N\bar{V}$ is the magnitude of the mean field $N\bar{V} = \langle U\rangle$, NW_0 is the field dispersion, and $W_0 = \int V^2(r)d\mathbf{r}$, that is, the value at zero of the potential correlator $W(r) = \int V(\mathbf{r}-\mathbf{r}')V(r')d\mathbf{r}'$. If $E - N\bar{V} < 0$, we have

$$\rho(E) = 2\pi\frac{2(2m)^{3/2}}{(2\pi\hbar)^3}\frac{8NW_0}{(E-N\bar{V})^2}\sqrt{|E-N\bar{V}|}\exp\left[-\frac{(E-N\bar{V})^2}{8NW_0}\right] \tag{6.27}$$

for high negative energies ($(E-N\bar{V})^2 \gg NW_0$). Consequently, deep in the "tail", $\rho(E)$ decreases exponentially. In the region of renormalized zero, the $\rho(E)$ energy is finite:

$$\rho(N\bar{V}) = 2\pi\frac{2(2m)^{3/2}}{(2\pi\hbar)^3}\Gamma(3/4)\frac{(2NW_0)^{1/4}}{(2\pi)^{1/2}}\,. \tag{6.28}$$

Given high positive energies, ρ assumes the regular root dependence but with an energy shift. Such a density of states is proper for strongly non-ideal systems.

It should be mentioned that Eq. (6.27) cannot describe the $\rho(E)$ tail correctly if the electron-atom potential depth is comparable to or exceeds the temperature $|\beta V| \geq 1$. In such cases, the $\rho(E)$ tail is formed by rare but high fluctuations in density, which may be strongly affected by interatomic interactions. Interatomic repulsion restricts the $\rho(E)$ tail from the side of high negative energies while attraction extends this tail.

The density of states of a classical electron in a field of potential scatterers were studied by a number of researchers and basic qualitative regularities were revealed. The most reliable results, however, can be obtained by way of numerical modeling using the Monte Carlo methods and molecular dynamics. The methods of modeling the problem in the dynamics of a classical electron were developed by Lagar'kov and Sarychev [398, 399]. An attraction potential with polarization asymptotics, a depth g and a rigid core with a radius δ was selected as the potential $V(r)$. The quantity $\delta = 3.3a_0$ is the gas-kinetic radius of a mercury atom. The parameters g were selected, which are equal to 0.18 eV. The negative ion Hg^- does not exist and, therefore, the potential $V(r)$ cannot be very deep.

Figure 6.8 illustrates the results of the calculations [398] of $\rho(\varepsilon)$ as a function of the relative electron energy $\varepsilon = E/g$ for different atomic densities. The $\rho(\varepsilon)$

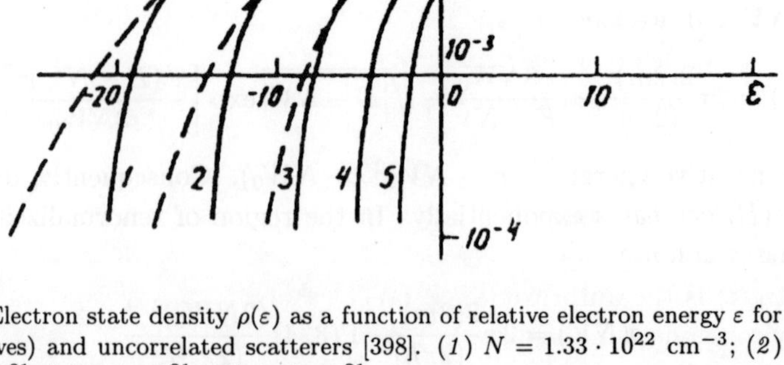

Fig. 6.8. Electron state density $\rho(\varepsilon)$ as a function of relative electron energy ε for correlated (solid curves) and uncorrelated scatterers [398]. (*1*) $N = 1.33 \cdot 10^{22}$ cm^{-3}; (*2*) $8.3 \cdot 10^{21}$; (*3*) $4.8 \cdot 10^{21}$; (*4*) $2.4 \cdot 10^{21}$; (*5*) $1.05 \cdot 10^{21}$.

dependences were constructed to allow for both the interatomic correlation and to ignore the latter. The scatterers were assumed to interact as the hard spheres of a radius δ_1. Interatomic repulsion limits clustering by cutting off the $\rho(\varepsilon)$ tail.

6.3.2. *Electron Mobility and Electrical Conductivity*

The linear response theory allows one to write the exact initial expressions for kinetic coefficients. In particular, as long as the linear relationship between the current and the field is valid (Ohm's law), the electrical conductivity is expressed in terms of the correlation function of currents.

Electrical conductivity, as given by the linear theory, is

$$\sigma_{\mu\nu}(\omega) = N_e e^2 \beta \int \langle \nu_\nu(0)\nu_\mu(t)\rangle e^{-i\omega t - \delta t} dt\,, \qquad \delta = +0\,. \tag{6.29}$$

Here, N_e is the concentration of current carriers, ω is the external field frequency, and $\langle \nu_\nu(0)\nu_\mu(t)\rangle$ is the autocorrelator of electron velocities. The averaging is performed over the electron distribution, velocities and position of scatterers:

$$\langle \ldots \rangle = A \int d\mathrm{v} e^{-m\nu_2\beta/2} \int d\mathbf{R}_1 \ldots d\mathrm{R}_\mathrm{N} \Omega^{-\mathrm{N}} \exp(-\beta u) \cdot (\ldots)\,,$$

where A is the normalization constant of averaging and U is the energy of interaction between scatterers. The latter should be included if the medium cannot be regarded as an ideal gas.

From a discussion of these formulas, it follows that the expression for the static electrical conductivity, $\sigma \equiv \sigma_{xx}(0)$, can be transcribed as

$$\sigma = (e^2/m) \int_{-\infty}^{\infty} N_e(E)\tau(E) dE\,. \tag{6.30}$$

Here, $\tau(E)$ is the autocorrelation time of electron with pre-assigned energy E. Under conditions of the electron travel in a medium of massive scatterers, its energy can be regarded constant. The value of time, $\tau(E)$, is given by the following expression:

$$\tau(E) = \int_{-\infty}^{\infty} \varphi_\mathrm{E}(t) dt\,, \qquad \varphi_\mathrm{E}(t) = \langle \nu_x(0)\nu_x(t)\rangle / \langle \nu_x(0)^2\rangle\,, \tag{6.31}$$

where $\varphi_\mathrm{E}(t)$ is the autocorrelation function of electron velocity and ν_x is the velocity component in the direction of an electric field. In Eq. (6.31), averaging is done over all possible atomic configurations at a fixed electron energy E.

In a rarefied gas, $\tau(E)$ is inversely proportional to the free path between two successive collisions of an electron with atoms. The integration range is restricted by positive energies because $\tau(E)$ is given by Eq. (6.31). Therefore, Eq. (6.30) changes into the Lorentz's formula. In a dense system, electrons with different energies make substantially different contributions to the electrical conductivity. We now turn to a discussion of these contributions by following Khrapak and Iakubov [400]. An ideal-gas approximation of free time of flight is only valid for very fast electrons. If $\beta(NW_0)^{1/2} \geq 1$, the bulk of conduction electrons are not free even if their energy is positive. They are constantly in the field of scatterers. This field is alternates between "wells" and "hills", with the most probable depths (heights) being close to $(NW_0)^{1/2}$ and space dimensions of the order of ℓ. The value of ℓ corresponds to the correlation length of the field of scatterers: it is close to the effective range of the potential $V(r)$. Therefore, the slow electrons $E < (NW_0)^{1/2}$ travel at velocities of $\nu \approx (Nw_0)^{1/4}m^{-1/2}$, which exceed thermal velocities. The electrons are scattered from density fluctuations. Apparently, τ is close to the time of flight of field correlation length, that is, it close to ℓ/ν.

The electrons of the $\rho(E)$ tail do not contribute to static conductivity because they correspond to low-mobility clusters. However, electrons with negative energies of a lower absolute value, ($|E| < (NW_0)^{1/2}$), may be conduction electrons. The point is that field fluctuations lead to the emergence of "percolation channels". The possibility of such channels permeating the entire macroscopic volume of the medium is a sharp, practically step, function (for more details, see [397]). The minimum energy at which electrons contribute to conductivity is referred to as the percolation energy $E_{\rm P}$. We now determine the $E_{\rm P}$ for a classical particle in an arbitrary pattern of potential energy $U(r)$. It is equivalent to such a minimum value of energy, at which it is still possible to find a region in space where $U(r) < E_{\rm P}$ and permeates the entire plasma volume. There is no analytical solution to the percolation problem. Numerical modeling shows that $E_{\rm P} \approx 0.3(NW_0)^{1/2}$. Consequently, along with electrons of positive energies, electrons with negative energies ($E > E_{\rm P}$) are conduction electrons. If $\beta(NW_0)^{1/2} \gg 1$, it is the electrons in the percolation energy range which make a major contribution to σ.

The most detailed information on electron dynamics is contained in the correlation function $\varphi_{\rm E}(t)$. Its qualitative form for electrons of different energies was also discussed by Khrapak and Iakubov [400]. It is only for fast electrons that $\varphi_{\rm E}(t) = \exp(-t/\tau)$, where $\tau = \nu^{-1}$ and ν is the frequency of successive electron-atom collisions. In the case of low positive energies, the correlator $\varphi_{\mathrm{E}}(t)$ is a non-Boltzmann one. It does not decrease exponentially, but does so under the Gaussian law $\varphi_{\mathrm{E}}(t) = \exp(-t^2/\tau^2)$, where $\tau = \ell/\nu$. Finally, the correlator of trapped electrons oscillates, $\varphi_{\mathrm{E}}(t) = \cos\omega_0 t$, where ω_0 is the natural frequency of electron oscillation in the potential well of cluster. This gives the autocorrelation time, τ, close to zero. The peculiarities of behavior of the velocity autocorrelator are well-defined in Fig. 6.9, which is extracted from [398]. These results were obtained via numerical modeling of the electron dynamics using the method of molecular dynamics. In Fig. 6.9, the autocorrelator $\varphi_\varepsilon(t)$ is given at different values of energy reckoned from the mean field level and measured in the units of the $V(r)$ potential depth. At positive energies, the electron is free and $\varphi_\varepsilon(t')$ decays monotonically. With a decrease in energy, the function $\varphi_\varepsilon(t)$ reaches a minimum at which $\varphi_\varepsilon < 0$. This is a result of the vibrational motion of electron. The electron appears as though trapped until it finds a passage (percolation channel) to a field of complex shape. In doing so, the autocorrelation time $\tau(E)$ is still substantially different from zero. With a further decrease in energy, the electron is trapped and $\tau(E)$ reaches zero.

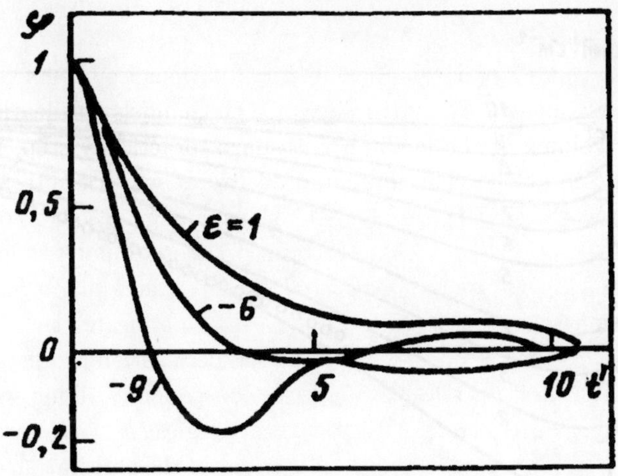

Fig. 6.9. Autocorrelation function of electron velocity versus dimensionless times t' for several values of relative energy ε [398], $N = 4.8 \cdot 10^{21}$ cm^{-3}.

Figure 6.10 illustrates the diffusion coefficient of electrons in the same medium, $D = \int_{-\infty}^{\infty} \langle \nu_x(0)\nu_x(t)\rangle_\varepsilon dt$. This quantity reaches zero at the energy value which is the percolation threshold $E = E_{\mathrm{p}}$.

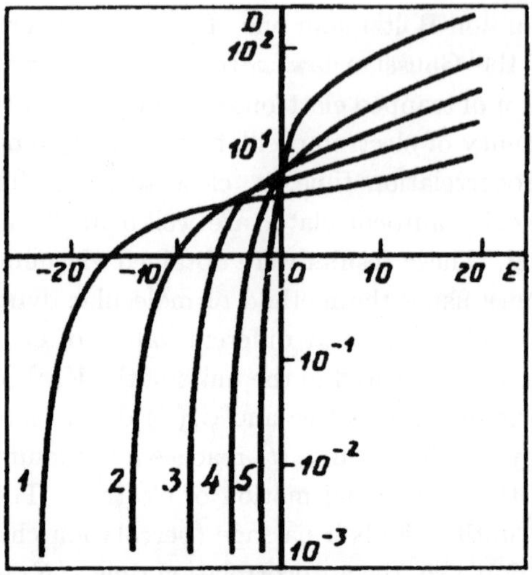

Fig. 6.10. Electron diffusion coefficient $D(\varepsilon)$ [398]. Designations are the same as in Fig. 6.9.

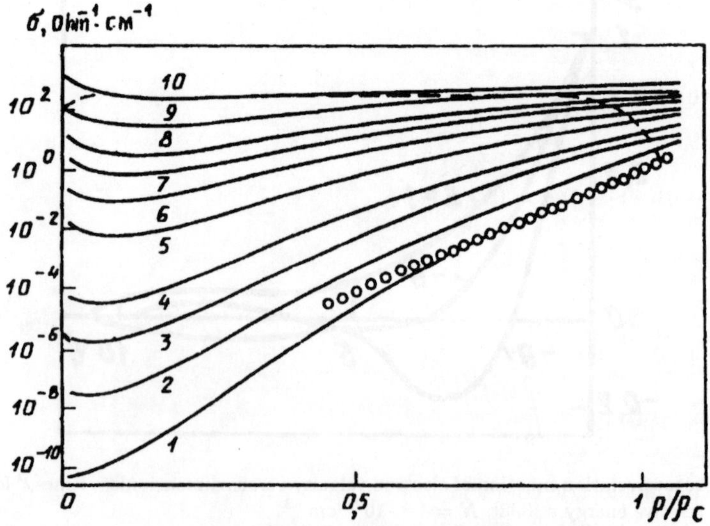

Fig. 6.11. Isotherms of electrical conductivity of mercury plasma [398]. Dots indicate the measurements [32], and curves the calculation at T = 1,823 K (*1*), 2,200 K (*2*), 2,600 K (*3*), 3,000 K (*4*), 4,000 K (*5*), 5,000 K (*6*), 6,000 K (*7*), 7,000 K (*8*), 10,000 K (*9*) and 15,000 K (*10*). The broken line defines the region with the degree of ionization $x \leq 0.1$.

Based on $\rho(E)$ and $D(E)$, Lagar'kov and Sarychev [399] calculated the conductivity of mercury plasma from, Eq. (6.30) in a wide range of temperatures and densities (Fig. 6.11). A good agreement is observed in both experiment and qualitative assumptions on the initial stage of "metal-dielectric" transition.

The answer to the question of the impact of quantum effects on mobility remains unclear.

6.4. The Thermoelectric Coefficient

In the presence of a temperature gradient, Ohm's law assumes the form of

$$\mathbf{F} + (1/\mathrm{e}) \cdot \nabla\mu = (1/\sigma) \cdot \mathbf{j} + S \cdot \nabla \cdot T\,. \tag{6.32}$$

The left-hand side includes $\nabla\mu$: the gradient of the chemical potential of electrons. The thermoelectric coefficient S appears on the right-hand side.

In Lorentz gas, the thermal e.m.f. is calculated analogously to the coefficient of electrical conductivity (see Section 6.1) [387]. The distribution function $f = f_0 + \delta f$, $f_0 = \exp[(\mu - E)/T]$, $E = m\nu^2/2$ is the electron energy and δf is a minor correction that is linear over the field and gradients of chemical potential and temperature. From the kinetic equation

$$-\frac{e\mathbf{F}}{m}\frac{\partial f_0}{\partial \mathbf{v}} + \mathbf{v}\frac{\partial f_0}{\partial \mathbf{r}} = -\nu(v)\partial f\,,$$

we derive

$$\delta f = \frac{f_0}{T\nu}\mathbf{v}\left(e\mathbf{F} + \nabla\mu - \frac{\mu - E}{T}\nabla T\right)$$

for δf. With $e\mathbf{F} + \nabla\mu = 0$, the thermal e.m.f. defines the current

$$\mathbf{j} = -S \cdot \nabla T = e \cdot \int \mathbf{v} \cdot \delta f \cdot d\mathbf{v}\,.$$

Finally, we derive

$$S = \frac{1}{eT}\left\{\mu - \int_0^\infty \frac{E^{5/2}dE}{\nu(E)}e^{-\beta E}\left[\int_0^\infty \frac{E^{3/2}dE}{\nu(E)}e^{-\beta E}\right]^{-1}\right\} \tag{6.33}$$

for the thermoelectric coefficient. In a weakly ionized plasma when $I \gg T$, the thermal e.m.f. is mainly defined by the temperature dependence of the degree of ionization. Because $\mu \approx T \cdot \ln(N_e\lambda^3)$, then in view of Saha's formula (6.2), we have

$$S = \frac{1}{eT}\left\{-\frac{I}{2} + \frac{1}{2}\ln(N_a\lambda^3) - \mathrm{const}\,T\right\} \approx -\frac{I}{2eT}\,. \tag{6.34}$$

Therefore, in the Lorentz approximation, the relationship between the coefficient of electrical conductivity and thermal e.m.f. is available:

$$S = -\frac{k_B T}{e} \cdot \frac{d \ln \sigma}{dt} ,$$

where k_B, the Boltzmann constant, is written explicitly. This expression is analogous to the known Ziman relation [234] for liquid metals:

$$S = \frac{\pi^2 k_B}{3e} \cdot \frac{T}{E_F} \cdot \left[\frac{d \ln \sigma}{d \ln E} \right]_{E=E_F} , \tag{6.35}$$

where E_F is the Fermi energy.

In a strongly ionized plasma, $\nu \sim E^{-3/2}$, and Eq. (6.33) yields

$$S = \frac{1}{e} \left(\frac{\mu}{T} - 4 \right) . \tag{6.36}$$

This expression is only valid if $Z \gg 1$. For small values of Z, inter-electron collisions affect the thermal e.m.f. and coefficient of electrical conductivity. A discussion of this problem may be found in [216].

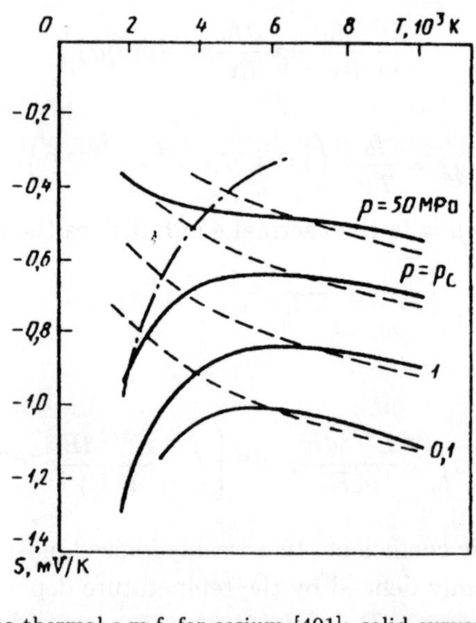

Fig. 6.12. Isobars of the thermal e.m.f. for cesium [401]: solid curves indicate the isobars of a partly ionized plasma; broken curves, asymptotics (6.36); dot-and-dash line, asymptotic (6.34).

Figure 6.12 gives the results of calculation by Redmer *et al.* [401] of the thermal e.m.f. of cesium on isobars. The procedure approximates that developed by researchers from Rostock to calculate the coefficient of electrical conductivity (see Section 7.3). The behavior of the asymptotics for fully ionized (6.36) and weakly ionized (6.34) plasma is also shown in Fig. 6.12. These asymptotics correlate well with the results of numerical calculation. Note that the calculation covers the $T \leq 3,000$ K range, in which the effect of non-ideality is still fairly small. At lower temperature (in the 2,000 K range), a sharp increase in S occurs due to the dielectric-metal transition (Fig. 4.2), from values of the order of $-1\ \mu$V/K to those of the order of $-1\ \mu$V/K.

We will now turn to a discussion of the thermal e.m.f. of non-ideal plasma. In a weakly non-ideal plasma, one must use the equation of ionization equilibrium. It contains the quantity $(I - \Delta I)$ instead of the ionization potential I, and allows for the presence of complex ions (4.32) in the plasma composition. We then assume, for example, that the ionization equilibrium on the subcritical isobars of metal vapors is mainly defined by the A_3^+ and A^- ions, and derive

$$S = [-I - (E_3 + E_5 - E_2)]/(2eT)\,, \tag{6.37}$$

where the designations of Table 4.9 are used. In Fig. 4.2, relations (6.36) and (6.37) are given.

As discussed in Section 4.2, the effect of non-ideality shows up on the supercritical isobars as a decrease in the ionization potential rather than as a variation of the composition of charged components of plasma. We then use formula (4.9) to derive

$$S = -(I - \Delta I)/2eT = [I - T(N/N^*)]/(2eT)\,. \tag{6.38}$$

Hence it follows that the quantity $S(T)$ must have a minimum value on the isobars. Indeed, at low temperature, S increases upon heating because ΔI decreases. At high temperature, however, $|S|$ decreases in inverse proportion to T. The existence of a minimum of $|S|$ was predicted by Alekseev *et al.* [25]:

$$S_{\min} \approx -10^{-3} \mu\text{V/K}\,.$$

Formula (6.34) may be transcribed as

$$S = -\Delta E/(2eT)\,, \tag{6.39}$$

where ΔE is the temperature coefficient of conductivity, $\Delta E = -2d\ln\sigma/d\beta$. The quantity ΔE is sometimes referred to as the "transport energy gap". Bolshov and Starostin [402] have demonstrated that the formula (6.35) is valid even if the interaction between electrons and disordered scatterers becomes very strong. One can use formulas (6.35) and (6.39) to obtain S if σ is known. In Fig. 6.13, this is done for cesium. At low temperature, the thermal e.m.f. of liquid cesium is proportional to temperature, $S \sim k_B T/eE_F$. This is because $d\ln\sigma/d\ln E_F$ is almost independent of temperature [387]. At high temperature, in accordance with (6.34), the thermal e.m.f. of a weakly ionized plasma is inversely proportional to T. The intermediate region, in which the thermal e.m.f. varies sharply, corresponds to the metal-dielectric transition.

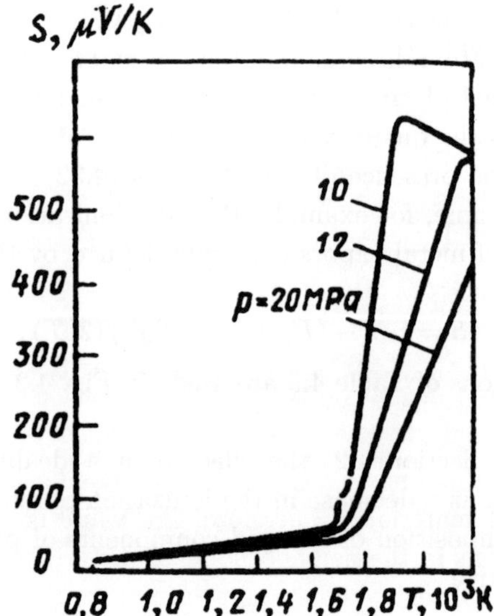

Fig. 6.13. Isobars of the thermal e.m.f. for cesium calculated in [25].

The pattern described above corresponds to the measurement results in Fig. 4.2. The subcritical isobars (2 and 6 MPa) are shown according to the theory of metal-dielectric transition discussed in Section 4.5 [220].

Lagar'kov and Sarychev [403] have studied the thermal e.m.f. using emolecular dynamics. The general expression is given by

$$S = (eT)^{-1}\left\{\mu - \int_{-\infty}^{\infty} \rho(E)\tau(E)E^{5/2}e^{-\beta E}dE \right.$$

$$\left. \times \left[\int_{-\infty}^{\infty} \rho(E)\tau(E)E^{3/2}e^{-\beta E}dE\right]^{-1}\right\}.$$

Therefore, the value of S can be calculated when $\rho(E)$ and $\tau(E)$ are given. Figure 6.14 shows the supercritical isobars of thermal e.m.f. for mercury. A good agreement is observed [34, 404] with experimental data. Lagar'kov and Sarychev [403] have observed that the pressure dependence of the position of the minimum thermal e.m.f. is slight.

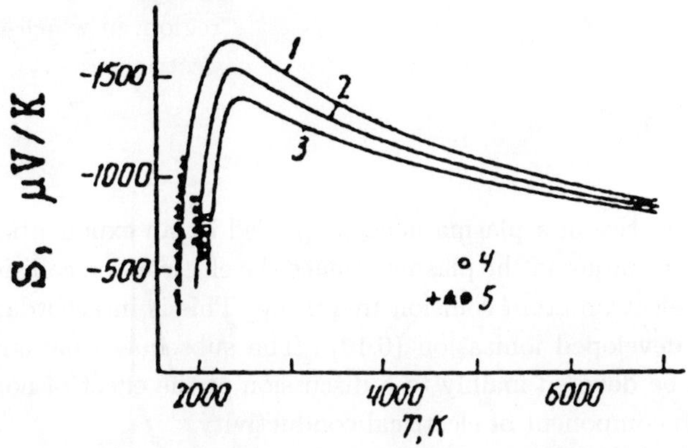

Fig. 6.14. Isobars of the thermal e.m.f. for mercury [403]. Calculation: *1*, $p = 2,200$ bar; *2*, 2,500 bar; *3*, 3,000 bar. Experiment: *4* [404]; *5* [34].

The thermal e.m.f. is known to be very sensitive to any slight variations in the parameters of materials. Therefore, the value of thermal e.m.f. may suffer drastic variations in the neighborhood of the critical point (at $\rho \approx \rho_c$). Neale and Cusack [405] registered such variations on the 1,700-bar isobar for mercury. The value of thermal e.m.f. changed its sign twice in the temperature range of 1,500C to 1,520C.

Note that many interesting problems associated with the behavior of thermal e.m.f. in the region of metal-dielectric transition are still not clarified.

Seven

Electrical Conductivity of Fully Ionized Plasma

A temperature rise in a plasma is accompanied by an exponential increase in the number of charges in the plasma. Hence the electron-ion collision frequency exceeds the electron-atom collision frequency. This is in accordance with the criterion of developed ionization (6.10). The subsequent paragraphs of this chapter will be devoted mainly to a discussion of the effect of non-ideality on the Coulomb component of electrical conductivity.

7.1. Kinetic Equations and the Results of Asymptotic Theories

Following the review by Iosilevskii *et al.* [9], we shall dwell briefly on the basic approaches to the calculation of kinetic coefficients for weakly non-ideal plasma. The starting point of the kinetic theory is Liouville's equation for the N-particle distribution function, which describes the time evolution of the ensemble of particles obeying Hamilton's equations. The integration over coordinates and particle momenta leads to the Bogoliubov, Born, Green, Kirkwood and Yvon (BBGKY) hierarchy of equations. It also enables one to find the expressions for transport coefficients in terms of time correlation functions.

The BBGKY hierarchy of equations is a set of integro-differential equations for S-particle distribution functions. Each contains a function of higher $(S+1)$-th order in the right side. To uncoupled this hierarchy, use is made of an approximation which ignores triple correlations. In addition, it includes plasma polarization, which indicates charge interaction with a large number of neighbors. This leads to the kinetic equation for distribution function [406]:

$$\frac{\partial f_1}{\partial t} = -\frac{\partial I(p_1)}{\partial p_1},$$

$$I(p_1) = \pi \int_{-\infty}^{\infty} \frac{d\omega}{2\pi} \int \frac{V^2(k)^2 \delta(\mathbf{k}\mathbf{v}_1 - \omega)\delta(\mathbf{k}\mathbf{v}_2 - \omega)}{|\varepsilon(k,\omega)|^2} \times \left[f(\nu_2)\frac{\partial f(\nu_1)}{\partial \nu_1} - f(\nu_1)\frac{\partial f(\nu_2)}{\partial \nu_2} \right] \frac{d\mathbf{r}}{(2\pi)^3} . \tag{7.1}$$

The dielectric permeability of plasma containing particles of diverse species,

$$\varepsilon(k,\omega) = 1 + \sum_a \frac{4\pi e_a^2}{k^2} \int \frac{d\mathbf{p}_a}{\omega - \mathbf{k}\mathbf{v}_a + i\delta} \mathbf{k} \frac{\partial f}{\partial p_a} , \tag{7.2}$$

is $\delta = +0$. $V(k) = 4\pi e^2/k^2$ is the Coulomb potential in Fourier representation. For fast ($\omega \gg \omega_{pe} = (4\pi e^2 N_e/m)^{1/2}$) processes $\varepsilon(\omega, k) = 1$, the screening does not have enough time to reach a steady-state value and the collision integral has the form proposed by Landau. This expression diverges at great and small interparticle distances. For slow ($\omega \ll \omega_{pe}$) processes, static screening is realized: $\varepsilon(0, k) = 1 + (k/k_{\mathrm{D}})^{-2}$. This eliminates divergence at great distances. In order to eliminate divergence at small distances, the collision integral was represented [407] as a combination of the Boltzmann I_{B} (infrequent strong collisions), the Landau I_{L}, and the Lenard-Balescu I_{LB} (6.44) collision integrals:

$$I = I_{\mathrm{B}} - I_{\mathrm{L}} + I_{\mathrm{LB}} . \tag{7.3}$$

In fact, this corresponds to the inclusion of the dynamic polarization of plasma in the Boltzmann collision integral.

Along with the approaches based on the BBGKY hierarchy of equations, the method of time Green functions is employed to write the kinetic equations for plasma. The emerging kinetic equations are written for advanced and retarded Green functions which define the particle density and probability of allowed states with pre-assigned momentum and energy. It is by using this method equipped with the diagram technique for classification and regrouping of terms in the series of perturbation theory (ring and ladder fragments are included) that Gould and DeWitt [408], as well as Williams and DeWitt [409], obtained a convergent equation, including the interparticle interaction, via screened Coulomb potential and which is valid in the first order in Γ. The emerging approximation is, in a sense, analogous to the combination (7.3).

The inclusion of collective effects in the Coulomb interaction was performed on the basis of the Focker-Planck equation, where the collision integral incorporates all moments of the distribution function. In such a procedure, the contribution made by close collisions is described by Boltzmann's integral and that made by distant collisions, by the Lenard-Balescu integral with exponential cutoff in the contributions made by close and distant collisions.

In order to discuss concrete results, we shall transcribe the expression (6.1) for electrical conductivity as

$$\sigma = e^2 N_e / m\nu_i \,,$$
$$\nu_i = (\pi^{3/2}/4\sqrt{2}) N_i \sqrt{T/m}\gamma_{\mathrm{E}}^{-1} (Ze^2/T)^2 \ln\Lambda \,, \tag{7.4}$$

where, in the roughest approximation, $\ln\Lambda = \ln(b_{\max}/b_{\min})$. The maximum impact parameter, $b_{\max}$, is provided by the screening length. The minimum impact parameter is the maximum of two lengths: the classical distance of the closest approach, e^2/T, and the thermal electron wavelength, $\lambda = (h/2\pi)$ $(2T/m)^{-1/2}$.

Spitzer's formula (6.8), which is derived for classical non-degenerate plasma, is only logarithmically exact. Spitzer's Coulomb logarithm may serve as the starting point for further refinement:

$$\ln\Lambda = \ln(3\Gamma^{-1}) \,, \qquad \Gamma = e^2/r_{\mathrm{D}}T \,, \qquad r_{\mathrm{D}} = (4\pi e^2 N_e/T)^{-1/2} \,.$$

The result of Williams and DeWitt [409], who used the collision integral (7.3), corresponds to the Coulomb logarithm with numerical correction:

$$\ln\Lambda = \ln(\Gamma^{-1}) - 2C + \ln 2 \,, \qquad C = 0.577 \,. \tag{7.5}$$

The subsequent correction is analogous to the relaxation correction to the ion mobility in the solutions of strong electrolytes [387]. The external electric field affects the correlation functions to bring about a three-dimensional strain of screening cloud. This reduces somewhat the coefficient of electrical conductivity because the electric field acting on a particle is compensated to some extent [410, 411]. As a result, Roepke [412] derived

$$\ln\Lambda = \ln\gamma^{-3/2} + 1.102 + (\sqrt{6}+\sqrt{3})^{-1}\gamma^{3/2}\ln\gamma^{-3/2} + \dots \,,$$

where $\gamma = e^2(4\pi N_e/3)^{1/3}T^{-1}$. The range of validity is limited by the inequalities $\gamma \ll 1$ and $\gamma^2\theta^{-1} \ll \ln\Lambda$, where $\theta = T/E_{\mathrm{F}}$.

The calculation of the coefficient of electrical conductivity for quantum non-degenerate plasma, using the Lennard-Balescu collision integral, leads to the expression

$$\ln\Lambda = \int_0^\infty \frac{dq}{q} \cdot \frac{S_{ii}(q)}{|\varepsilon(q)|^2} \cdot \exp\left(-\frac{(h/2\pi)^2 q^2}{8mT}\right) . \tag{7.6}$$

The use of Debye-Hüeckel static expressions for $S_{ii}(q)$ and $\varepsilon(q)$ (5.6) leads to the series whose first terms are [413]

$$\ln\Lambda = \int_0^\infty q^3(q^2+2)^{-1}(q^2+1)^{-1}\exp(-\eta^2 q^2/4)$$
$$= \ln\eta^{-1} - (C - \ln 2)/2\,, \tag{7.7}$$

where $\eta = \lambda/r_D$. The subsequent terms of this series are given in [414].

Not only are the resultant expressions (7.5) and (7.7) logarithmically exact, but their extra-logarithmic terms are exact as well. For a classical plasma ($T \ll$ Ry), (7.5) is valid and for a quantum plasma, (7.7). The solution derived by Williams and DeWitt [409] joins (7.5) and (7.7):

$$\ln\Lambda = \ln\eta^{-1} - 2C + \ln 2 - \frac{1}{2}\exp(z)E_i(z)\,,$$

where $E_i(z) = \int_z^\infty dt\, t^{-1}\exp(-t)$ and $z = \exp(2C)(e^2/2\lambda T)$. The high-temperature limit corresponds to the limit when $z \ll 1$, and the low-temperature limit corresponds to that with $z \gg 1$.

Later, the results of [408, 409] were improved by Rogers *et al.* [415].

7.2. Electrical Conductivity Measurement Results

Electrical conductivity is the most revealing and readily observed characteristic of plasma which governs the dissipative heating of the latter and its interaction with the electromagnetic field. Researchers are further attracted by the relatively simple and well-developed registration techniques, as well as by the possibility of performing electrophysical measurements under the most diverse experimental conditions. In the early experiments reviewed by Asinovsky *et al.* [416], the non-ideality parameters attained did not exceed $\Gamma \approx 0.2 - 0.3$. Meanwhile, the measured values of electrical conductivity were close to the results of calculations performed on the basis of kinetic equations. Nevertheless, even at these values of Γ, slight but systematic divergences of the measured and calculated values were observed.

In the early 1970s, various researchers performed a series of studies on the electrical conductivity of the plasma of inert gases, mainly in a plasma of arc discharges in the temperature range of $(1\text{–}10)\cdot 10^4$ K and charged particle concentration range of $10^{16} - 10^{18}$ cm^{-3}. The results of those measurements demonstrated appreciable deviations of electrical conductivity from the Spitzer values, although the non-ideality parameter was still small ($\Gamma \leq 0.1$). It turns out that for the bulk of experimental data, an agreement with calculation results is attained if one allows for the fact that the plasma is not fully ionized and collisions with atoms may greatly reduce the electrical conductivity. This can be done either with the aid of Eq. (6.11) or by regular methods. Figures 7.1 and 7.2 present the results of measurements of the electrical conductivity for argon plasma and the calculation results by Devoto [417].

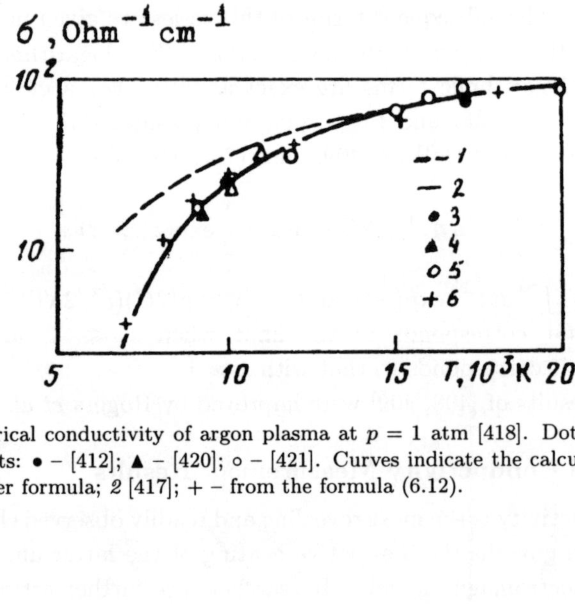

Fig. 7.1. Electrical conductivity of argon plasma at $p = 1$ atm [418]. Dots mark the measurement results: • – [412]; ▲ – [420]; ○ – [421]. Curves indicate the calculation results: *1*, from the Spitzer formula; *2* [417]; + – from the formula (6.12).

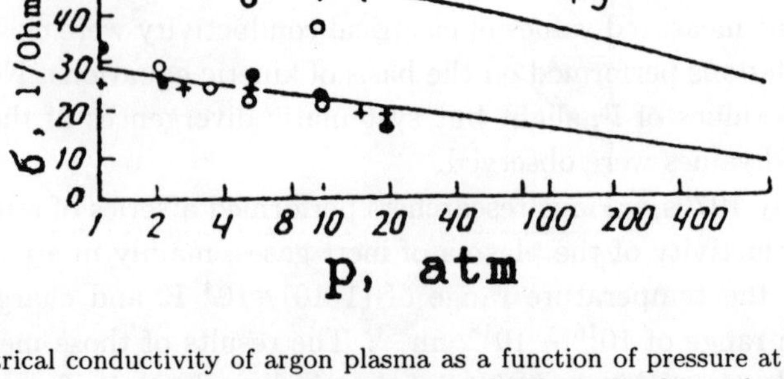

Fig. 7.2. Electrical conductivity of argon plasma as a function of pressure at different temperatures. Measurement results marked by dots: ○ – [419]; • – [422]; + – [423]. Calculation results are shown by curves [417].

Even at Γ somewhat greater than 0.1, the inclusion of electron collisions did not eliminate the divergence of the observed values of electrical conductivity from the calculated values obtained by way of solving kinetic equations. The typical experimental conditions [95, 424–426] correspond to pressures in the discharge, $p = 10 - 20$ atm and $T = 10^4$–$1.8 \cdot 10^4$ K (Fig. 7.3). The inclusion of collisions with atoms (curve *2*) leads to an agreement between theory and experiment in the low-temperature range. However, one can see that an increase in temperature causes an appreciable divergence. Curve *3* is plotted with due regard for collision complexes and it agrees well with the experimental data. The model of Vorob'ev and Khomkin [427] introduces new quasi-particle states into the classification of electron-ion interactions, namely, quasi-bound states and collision complexes. They show up when Γ is close to unity and enable one to describe the decrease observed in the coefficient of electrical conductivity. In [428], this theory is outlined and in [429], the role played by electron-ion collisions is discussed.

An analogous situation is observed in xenon plasma (Fig. 7.4).

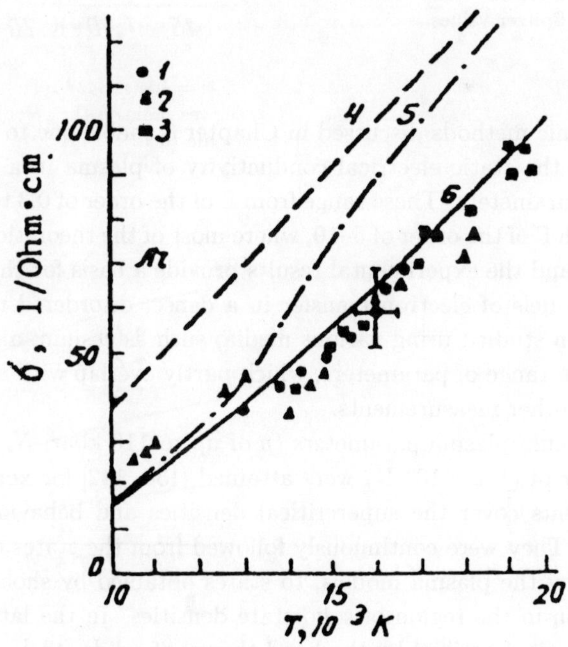

Fig. 7.3. Electrical conductivity of argon plasma [69]. Experiment: *1* [92]; *2* [93]; *3* [94]. Theory: *4*, σ_{sp}; *5*, inclusion of collisions with atoms; *6*, inclusion of collision complexes.

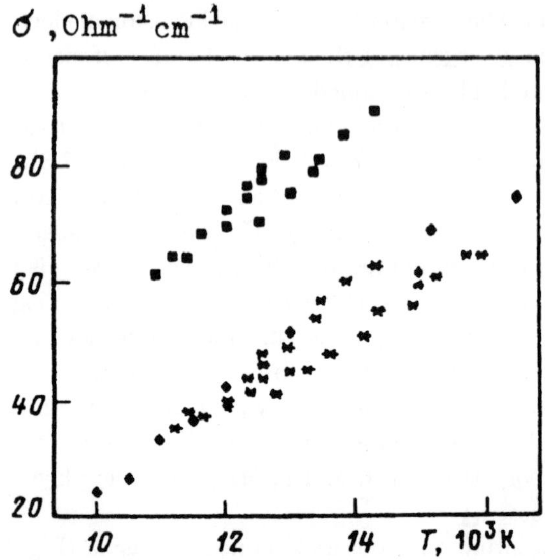

Fig. 7.4. The coefficient of electrical conductivity of xenon plasma [430]. Experiment: ♦ [425]; ∗ [430]; ■ Spitzer values.

The dynamic methods discussed in Chapter 3 enable one to measure [138, 142, 431–433] the static electrical conductivity of plasma in a wide range of non-ideality parameters. These range from Γ of the order of 0.3 to the region of extremely high Γ of the order of 5–10, where most of the theoretical approximations diverge and the experimental results provide a basis for the construction of physical models of electron transfer in a dense, disordered medium. This range has been studied using various media, such as cesium, air, neon, argon and xenon in a range of parameters, which partly overlap with each other and with those of other measurements.

The maximum plasma parameters (p of up to 110 kbar, $N_e \leq 10^{21}$ cm^{-3}, T of the order of $(1–2) \cdot 10^4$ K) were attained [138, 142] for xenon (Fig. 3.9). The experiments cover the supercritical densities and behavior of electrical conductivity. They were continuously followed from the states of low density, as described by the plasma models, to states obtained by shock compression of liquid xenon in the region of solid-state densities. In the latter, the shock compressibility is described by the band theory of solids [434].

Figure 7.5 shows the values of "reduced" dimensionless electrical conductivity:

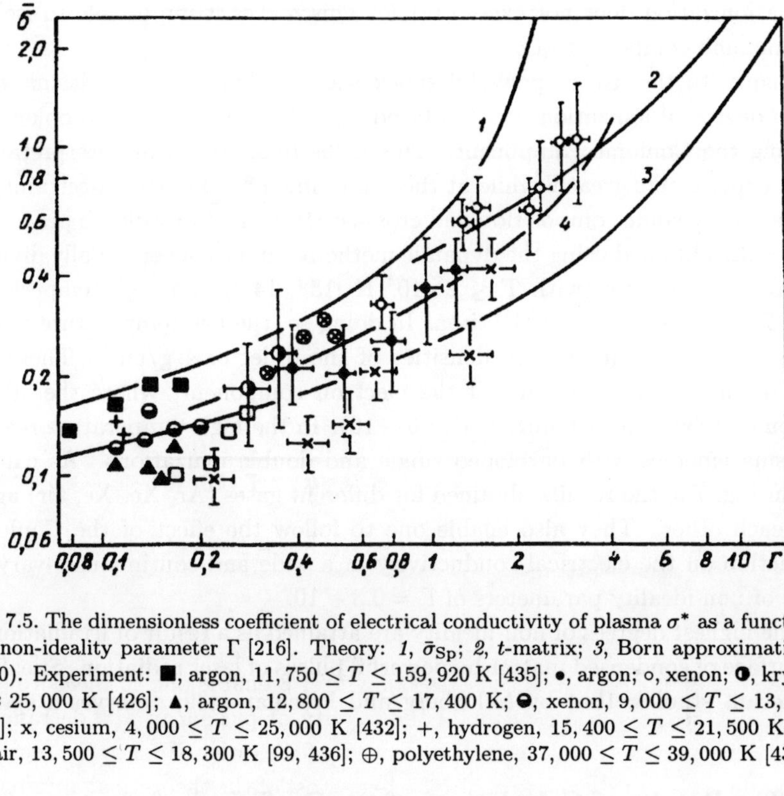

Fig. 7.5. The dimensionless coefficient of electrical conductivity of plasma σ^* as a function of the non-ideality parameter Γ [216]. Theory: *1*, $\bar{\sigma}_{\text{Sp}}$; *2*, t-matrix; *3*, Born approximation; *4*, (7.10). Experiment: ■, argon, $11,750 \leq T \leq 159,920$ K [435]; •, argon; ∘, xenon; ◑, krypton, $T \approx 25,000$ K [426]; ▲, argon, $12,800 \leq T \leq 17,400$ K; ⊖, xenon, $9,000 \leq T \leq 13,700$ K [424]; x, cesium, $4,000 \leq T \leq 25,000$ K [432]; +, hydrogen, $15,400 \leq T \leq 21,500$ K [435]; □, air, $13,500 \leq T \leq 18,300$ K [99, 436]; ⊕, polyethylene, $37,000 \leq T \leq 39,000$ K [437].

$$\bar{\sigma} = e^2\sqrt{m}\sigma/T^{3/2}\,. \tag{7.8}$$

The resulting combination of experimental data (Fig. 7.5) definitely points to an under-estimation of the measured values of electrical conductivity as compared to the Spitzer's theory. Meanwhile, more stringent theories predict an increase in conductivity with a rise in the Coulomb interaction as compared to Eq. (7.4). The existing quantitative divergence between different groups of experiments in Fig. 7.5 is due to the physical peculiarities of the high-temperature plasma behavior, as well as to the actual discrepancy of the primary data and the difficulties in isolating the Coulomb component in a weakly ionized plasma. The latter is most characteristic in experiments with alkali metals [44, 432, 437], where the contribution by neutrals to the collision frequency is especially great in view of the large cross-sections of electron-atom scattering. Thus, at maximum pressures in the experiment [50], the degree of plasma ionization does not exceed 0.1%. This makes it impossible to isolate the Coulomb component of σ.

Thanks to the use of powerful shock waves (Chapter 3), a plasma with a high degree of ionization was produced [138, 143] without any problems in isolating the Coulomb component. This helps to simplify the interpretation of the experiments greatly while at the same time reducing the uncertainty in the thermodynamic composition and cross-sections of non-Coulomb processes. The results obtained using the dynamic methods can be conventionally divided into low-temperature (with $T \leq 2 \cdot 10^4$ K [138, 142]) and high-temperature (with $T > 2 \cdot 10^4$ K [143, 431]) data. In doing so, the low-temperature points correspond to extremely high densities of the order of 4 g/cm^3. These are close to the degeneracy limit of the electron component, where the strong Coulomb interaction is realized ($\Gamma = 6 - 10$). In the high-temperature region, a plasma emerges with developed single and double ionizations. As can be seen in Fig. 7.5, the results obtained for different gases (Ar, Xe, Xe, air) agree with each other. They also enable one to follow the effect of the Coulomb interaction on the electrical conductivity in a wide and continuously varying range of non-ideality parameters of $\Gamma = 0.3 - 10$.

The highest degrees of non-ideality are attained as a result of irradiation of the surface of condensed matter by powerful pulses of laser radiation. Speaking of relevant studies, that by Milchberg and Freeman [439] deserves a special note.

7.3. The Results of Calculations of the Coefficient of Electrical Conductivity

The asymptotic expressions for $\ln \Lambda$, given in Section 7.1, are justified only on the condition that $\ln \Lambda \gg 1$ and, when they go into infinity (for example, $\sigma_{\mathrm{Sp}} \to \infty$ at $\Gamma \to 3$), they lose their meaning. The most simple convergent value for σ may be derived if, in order to avoid formal divergences, one uses

$$\ln \Lambda = \frac{1}{2} \ln(1 + b_{\max}^2 / b_{\min}^2) , \qquad (7.9)$$

where $b_{\max}$ and $b_{\min}$ are the maximum and minimum impact parameters, respectively. Of course, this expression fails to solve the problem set, that of constructing an expression for σ, which would turn into correct limiting formulas and describe well the available experimental data.

In the case of strongly non-ideal plasma, it is important to include the effect of ion correlations on electron scattering. For this purpose, Gryaznov *et al.* [440] proposed a model based on Ziman's approximation. The reference expression is the expression for conductivity obtained in a Lorentz approximation as discussed in Section 6.1:

$$\sigma = \frac{4e^2T^{-3/2}}{3\sqrt{\pi}m_e}\frac{2}{\lambda^3}\int_0^\infty E^2\nu_i(E)^{-1}\left(-\frac{\partial f_0}{\partial E}\right)dE\,. \tag{7.10}$$

Here, $f_0 = \{1+\exp[(E-\mu)/T]\}^{-1}$ is the electron distribution function and μ is the chemical potential related to the electron concentration, N_e, by

$$N_e = (4/\sqrt{\pi})(2\pi\hbar^2/m)^{-3/2}\int\sqrt{E}f_0(E)dE\,.$$

The following expression was proposed for the collision frequency:

$$\gamma_i(E) = \sqrt{\frac{2}{\pi}}\frac{\pi e^4}{E^{3/2}}N_e\int_0^{q_m}\frac{q^3}{(q^2+q_{\mathrm{D}}^2)^2}S(q)dq\,. \tag{7.11}$$

Here, the quantity $q_m = 2E/e^2$ ensures for Eq. (7.10), with an accuracy to electron-electron scattering, Spitzer's asymptotics at $e^2/\hbar v \gg 1$ (in this limit, $S(q) = 1$). Restricting ourselves to a Debye approximation for the ion correlation function $g(r)$,

$$g(r) = 1 - (e^2/rT)\exp(-q_{\mathrm{D}}r)\,, \tag{7.12}$$

and substituting it in the expression for a structure factor, we derive

$$I = \frac{1}{2}\left[\ln(1+\alpha) - \frac{\alpha}{\alpha+1} - \frac{\alpha^2}{2(\alpha+1)^2}\right], \qquad \alpha = (q_m r_{\mathrm{D}})^2 \tag{7.13}$$

for the integral in Eq. (6.64). In the limit of weakly non-ideal plasma ($\alpha \gg 1$), this expression changes to the regular Coulomb logarithm, $I = 4\ln(\alpha^2/2)$. For strongly non-ideal plasma ($\alpha \ll 1$), the expression for electrical conductivity proves finite and contains no non-physical divergences. However, it describes well the low-temperature experiment at up to extremely high values of Γ (cf., Fig. 7.5).

Apparently, the most systematic approach in constructing interpolation expressions is based on approximation (7.3). It consists of the use of the unscreened t-matrix approximation, the subtraction of the unscreened Born approximation, and the addition of the Born approximation with dynamic screening. In addition, the electron-atom collisions and inter-electron interactions must be included. The result of such treatment would adequately describe exact limiting expressions. Kraeft *et al.* [216] presented the results of extended studies and gave the results of calculations in a wide range of conditions.

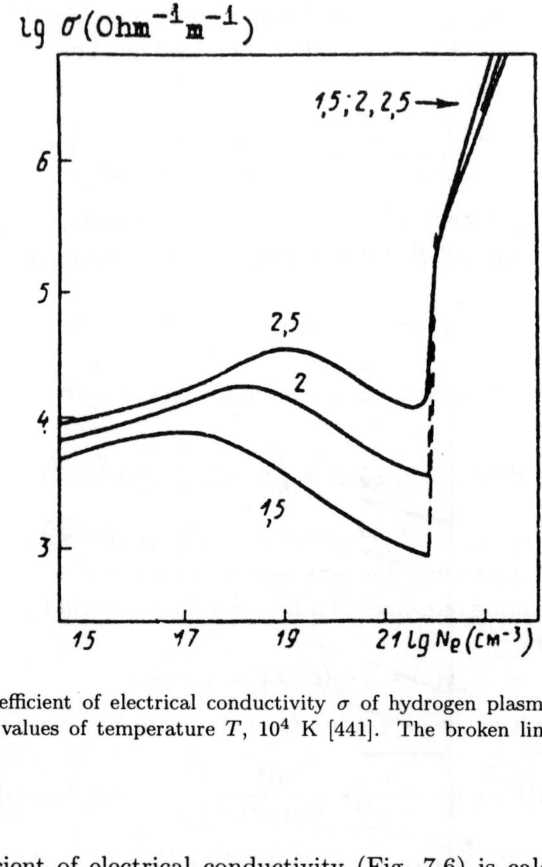

Fig. 7.6. The coefficient of electrical conductivity σ of hydrogen plasma as a function of N_e for different values of temperature T, 10^4 K [441]. The broken lines indicate Mott's transition.

The coefficient of electrical conductivity (Fig. 7.6) is calculated in a second Born approximation with dynamic screening. At low density, in order to make an effective transition to a t-matrix approximation, the thermal wavelength has been replaced (at the place where it acts as the cutoff length) by the Landau length. At low and high density, σ has the Lorentz and Ziman asymptotics. The equation of ionization equilibrium, whose solution gave the electron number density, describes ionization by pressure. The broken curve corresponds to Mott's transition which, in the approximation used, occurred in a jump rather than continuously. A minimum value of the coefficient of electrical conductivity is observed in the region of Mott's transition due to electron-atom scattering.

Ziman's formula may be derived from (7.10) and (7.11) with due regard for the fact that under strong degeneracy, $df_0/dE = \delta(E - \varepsilon_F)$ and the maximum transferred momentum is equal to $2(h/2\pi)k_F$. Thus,

$$\sigma_{\rm Z} = \frac{\varepsilon_{\rm F}^{3/2}}{\pi\sqrt{2}Ze^2\sqrt{m}}\left[\int_0^{2k_{\rm F}} dq q^3 S(q)\left|\frac{1}{q^2\varepsilon(q)}\right|^2\right]^{-1},$$

where $\varepsilon(q)$ is the dielectric permeability and $\varepsilon_{\rm F} = (h/2\pi)^2 k_{\rm F}^2/2m$.

The importance of inter-electron interactions is maximum under conditions of low number density of charged particles, when it is defined by the Spitzer factor $\gamma_{\rm E}$. The part played by these interactions diminishes as non-ideality increases. In the case of strong degeneracy, it is suppressed by the effect of Pauli's principle [442] (Fig. 7.7).

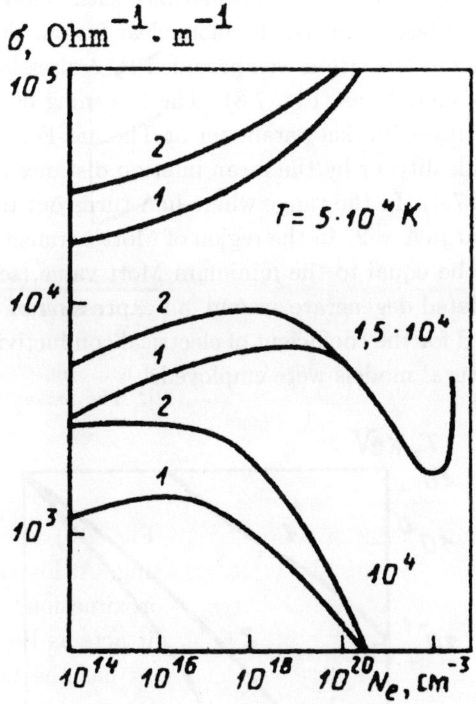

Fig. 7.7. The coefficient of electrical conductivity σ of hydrogen plasma calculated (*1*) in view of and (*2*) disregarding the inter-electron interactions [442].

In Fig. 7.5, the experimental data are compared with the results of calculations in both statically screened t-matrix approximation *2* and Born approximation *3*. At small values of Γ, curve *2* changes to Spitzer curve *1*.

The calculations performed by Kraeft *et al.* [216] may be classified as wide-range calculations of the coefficient of electrical conductivity. Such methods were called into being by application requirements. Under conditions of great pulsed energy contribution, the matter may pass through the entire range of non-ideality parameters. Ichimaru and Tanaka [414] have tabulated the Coulomb logarithm, given by expression (7.6), in a very wide range of parameters. In a plasma with an arbitrary degree of degeneracy, the exchange-correlation correction with Lindhardt's dielectric permeability was included and the value of the structure factor was borrowed from the solution of hypernetted chain equations (see Section 5.1). The non-ideality parameter $\gamma = Ze^2T^{-1}(4\pi N_i/3)^{1/3} \leq 2$.

Lee and More [443] performed calculations with reference to the conditions realized during laser-induced thermonuclear fusion. Simple calculations, in which the Coulomb logarithm was provided by expression (7.9), covered a very wide range of conditions (Fig. 7.8). The screening distance in [443] was provided by the Debye-Hueckel parameter or Thomas-Fermi radius in the region of weak non-ideality, or by the mean interion distance in the range where $\gamma \geq 1$ (see Section 7.4). In the range where $\ln\Lambda$ turns out to be less than two, it was assumed that $\ln\Lambda = 2$. In the region of Mott's transition, the coefficient σ was assumed to be equal to the minimum Mott value (see Section 4.5). In the strongly correlated degenerate system, an expression of the type of Ziman expression was used for the coefficient of electrical conductivity. Therefore, the most simple structural models were employed.

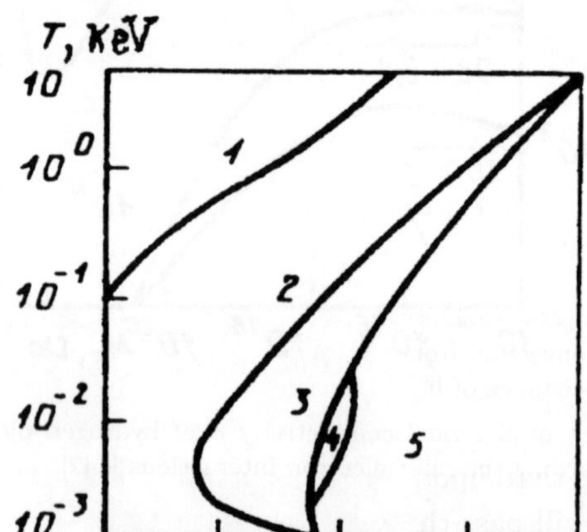

Fig. 7.8. The temperature and density regions treated in [443]: *1*, Debye-Hueckel and Thomas-Fermi screening; *2*, screening over the average interion distance; *3*, $\ln\Lambda = 2$; *4*, $\sigma = \sigma_{\min}$; *5*, Ziman region.

Under the conditions of strong non-ideality, the coefficient of electrical conductivity may be calculated from the expressions of the linear response theory. According to this formalism, the average values of the quantities observed for a system subjected to external perturbations are expressed in terms of time correlation functions (for example, such as formula (6.29)). This enables one to calculate σ [444, 445] using the numerical methods of molecular dynamics.

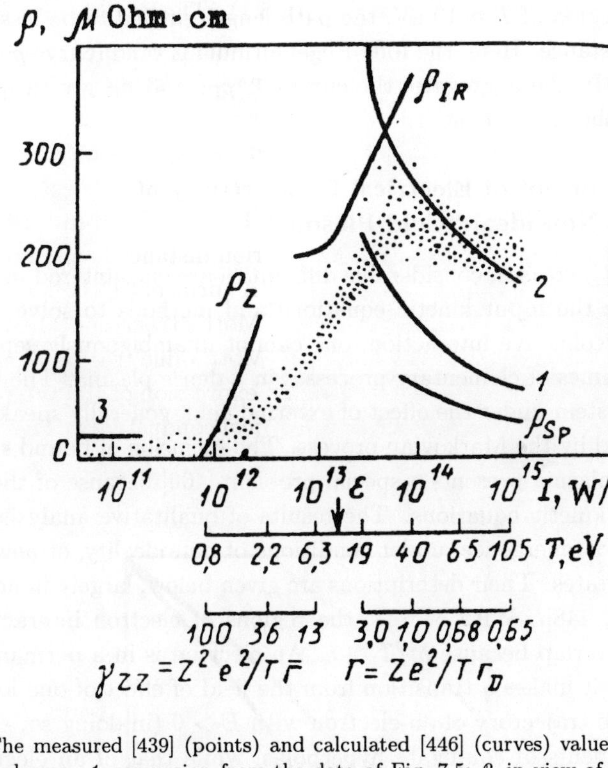

Fig. 7.9. The measured [439] (points) and calculated [446] (curves) values of ρ for a dense aluminum plasma: *1*, conversion from the data of Fig. 7.5; *2*, in view of the non-Coulomb scattering; *3*, specific resistance of liquid aluminum.

One can assume that upon isochoric heating from the melting point to 100 eV, the matter will pass through the entire set of non-ideal plasma states from liquid-metal to ideal plasma. Milchberg and Freeman [439] constructed the isochore of specific resistance ρ (Fig. 7.9) from a treatment of the results of measurements of the coefficient of reflection of radiation from the aluminum surface heated by a powerful laser pulse. During the period of 400 fs, the matter does not have enough time to expand and retains the density of solid aluminum. In [446], the measured values were compared with those derived using the methods of wide-range description of ρ. The effect of non-Coulomb scattering (see Section 7.5) from the core of complex Al^{+3} ion proves to be of extreme importance. The resistance of degenerate plasma is described by curve *2* in view of the non-Coulomb scattering from a core of radius $R_c = 1.1a_0$. In the degeneracy region, the ρ_Z curve is constructed by Ziman's formula in view of the scattering by Ashcroft's pseudopotential with the core radius $R_c = 1.1a_0$. In the region of $T \approx 19$ eV, the path length proved to be less than the interparticle distance. Here, the Ioffe-Regel formula is valid (curve $\rho_{\rm IR}$): $\sigma_{\rm IR} = (N_e e^2/m)\bar{r}/v$. In the aggregate, the curves *2*, $\rho_{\rm IR}$ and ρ_Z give a qualitative description of the experiment.

7.4. The Coefficient of Electrical Conductivity of Strongly Non-ideal "Cold Plasma"

As non-ideality increases, considerable difficulties are encountered in attempts to substantiate the input kinetic equations and methods to solve them. In view of strong collective interaction, one cannot unambiguously separate the characteristic times of elementary processes in a dense plasma. The time evolution of the system under the effect of external field, generally speaking, is no longer described by the Markovian process. The inclusion of bound states in a partly ionized plasma presents a special problem [406] because of the absence of appropriate kinetic equations. The results of qualitative analysis point to the possibility of emergence, under conditions of non-ideality, of new types of quasi-particle states. Their descriptions are given below, largely in accordance with [206, 447, 448]. With $\gamma > 1$, the regions of electron interaction with different ions overlap because $e^2/T > \bar{r}$. An electron is in a permanent state of collision and it makes a transition from the field of effect of one ion to that of another. The trajectory of an electron with $E > 0$ (in doing so, $e^2/E > \bar{r}$) consists of segments of conjugate hyperbolas, while that of an electron with $E < 0$ consists of conjugate segments of ellipses ($e^2/|E| > \bar{r}$). Of course, no significant differences are observed between these two types of trajectories.

In view of the foregoing, it is obvious that the electron density of states $\rho(E)$ must vary relatively little for low energy values when $|E| \leq e^2/\bar{r}$ and correspond qualitatively to curve *1* in Fig. 7.10. For high energy values, when $E \gg \gamma T$, the density of states is close to (6.22). For negative energy of high absolute values, $\rho(E)$ reaches the asymptotic of quasi-classical density of states for a hydrogen-like atom, $\rho(E) = N_i \mathrm{Ry}^{3/2} Z^3/|E|^{5/2}$, when $|E| \gg \gamma T$. The estimation of $\rho(E)$ in the neighborhood of zero energy may be performed with

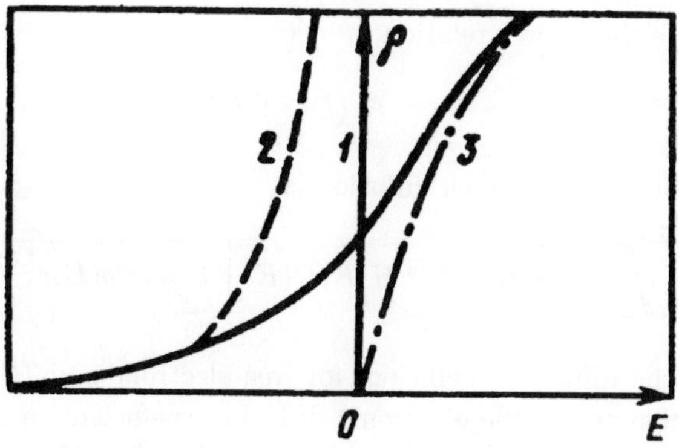

Fig. 7.10. The density of electron states: *1*, density of states in a non-ideal plasma; *2*, density of states of an isolated atom; *3*, density of states of free electrons.

the aid of formula (6.25) if one uses the distribution of potential energy of an electron in the field of the nearest ion:

$$P(U) = 4\pi N_i \int_0^\infty R^2 dR \exp\left(-\frac{4\pi}{3} N_i R^3\right) \delta\left(U + \frac{Ze^2}{R}\right). \tag{7.14}$$

Indeed, the probability of finding the nearest ion at the distance $F(R) = 4\pi^2 N_i \cdot \exp[-4\pi N_i \cdot R^3/3]$ from the electron decreases exponentially at $R > \bar{r}$. We use (7.14) to derive

$$g(E) = 2\pi \frac{2(2m)^{3/2}}{h^3} \Gamma(5/6) \frac{\sqrt{Ze^2/\bar{r}}}{2} \left[1 + \frac{E\bar{r}}{2Ze^2} \Gamma(7/6)/\Gamma(5/6)\right] \tag{7.15}$$

from (6.25). Thus, for low values of energy, the density of states is weakly dependent on energy.

The situation approaches that discussed in Section 6.4 for a weakly ionized dense plasma. As in the latter case, the importance of the energy E_n increases, at which the classically accessible regions of motion for negative energy electrons overlap for the first time. The criterion of overlapping of randomly arranged spheres with radius R at concentration N_i has the form $1 - \exp(-4\pi N_i R^3/3) = 0.29$ [397]. This yields the level of percolation energy

$$E_n = -e^2/(0.7\bar{r}). \tag{7.16}$$

Electrons with $E > E_n$, as well as free electrons, are conduction electrons. However, even if the energy, though exceeding E_n, is close to it, these electrons do not yet possess all the characteristics of free electrons. The mobility is lesser.

We will write the known relation

$$\sigma = e^2 \cdot N_e \langle D(E) \rangle / T \,. \tag{7.17}$$

The mean coefficient of electron diffusion is

$$\langle D(E) \rangle = \int_{E_n}^{\infty} dE \rho(E) e^{-E/T} H(E) \tilde{D}(E) \left[\int_{E_n}^{\infty} dE \rho(E) e^{-E/T} \right]^{-1} .$$

Here, $\tilde{D}(E)$ is the diffusion coefficient for free electrons and $H(E)$ takes into account the difference of the electronic diffusion coefficient in the vicinity of the percolation threshold $D(E)$ from $\tilde{D}(E)$. The function $H(E)$ may be determined very precisely with the aid of the effective medium theory [449]. This function, however, has evident properties as well. It is zero at point $E = E_n$, and increases with E to unity when $E \geq 0$ is reached.

Under conditions of strong non-ideality and in view of the properties of $H(E)$, we derive

$$\langle D(E) \rangle = \tilde{D}(E_n) \left. \frac{dH}{dE} \right|_{E=E_n} \cdot T \approx \tilde{D}(E_n) \frac{T}{E_n} \,. \tag{7.18}$$

In order to calculate the diffusion coefficient $\tilde{D}(E_n) = v(E_n)\bar{r}$, one must know the mean velocity of an electron with energy close to E_n. It is believed that during the transition from one potential well to another, much of the time is spent by the electron passing through the saddle point. In the saddle point region, the kinetic energy of the electron is not high, being of the order of T (there is no other energy scale). Thus $\tilde{D}(E_n) \approx v_T \bar{r}$. We substitute the quantities written in (7.17) to derive (accurate to the numerical factor), finally, the following expression for the coefficient of electrical conductivity:

$$\sigma \sim \omega_p / \sqrt{\gamma} \,, \tag{7.19}$$

where $\omega_p = \sqrt{4\pi N_e e^2/m}$ is the plasma frequency and N_e is the number density of conduction electrons, that is, electrons with energy $E \geq E_n$.

The question of the coefficient of electrical conductivity for such a system may hardly be regarded as closed. On the contrary, if one assumes that an electron passes through the saddle point very quickly, then $v(E_n) \approx \sqrt{|E_n|/m}$. Thus, $\sigma \sim \omega_p$. The latter expression was proposed by Kurilenkov and Valuev [450] on entirely different grounds. According to them, the electron dynamics are defined by scattering from thermal plasma vibrations.

7.5. The Electrical Conductivity of High-temperature Non-ideal Plasma. The Ion Core Effect

Non-ideal plasma experiments are characterized by relatively low ($T < 3 \cdot 10^4$ K) temperatures. This is because they are directed at attaining a developed Coulomb non-ideality which decreases with a rise in temperature. The use of cumulative shock tubes and the effects of shock wave reflection from obstacles enable one to advance [143] into the region of extremely high temperatures, and to obtain a highly heated, multiply ionized plasma with a developed Coulomb non-ideality of the order of 1–5 (Fig. 7.11). The electrophysical properties of such plasma turned out to be very unexpected [143, 431]. This is because they point to the non-existence of similarity of the Coulomb component of non-ideal plasma: the dimensionless electrical conductivity of high-temperature plasma turns out to be less than that of low-temperature plasma at the same values of the Coulomb non-ideality Γ.

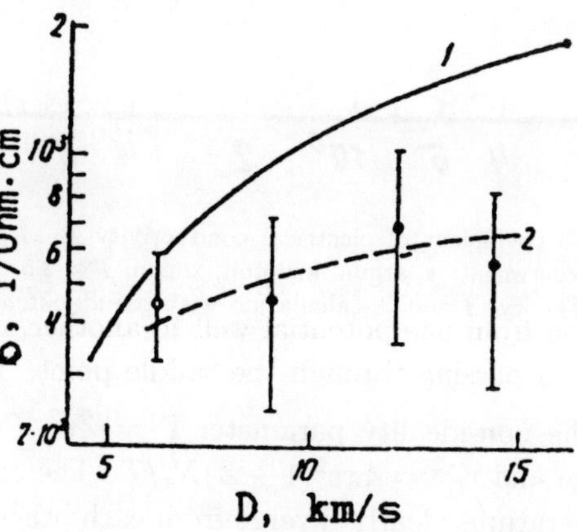

Fig. 7.11. Electrical conductivity as a function of shock wave velocity in xenon at $p_0 = 1$ bar. Experiment: • [143]; ○ [138]. Calculation: *1*, from Eq. (6.8); *2*, account for non-Coulomb corrections.

The dimensionless coefficient of electrical conductivity for a high-temperature plasma is less than that for a low-temperature plasma at the same values of Γ (see Figs. 6.6 and 7.12). Figure 7.12 yields the quantity

$$\bar{\sigma} = (1+\bar{Z})^{-1/2}\gamma_{\rm E}^{-1}(\bar{Z})\frac{\sigma}{\omega_p}\Gamma \tag{7.20}$$

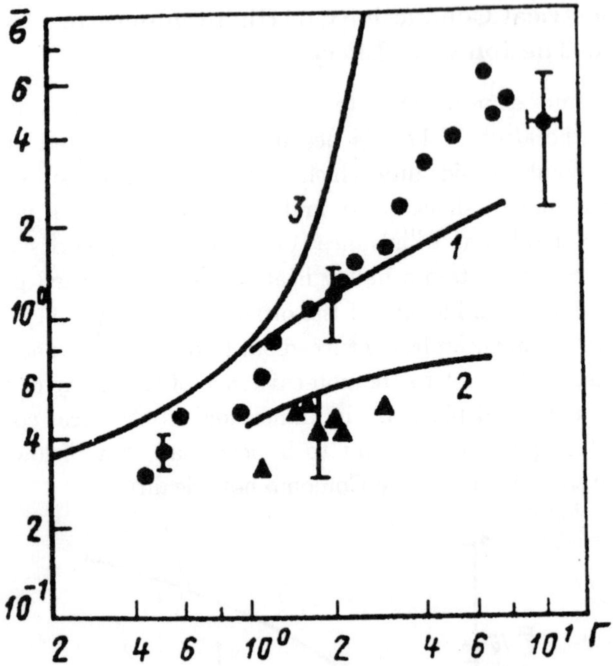

Fig. 7.12. The reduced coefficient of electrical conductivity @ as a function of the non-ideality parameter. Experiment: •, argon, krypton, xenon, $T \approx 2.5 \cdot 10^4$ K [431]; Δ, xenon, $T \approx 7 \cdot 10^4$ K [143]. Theory: *1* and *2*, calculation with pseudopotential (7.21) at the same temperatures; *3*, $\bar{\sigma}_{\mathrm{Sp}}$.

as a function of the non-ideality parameter $\Gamma = \bar{Z}e^2/Tr_{\mathrm{D}}$, where $\bar{Z}$ is the averaged ion charge and $r_{\mathrm{D}}^{-2} = 4\pi e^2(1+\bar{Z})N_e/T$. The experimental data at low and high temperatures clearly diverge from each other.

Indeed, in view of the long-range character of the Coulomb potential, the greatest contribution to the transport coefficient at moderate temperatures is made by electron scattering from great impact parameters. This justifies the use of semi-qualitative inclusion of close collisions by introducing various forced cutoffs [451–452]. As the temperature rises, the amplitude of Coulomb scattering, $\bar{Z}e^2/T$, turns out to be comparable with the proper size of ions, r_i. Hence the high energy conduction electrons may approach the nucleus rather closely on their scattering, where the interaction potential is no longer purely Coulombian and appears distorted by inner electron shells. The estimates [142, 453] made for rarefied plasma indicate that this effect starts to show against the background of Coulomb scattering at only extremely high temperatures of the order of $2 \cdot 10^6$ K. An increase in the plasma density causes a strengthening of the Coulomb interaction screening and, consequently, an increase in the importance of close collisions in the plasma. This in turn makes it possible to manifest the effect of non-Coulomb scattering of electrons from small impact parameters found in the region of lower temperatures accessible to experiment, $T \geq 5 \cdot 10^4$ K.

Mintsev *et al.* [143] measured the electrical conductivity of a dense xenon plasma in a previously uninvestigated high-temperature range of parameters characterized by developed ionization ($\bar{Z} \approx 3$) and strong interparticle Coulomb interaction ($\Gamma \geq 1$). In the experiments, explosion cumulative plasma generators was used. Linear explosion systems were also used to prepare a highly heated plasma in reflected shock waves. The experiments involved the use of xenon due to its high molecular weight, which makes it effective to heat this gas by shock waves. Hence a great number of bound electrons will cause a marked distortion to the Coulomb potential at close distances. The cumulative systems allow the preparation of a plasma with $p \approx$ 5–35 kbar, $T \approx (5\text{–}10) \cdot 10^4$ K, $\rho \approx 0.05\text{–}0.5$ g/cm^3, $N_e \approx (0.4\text{–}3) \cdot 10^{21}$ cm^{-3}, $\Gamma \approx 1.1\text{–}2.6$ and $\bar{Z} \approx 2\text{–}3$. The plasma parameters behind the reflected shock wave amounted to $p \approx$ 40–80 kbar, $T \approx (3\text{–}8) \cdot 10^4$ K, $\rho \approx 0.3\text{–}2$ g/cm^3, $N_e \approx (2\text{–}6) \cdot 10^{21}$ cm^{-3}, $\Gamma \approx 2\text{–}5$ and $\bar{Z} \approx 2$.

In the region of low temperatures, the data obtained agree well with results obtained previously (Fig. 7.5). However, the electrical conductivity of plasma increases with the shock wave velocity at a much slower rate than one could expect on the basis of conventional plasma models of electron transfer.

The potential of electron-ion interaction at small distances is stronger than $\bar{Z}e^2/\bar{r}$, which leads to an increase in the scattering cross-section as compared to the Coulombian one and, consequently, to a relative decrease in electrical conductivity.

To interpret the experimental data, use was made of a model employed to study the effect of non-Coulombian behavior of electron scattering at small impact parameters on the electron mobility. The conductivity was calculated in the approximation (6.12) while the transport cross-sections of scattering used in the calculations were found numerically using the partial wave method. The electron-ion interaction was described by the effective two-body potential [453], which allows for the presence of ion core,

$$V_{ei}(r) = -\left[\frac{Z_j}{r} - \frac{(Z - Z_j)}{r}\exp(-Br)\right]\exp\left(-\frac{r}{r_{\rm D}}\right), \quad B = \frac{1.8Z^{4/3}}{(Z - Z_j)a_0}, \tag{7.21}$$

where Z, Z_j are the nuclear and ion charges and $r_{\rm D}$ is the Debye screening distance. At small values of r, the potential (6.71) coincides with the Thomas-Fermi potential and at $r \gg B^{-1}$, it changes to the Debye potential. This potential was used in the numerical solution of Schrödinger's equation for the radial part of the wave functions.

The results of these calculations for the pseudopotential (7.21) in the region of electrons with an energy of the order of 0.7 a.u. make their main contribution to the electrical conductivity of plasma at $T = 7 \cdot 10^4$ K. The presence of interaction is stronger than $(Z_j e^2/r)$ and is approximately twice that of the scattering.

In accordance with the model employed for $T = 2.5 \cdot 10^4$ and $7 \cdot 10^4$ K, the results of the calculation of the electrical conductivity are marked in Fig. 7.11 with a broken line. It can be seen that the ion core model permits an intelligent description of the effect of "separation" of the isotherms of electrical conductivity of a highly heated non-ideal plasma.

Eight

The Optical Properties of Dense Plasma

8.1. Optical Properties

Optical properties are of considerable interest from the standpoint of plasma physics. This is because they enable one to follow the effect of non-ideality on the dynamics and energy spectrum of electrons in a dense, disordered medium. The plasma radiation bears information on the temperature and concentration of particles, elastic and inelastic collisions, and ionization and recombination processes. Thus far, extensive data have been obtained on the optical properties of rarefied plasma where the elementary processes are readily separated. The theoretical models developed also provide an exhaustive description of numerous experiments (Section 7.1). As the density increases, the optical consequences of non-ideality are registered before the corresponding variations of the thermodynamic and transport properties take place. They also show up as the spectral line shift and broadening, as well as the shift of photoionization continua (Section 7.2).

With a further increase in non-ideality, the effects do not change in their behavior but only increase quantitatively ("spectroscopic stability"). It is only under very high density that a marked rearrangement of the electron energy structure occurs in plasma (primarily in the threshold region of the spectrum). This is described in various, and often still controversial, models. Numerous and clearly defined manifestations of the electron spectrum in a highly compressed plasma have been registered. This plasma emerges as a result of the contributions of great pulsed energy under laser irradiation of condensed matter in pinched electric discharges, as well as in a number of dynamic experiments. It was also revealed in static experiments that upon reaching the near-critical values of density, a qualitative variation of the optical spectra of metal vapors takes place.

We will introduce the basic concepts of the radiation theory. The spectral absorptivity k_ν is determined in terms of attenuation dI_ν, which is experienced by the radiation intensity I_ν whilst passing through a layer of matter with a thickness dl:

$$dI_\nu = -k_\nu I_\nu dl\,. \tag{8.1}$$

Under thermodynamic equilibrium, k_ν is related to the radiation intensity I_ν by Kirchhoff's law:

$$I_\nu = k_\nu B_\nu(T)\,, \qquad B_\nu(T) = 2\pi h\nu^3 c^{-2}\left[\exp\left(\frac{h\nu}{T}\right) - 1\right]^{-1}, \tag{8.2}$$

where $B_\nu(T)$ is the Planck's function. The quantity $I_\nu d\nu$ defines the energy emitted by volume dV per unit time in a unit solid angle in the frequency interval $d\nu$.

These quantities are used to express the divergence of the radiant energy flux q_R, which characterizes the energy losses (or release) due to radiation. In two limiting cases, that is, with high and low optical density, ∇q_R is written in a simple manner.

If the plasma dimension 1 exceeds the radiation path length k_ν^{-1}, $k_\nu l \gg 1$ in the entire frequency interval covering the spectral region in which the bulk of radiant energy is transferred, the radiative energy transfer may then be described in the approximation of radiant thermal conductivity,

$$q_\mathrm{R} = -\frac{16}{3}\sigma_\mathrm{SB} l_\mathrm{R}(\mathbf{r}) T^3(\mathbf{r}) \nabla T(\mathbf{r})\,,$$

$$l_\mathrm{R}(\mathbf{r}) = \int k_\nu^{-1}(\mathbf{r}) B_\nu(\mathbf{r}) d\nu \Big/ \int B_\nu(\mathbf{r}) d\nu\,, \tag{8.3}$$

where σ_SB is the Stefan-Boltzmann constant and l_R is the Rosseland length, which is the mean photon path length.

Conversely, in the limiting case of an optically thin medium, the divergence of energy flux is written easily:

$$\nabla q_\mathrm{R} = 4\pi\sigma_\mathrm{SB} k_{Pl}(\mathbf{r}) T^4(\mathbf{r})\,,$$

$$l_{Pl}(\mathbf{r}) = \int k_\nu(\mathbf{r}) B_\nu(\mathbf{r}) d\nu \Big/ \int B_\nu(\mathbf{r}) d\nu\,, \tag{8.4}$$

where the quantity k_{Pl} is referred to as the mean Planck absorptivity.

Although both limiting cases are seldom realized in their pure form, the representation of the results of calculations of radiative properties in the form of density and temperature dependences of mass absorption cross-sections yields important information. Figure 8.1 gives the Rosseland and Planck mass absorption cross-sections as functions of the atomic nuclear charge.

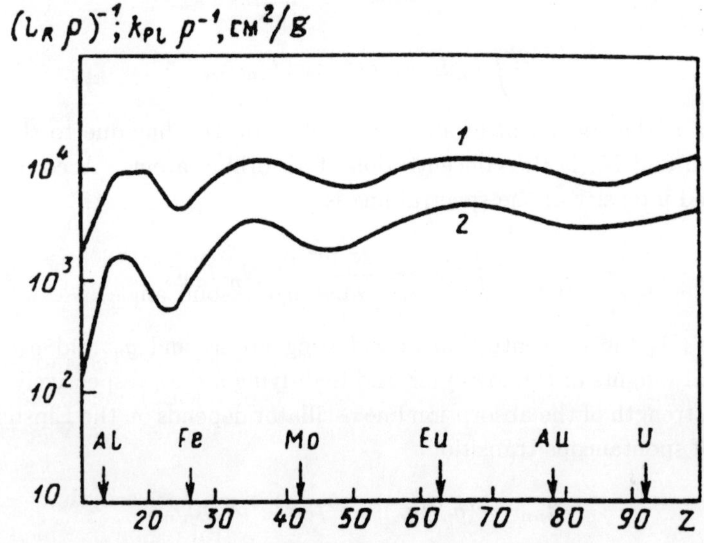

Fig. 8.1. The (*1*) Planck and (*2*) Rosseland mass absorption cross-sections as functions of the atomic nuclear charge [464]; $T = 100$ eV, $\rho = 0.1$ cm^2/g.

8.2. Basic Radiation Processes in Rarefied Atomic Plasma

The radiation processes in ideal plasma have been fairly well studied (cf., [454–456]). We will only present here the basic information along with some reference formulas.

Depending on the type of the corresponding optical spectra, the electron radiation processes are divided into two groups:

(i) Bound-bound transitions in atoms. They provide a series of spectral lines which converge at the photoionization thresholds. Near the threshold, the convergence of lines occurs. As a result, the position of the threshold shifts to the long-wave side (Section 8.3).

(ii) Bound-free and free-free transitions. The photoionization and bremsstrahlung processes defining the continuous spectrum manifest little dependence on frequency and experience sharp discontinuities at frequencies which are equal to the threshold frequencies. Such a division is not absolute. For example, strongly broadened spectral lines overlap and are superimposed on the continuous spectrum. These effects are especially strong in a strongly non-ideal plasma, that is, high-pressure plasma.

The integrated intensity of spectral line depends on the oscillator strength $f_{nn'}$:

$$\int k_n d\nu = (\pi e^2/mc) f_{nn'} N_n ,$$

where k_ν is the coefficient of absorption in a spectral line due to the $n \to n'$ transition and N_n is the concentration of absorbing atoms. Accordingly, the integrated intensity of the spectral line is

$$I = \int I_\nu d\nu = \frac{2\pi h e^2}{m\lambda^3} \frac{g_n}{g_{n'}} N_{n'} f_{nn'} ,$$

where N_n is the concentration of radiating atoms and g_n and $g_{n'}$ are the statistical weights of the low-lying and high-lying states, respectively.

The strength of the absorption line oscillator depends on the Einstein probability of spontaneous transition,

$$f_{nn'} = (g_{n'}/g_n)(mc^3/8\pi^2 e^2 \nu^2) A_{n'n} ,$$

which is related to the matrix element of the dipole moment of an atomic system. The absorbing and radiating states are actually characterized by several quantum numbers such that the subscript n characterizes their entire required set. The fullest tables for $f_{nn'}$ and $A_{nn'}$ are provided by Wiese *et al.* [457]. The quasi-classical Kramers formula for hydrogen-like atom appears useful. The transition is considered between states with the main quantum numbers k and m averaged over the remaining quantum numbers. For such a transition,

$$f_{km} = \frac{32}{3\pi\sqrt{3}} \frac{1}{k^3 m^3} (k^{-2} - m^{-2})^{-3} = \frac{1.96}{k^5 m^3} \left(\frac{E_k - E_m}{\mathrm{Ry}} \right)^{-3} ,$$

where E_k and E_m are the binding energies of the m-th and k-th levels. The error of f_{km}, when calculated using this formula for a hydrogen atom, does not exceed 30% for all values of k and m.

The line spectral intensity depends fully on the frequency dependence of the absorption coefficient. This is because the Planck function varies slightly within the line. The ν independence of k is defined by the behavior of line broadening. In a rarefied plasma, it is defined by radiation damping and the Doppler effect. The broadening in a non-ideal plasma, which is due mainly to interparticle interactions, will be discussed below.

Continuous absorption and emission spectra are formed as a result of electron transitions from the bound to free state and *vice versa* (bound-free and free-bound transitions), and as a result of free-free transitions. Bound electrons are found in the atom while the free electron may be in transit through the field of atom or ion. Accordingly, for the emission of radiation, a distinction is made between free-bound transitions in the field of ion (recombination radiation),

$$\mathrm{A}^+ + e \to \mathrm{A} + h\nu\,,$$

and the two types of free-free transitions: bremsstrahlung in the field of ion,

$$\mathrm{A}^+ + e(mv_1^2/2) \to \mathrm{A}^+ + e(mv_2^2) + h\nu\,, \quad h\nu = mv_1^2/2 - mv_1^2/2\,,$$

and in the field of atom,

$$\mathrm{A} + e(mv_1^2/2) \to \mathrm{A} + e(mv_2^2) + h\nu\,.$$

In the case of absorption, a distinction is made between bound-free transitions in the field of ions (phototonization of atoms),

$$\mathrm{A} + h\nu \to \mathrm{A}^+ + e\,,$$

and free-free transitions in the fields of atom and ion.

If an atom has a bound state with an electron, one should add the photodetachment of electron (in the case of absoption) to the list of processes outlined above:

$$\mathrm{A}^- + h\nu \to \mathrm{A} + e\,,$$

as well as the photoattachment (in the case of radiation)

$$\mathrm{A} + e \to \mathrm{A}^- + h\nu\,.$$

As a rule, the resultant continuous spectra represent the superposition of several continua due to individual processes. The number of such resuperposed continua may be considerable. Hence the determination and, especially, analysis of the resultant spectrum may prove to be a non-trivial problem.

In a plasma with developed ionization, the greatest contribution to the continuous spectrum intensity is made by the free-free transitions of electrons in the fields of ions. The absorption coefficient, including the correction for stimulated radiation, is given by the Kramers' formula:

$$k_\nu = \frac{2\sqrt{2}e^6 Z^2}{3\sqrt{3}\pi c\hbar m^{3/2}\sqrt{T}} \frac{g}{\nu^3} N_e N_i (1 - e^{-\nu/T}) , \tag{8.5}$$

where Z is the ion charge and g is the Gaunt factor varying, as a rule, in the 1.1–2 range. The bremsstrahlung contribution to the total intensity of continuum prevails at $h\nu \leq 0.7T$ (cf., Fig. 8.2). At higher frequencies, photorecombination radiation is significant.

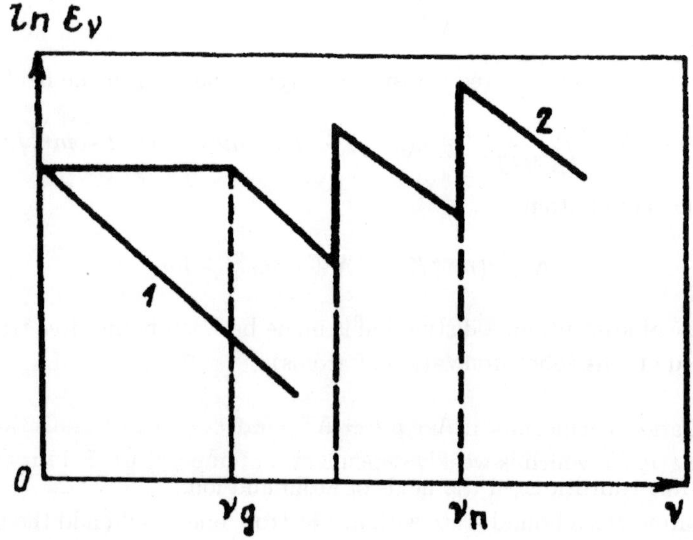

Fig. 8.2. The spectral intensity of continuous (*1*) bremsstrahlung and (*2*) recombination spectra.

In photoionization, the absorption coefficient is calculated by summation of the products of cross-sections of photoabsorption, $q_n(\nu)$, by various states n multiplied by the populations of these states, N_n:

$$k_\nu(T) = \sum_n q_n(\nu) N_n(T) . \tag{8.6}$$

Photoionization is a threshold process: for the n-th state, $q_\nu = 0$ if $\nu < \nu_n$, and $q_n = {}_n(\nu)$ if $\nu > \nu_n$, where ν_n is the threshold frequency defined by the bonding energy of the n-th state (cf., Fig. 8.2). The cross-section of absorption by a hydrogen-like atom in the n-th state is given by the Kramers' formula:

$$q_n(\nu) = \frac{64\pi^4}{3\sqrt{3}} \frac{me^{10}Z^4}{ch^6n^5} \frac{1}{\nu^3} . \qquad (8.7)$$

The sum (8.6) can be calculated because the atomic level populations are known in the equilibrium plasma. The summary absorption coefficient depends non-monotonically on frequency. This is because given an increase in frequency, new levels begin to contribute to the sum (8.6) starting from their threshold frequencies. At low frequencies, however, the jumps in k_ν are small and the summation may be replaced with an integration. This leads to the Kramers-Unsöld formula. In doing so, it is convenient to add up the expression for k_ν, which is due to photoionization and expression (8.5). In this manner, we obtain the summary coefficient of absorption which allows for both bound-free and free-free transitions:

$$k_\nu = \frac{2\sqrt{2}e^6 Z^2 g}{3\sqrt{3\pi}c\hbar m^{3/2}T^{1/2}} N_e N_1 \frac{1}{\nu^3} \mathrm{e}^{h\nu/T}\left(1 - \mathrm{e}^{-h\nu/T}\right), \quad \nu \le \nu_g . \qquad (8.8)$$

The value of the boundary frequency depends on the bonding energy of the lower excited state of the atom. The photoionization of the latter is included in Eq. (8.8).

In order to describe the plasma of complex atoms and ions, Biberman and Norman [454] developed an approximate calculation technique. It enables one to derive unified and fairly simple formulas to calculate the absorption and radiation coefficients in both free-free and bound-free transitions. The function $\xi(\nu, T)$, which is weakly dependent on temperature, is introduced and calculated. It allows for a non-hydrogen likeness of complex atoms. The final expressions have the form

$$k_\nu = \begin{cases} \dfrac{2\sqrt{2}e^6 Z^2 g}{3\sqrt{3\pi}c\hbar m^{3/2}T^{1/2}} N_e N_i \dfrac{1}{\nu^3} \mathrm{e}^{h\nu/T} \xi(\nu) \exp[(h\Delta\nu - \Delta I)/T] \,, \; \nu \le \nu_g & (8.9) \\ k_\nu(\nu_g) \dfrac{\xi(\nu)}{\xi(\nu_g)} \left(\dfrac{\nu_g}{\nu}\right)^3 + \sum\limits_{n<n_g} q_n(\nu) N_n(T) \,, \qquad \nu \ge \nu_g . & (8.10) \end{cases}$$

The factor $\exp[(h\Delta\nu - \Delta I)/T]$ is due to two effects: the merging of spectral lines near the continuous spectrum boundary and the decrease in the atomic ionization potential ΔI. The latter has been discussed in detail in Chapters 4 and 5. The value of the shift in the continuous spectrum boundary, $\Delta\nu$, is discussed in Section 8.3. The sum in (8.10) is calculated in view of all the states that are not included in the integral formula, $h\nu_n > h\nu_g$. This expression defines the frequency ν_g, which is referred to as boundary.

Using the equation of ionization equilibriums, the absorption coefficient k_ν can be related to the atomic concentration N_a rather than to N_eN_i. Instead of Eq. (8.9), we have

$$k_\nu = \frac{32\pi^2 e^6}{3\sqrt{3}ch^4} N_a \frac{\Sigma^+}{\Sigma}\frac{Z^2}{\nu^3} T \exp^{(-I/T)} \exp\left(\frac{h\nu + h\Delta\nu}{T}\right) \xi(\nu, T)\,, \nu \le \nu_g \tag{8.11}$$

$$\tag{8.12}$$

where Σ^+ and Σ are the statistical sums of ion and atom.

At low degrees of ionization, the continuous spectrum can be defined by the free-free transitions of electrons in the field of atoms. The probability of free-free radiative transitions in a certain approximation is related to the cross-section of elastic electron-atom scattering. Zel'dovich and Raizer [458] have shown that the differential effective radiative cross-section in the $d\nu$ frequency range, upon the collision of an electron having a velocity ν with an atom, takes the form of

$$dq_\nu = (8e^2\nu^2/3c^3h\nu)qd\nu\,, \tag{8.13}$$

where q is the transport cross-section of electron-atom scattering.

If q does not depend on the energy of the electron, the absorption coefficient may be estimated by the formula

$$k_\nu = e^2(\pi mc\nu^2)^{-1} N_e N_a \langle q(mv^2/2)v\rangle\,, \tag{8.14}$$

where $\langle\ldots\rangle$ implies an averaging over the Maxwellian distribution. Bremsstrahlung on atoms predominates over bremsstrahlung on ions if the following condition is satisfied:

$$\frac{1.5\cdot 10^{13} T^{-5/4}}{q\sqrt{N_a}} \exp\left(-\frac{I}{2T}\right) \ll 1\,, \tag{8.15}$$

where T is in eV, the cross-section is in 10^{-16} cm^2, and the number density of atoms is in cm^{-3}. Therefore, the intensity of bremsstrahlung on an atom prevails in a weakly ionized plasma.

In the plasma of a number of chemical elements, an important part is played by the photoattachment processes and photodetachment from negative ions in the formation of continuous optical and ultraviolet spectra. We now compare the coefficients of absorption, which are due to the photodetachment k_ν^-, with the electron transitions in the fields of ions. We shall write the first coefficient as

$$k_\nu^- = N^-\bar{q}(\nu)\,,$$

where N^- is the negative ion concentration and $q(\nu)$ is the photodisintegration cross-section. The second coefficient, k_ν^+, is given by Eq. (8.11). Thus [459],

$$k_\nu^-/k_\nu^+ = [3\sqrt{3}ch^4(128\pi^3\sqrt{2\pi}e^6m^{3/2})^{-1}]N_eT^{-5/2}(h\nu)^3 \\ \times q(\nu)\xi^{-1}(g^-/\Sigma^+)\exp\{[I+E-h(\nu+\Delta\nu)]/T\}, \quad (8.16)$$

where g^- is the statistical weight of the negative ion and E is the bonding energy of an electron in the latter. The relation (8.15) is at its maximum near $\nu \approx 3T/h$. Upon substituting the characteristic values in Eq. (8.15), we derive

$$k_\nu^-/k_\nu^+ \approx 10^{-26}N_e\sqrt{T}\exp[(I+E)/T].$$

Hence it follows that in the plasma of elements with the ionization potential $I \geq \text{Ry}$, the continuum due to negative ions contributes greatly to the continuous spectrum.

8.3. The Effect of Weak Non-ideality: Spectral Line Broadening and Shift; Photoionization Threshold Shift

The influence of weak interparticle interaction on the radiative properties of plasma reduces the latter to the well-known effects of spectral line broadening and shift. The photoionization threshold also shifts to lower energies. In view of the ample literature on these subjects [460–463], we shall only discuss them briefly.

In a dense plasma, both broadening and shift are caused by the interaction between a radiating atom or ion and the surrounding particles. In doing so, the interaction between the radiator and plasma electrons and ions is of great importance.

The interaction with neutral atoms which are of the same type as the radiating system may lead to resonance broadening if one of the states in the line has an optically allowed transition to the ground state. The interaction with atoms of another type leads to the van der Waals broadening. Both types of broadening in a plasma with developed ionization (at degrees of ionization of ≥ 0.01) are less important than the Stark broadening.

For the actual calculation of pressure broadening, use is made of impact (collision) and quasi-static approximations. The quasi-static approximation is applicable when the perturbing particles are moving relatively slowly. Hence the perturbation over the period of time which is of interest to us is practically constant. In the other limiting case, the radiating system does not experience any perturbations most of the time except for short impacts which are well separated in time. This approximation is applicable in describing the perturbation of an atom by electrons. It is valid when the time lapse between the collisions is much longer than that of the collision proper, that is,

$$N_a^{1/3}/v \ll (N_e q v)^{-1},$$

where v is the electron velocity and q is the cross-section of an electron-atom scattering. For the long-range Coulomb interaction, which is responsible for line broadening in plasma, the collisions are not separated in time. This is especially true for the conditions of non-ideality. Therefore, the impact approximation is only valid when the interaction is weak on the whole.

The problem of calculating the line broadening through the interaction of ions and electrons in a weakly non-ideal plasma is reduced to the following [462]. The electric field generated by ions is assumed to be constant and the Stark broadening is determined for an atom in this field. The broadening of each Stark component is then calculated within the impact approximation. After that, the resultant distribution of intensity is averaged over all the possible values of intensity of the ion microfield.

Collisional broadening by electrons produces the dispersion line contour

$$k_\nu = k_0[1 + ((\nu - \nu_0) - \Delta)^2/(\gamma/2)^2]^{-1}.$$

Here, γ is the line width and Δ is its shift. The absorption coefficient at the center of the line is $k_0 = (8\pi)^{-1}\lambda^2(g_2/g_1)\ \mathrm{A}_{21}N_1(2/\pi\gamma)$, where λ is the radiation wavelength and N_1 is the number density of absorbing atoms.

The shift and width are expressed in terms of the amplitude of elastic forward scattering, $f(0)$:

$$\Delta = -\frac{h}{m}N_e\,\mathrm{Re}[f(0)]_{a\nu}, \qquad \gamma = \frac{h}{m}N_e\,\mathrm{Im}[f(0)]_{a\nu} = \frac{1}{2}N_e(vq_{\mathrm{tot}})_{a\nu}.$$

The latter expression includes the averaged scattering frequency; q_{tot} is the total scattering cross-section. Averaging is performed over the electron energies and impact parameters. The maximum impact parameter is limited by the screening distance. This is important from the standpoint of the shift, but has little effect on the width as long as the transition frequency is much higher than the frequency of the plasma.

The spectral lines play an important part in the spectral and integral characteristics of dense plasma. Figure 8.3 illustrates the spectral absorptivity of air plasma [465]. The broken line indicates the contribution made by continua. As the density increases, the lines broaden considerably and merge to

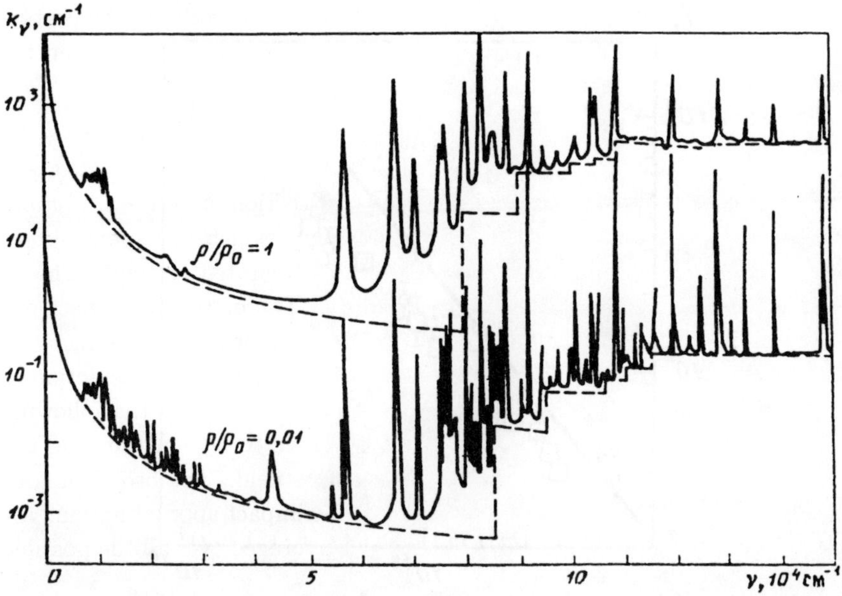

Fig. 8.3. The absorption coefficient of air plasma for $T = 2.2 \cdot 10^4$ K and two values of relative density ρ/ρ_0; ρ_0 is the normal density of air [465].

form quasi-continua, thus making a sizable contribution to the characteristics of the integral.

The methods of calculating the absorption coefficient in lines involve the use of the Holtsmark distribution of ion microfields and the impact approximation for electrons. That is, the effects of strong correlation are ignored while other simplifying elements are employed which correspond to the conditions of weak non-ideality. No systematic studies have yet been performed on the effect of strong non-ideality on the spectral lines.

Figure 8.4 gives the dependence of the shift in the spectral line of xenon $\lambda = 467.1$ nm on the electron number density. Guenter *et al.* [468] have carried out measurements in a plasma of pulsed electric discharge and in an adiabatically compressed plasma of up to $N_e = 3 \cdot 10^{18}$ cm^{-3}. It is only for the maximum values of N_e that one can observe some deviations of the dependence ΔN_e from the linear dependence $\Delta \sim N_e$, which is characteristic of ideal plasma with lines broadened by the quadratic Stark effect. Sechenov and Novikov [470] have performed measurements for the same line in shock-compressed xenon of up to $N_e = 10^{19}$ cm^{-3}, $T = 1.4 \cdot 10^4$ K. These are

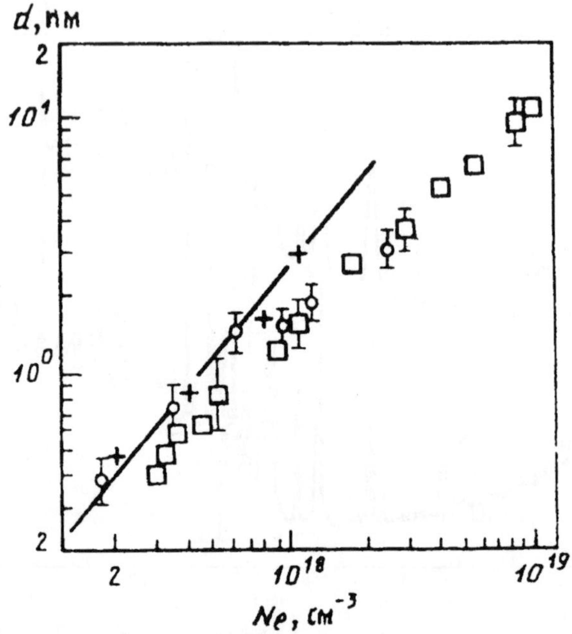

Fig. 8.4. The shift in the xenon line of 467.1 nm. Measurement results: curve, extrapolation of the data of [466, 467]; + [469]; ∘ [468]; □ [470].

the maximum parameters reached in xenon, which are discussed below (see Section 8.4).

The interparticle interaction leads to a shift in the boundary of continuous spectrum. This is interpreted as a decrease in the ionization potential. In addition, the highest states of discrete spectrum broaden strongly. In other words, an actual decrease in the ionization potential occurs and, in addition, the broadening of higher levels leads to an apparent decrease in the ionization potential. In spectrograms, the relative importance of these two effects is hard to identify. Therefore, they are combined and one speaks, bearing in mind both effects, of the transformation of lines into a continuous spectrum and of the respective shift in the ionization threshold.

In order to estimate the threshold shift, one usually proceeds from qualitative considerations. The photoionization spectrum shifts towards the red up to the position of the line where the total Stark width is equal to the distance of the nearest line in the series. The Inglis-Teller criterion is based on this principle, which gives

$$n^{Z-1} = (1/2)Z^{3/5}(N_i a_0^3)^{-2/15}$$

for the main quantum number of the last line, and for the boundary shift,

$$\Delta E^{Z-1} = Z^2\mathrm{Ry}/(n^{Z-1})^2 = 4Z^{4/5}(N_i a_0^3)^{-4/15}\mathrm{Ry}\,. \qquad (8.17)$$

The models which treat the shift in the photoionization threshold as a result of spectral line overlap are sufficiently valid only for the plasma of hydrogen, whose lines are closely arranged and strongly broadened due to the linear Stark effect. Even for ionized helium having a broad but wider (by a factor of four) spaced lines, such treatment is not valid. This is even more true of the spectral series of complex atoms where the line broadening is much weaker.

Therefore, a number of other methods have been proposed to estimate the shift in the photoionization threshold. Armstrong [471] has equated the inter-level energy distance $e^2(a_0 n^{*^3})^{-1}$ with the level width given by the asymptotic Baranger formula which describes broadening by electron impacts. This yields

$$\omega_{n^*} = 8\pi^2(e^2\alpha a_0^2 N_e)(me^2/2\pi T)^{1/2}(n^*)^4\,,$$

where $\alpha = e^2/hc$ and n^* is the effective main quantum number of the level with the bond energy $E_{n^*} = \mathrm{Ry}/n^{*^2}$. The shift

$$\Delta E = 4.71 \cdot 10^{-6} N_e^{2/7} T^{-1/7}\,, \qquad (8.18)$$

where T and ΔE are expressed in eV and N_e, in cm^{-3}.

In the spectra of complex atoms, the lines of different spectral series overlap. The spectra of the inert gases of argon, krypton, xenon and neon have a dense sequence of terms starting from the bond energy of 1 eV. The distance between the levels in this sequence is 0.01–0.05 eV. By introducing some averaged distance among the levels of dense sequence, δE, one can compare this distance with the width of the given higher level. The levels overlap if their bond energy is less than ΔE, which is defined by the expression [472]:

$$\Delta E = 3.5 \cdot 10^{-10}(N_e/\sqrt{T}\delta E)^{1/2}\,. \qquad (8.19)$$

Figure 8.5 gives the values of ΔE derived in experiment for argon and xenon as a function of N_e with $T_e = 0.8$–1.5 eV.

The behavior of the near-threshold Balmer spectral region is shown in Fig. 8.6. The higher terms in the Balmer series are already absent: they have been transformed into continuum (the undisturbed threshold is indicated in Fig. 8.6 by an arrow). The shift ΔE increases with N_e.

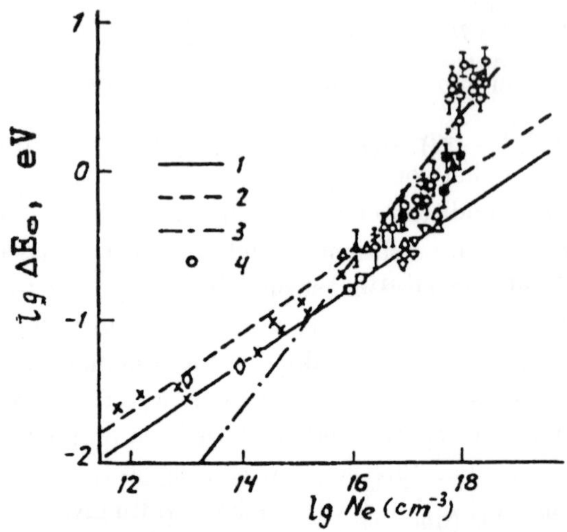

Fig. 8.5. The shift in the photoionization boundary as a function of the electron number density [472]. Theory: *1*, Inglis-Teller formula; *2*, (8.18), $\delta = 0.02$ eV; experimental data are borrowed from [473].

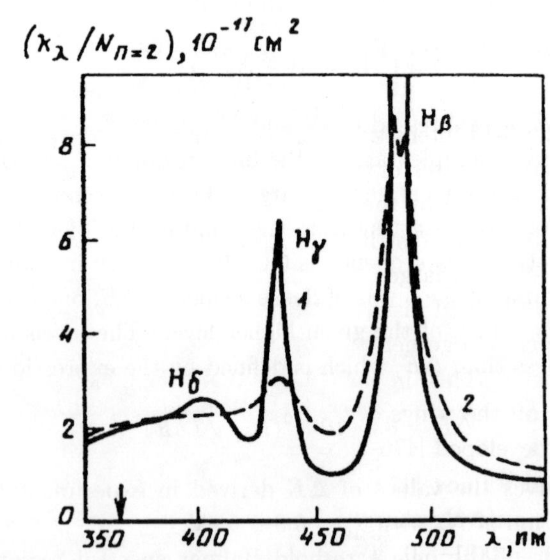

Fig. 8.6. The absorption coefficient of the Balmer series per single absorbing atom as a function of wavelength λ [468]. Measurement results: *1*, $N_e = 1.7 \cdot 10^{17}$ cm^{-3}, $T = 1.6 \cdot 10^4$ K, $\gamma = 0.07$; *2*, $N_e = 8.4 \cdot 10^{17}$ cm^{-3}, $T = 2.22 \cdot 10^4$ K, $\gamma = 0.09$.

In conclusion, we emphasize on the conditionality of the concept of "optical shift". At any frequency that less than the threshold frequency, an atom may radiate or absorb a quantum in both bound-bound and bound-free transitions. One can see from Fig. 8.6 that the optical shift, that is, the distance from the shifted photoionization threshold to the first line discernible in the spectrogram, is a conditional quantity dependent on subjective factors.

8.4. The Microfield Distribution Function in Non-ideal Plasma

As a result of the spontaneous violation of the electroneutrality of plasma on scales of the order of the screening distance, microfields which fluctuate in time occur in plasma [474, 475]. The microfield distribution function is used to calculate the optical properties of line broadening and shift, as well as the shift in the photoionization thresholds. Microfields are divided into two groups: low-frequency microfields (due to ions) and high-frequency microfields (due to electrons).

The distribution function for ion microfields is usually calculated within a one-component system of $\boldsymbol{N}$ ions against a uniform neutralizing electron background:

$$P(\mathbf{F}) = \int d\mathbf{R}_1 \ldots d\mathbf{R}_N \delta \left[\mathbf{F} - \sum_{i=1}^{N} \mathbf{F}(\mathbf{R}_j) \right] W(\mathbf{R}_1 \ldots \mathbf{R}_N)/\Omega^{\boldsymbol{N}} . \qquad (8.20)$$

Here, $\mathbf{F}(\mathbf{R}_j)$ is the field intensity at the sampling point developed by the jth ion, $\mathbf{R}_j$ are the co-ordinates of ions, Ω is the system volume, and W is the Gibbs' probability defined by the potential energy of charge interaction.

The known Holtsmark distribution of the ion microfield was derived for systems of uncorrelated charges, in which $W = 1$. In a non-ideal plasma, the correlation affects the field distribution. A strong interion repulsion prevents ions from coming close together, thereby reducing the probability of the emergence of strong fields. This, in turn, must affect the spectra, by reducing the shift and cutting off the wings of the spectral lines. A number of theoretical approaches were developed [476–479] to include the correlation.

Kurilenkov and Filinov [479, 480] have calculated the microfield correlation functions at positively charged, neutral and negatively charged points of a plasma with $\gamma = 0.01$–5.0, where $\gamma = e^2(N_i + N_e)^{1/3}/T$. The classical expression (8.19) was generalized to the quantum case using the interaction pseudopotentials which enable one to include the quantum effects when the particles are drawing together (see Section 5.3). The Monte Carlo method was used in further calculations. The microfield distribution functions are given in Fig. 8.7. The correlation indeed leads to a reduction in the probability of the emergence of strong fields.

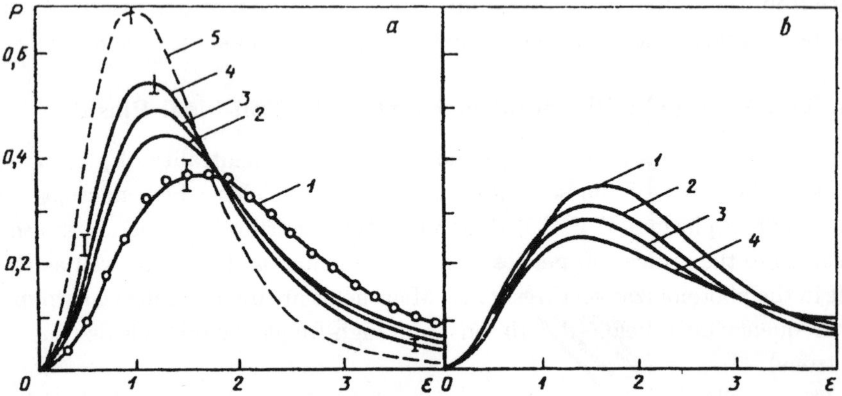

Fig. 8.7. The ion microfield distribution function $P(\varepsilon)$ at (*a*) positively charged point and (*b*) neutral point [479]: *1*, Holtsmark distribution; distributions in a non-ideal plasma; *2*, $\gamma^3 = 0.01$; *3*, $\gamma^3 = 0.1$; *4*, $\gamma^3 = 1.0$; *5*, $\gamma^3 = 5.0$.

In the experiments of [152, 468, 470], deviations were observed from the linear dependence of the shift in the xenon and argon spectral lines on the electron number density. In xenon, weak deviations were recorded at $N_e \approx 3 \cdot 10^{18}$ cm^{-3} (see Fig. 8.4). Kurilenkov and Filinov [480] studied an argon spectrum which corresponds to $4s - 4p$ transitions at $N_e \leq 10^{19}$ cm^{-3}, $T \leq 1.7 \cdot 10^4$ K, $p \leq 17$ MPa. The generation of plasma and the recording of its thermodynamic parameters were performed during compression and heating in the shock front. For all lines investigated in the 730–780 nm range, the extrapolation of ideal plasma calculations to the region of attained parameters showed that the observed shifts are four to five times less than the calculated values (Fig. 8.8). It is not improbable that the observed deviations towards reduced shifts reflect the tendency of microfield reduction with an increase in non-ideality.

When observing the radiation in spectral lines, the plasma microfields are responsible for yet another qualitative effect. The potential field of atomic residue is added to the field produced by free plasma particles. This field is uniform on an atom scale. However, as a result, a three-dimensional potential

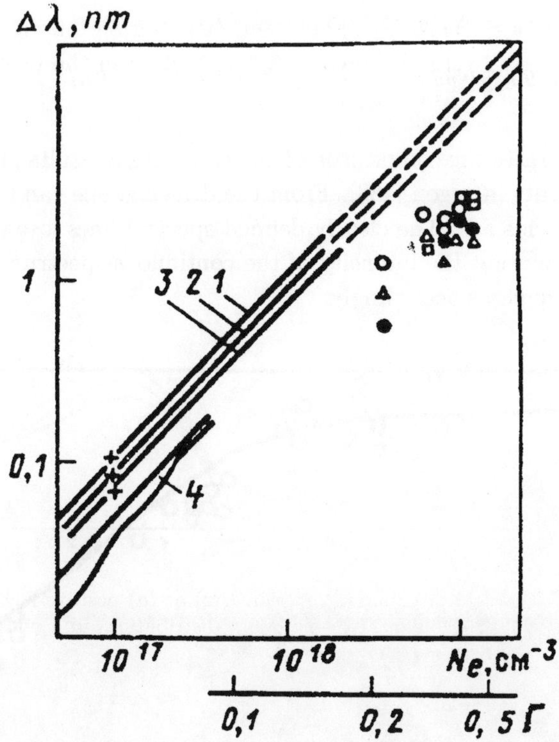

Fig. 8.8. The shift of some $4s$–$4p$ lines of argon as a function of N_e [480]. Theory: *1–3*, for 750.4, 772.4, 763.5 nm [461]. Experiment: +, 750.4, 751.5, 772.4 nm [481]; ∘, 750.4, 751.5 nm; □, 738.4 nm; •, 763.5 nm; △, 772.4 nm [480]; *4*, experimental points in the 763.5 nm range [482].

surface is formed. In doing so, a potential barrier emerges for the electron in the atom. The electron may overcome it by either thermal activation or tunneling. Highly excited states in the threshold region may be washed out fully in the given recorded microfield. The populations of other states, which still manifest themselves in the experiments as discrete, may be reduced because of a decrease in their partition function.

The destruction of bound states by plasma microfields was investigated beginning with the study by Unzoeld. The decay of bound state occurs when the values of the field intensity $\mathbf{F}$ exceed some critical value $\mathbf{F}^*$. The probability of the existence of this bound level is $\omega = \int_0^{\varepsilon^*} \rho(\varepsilon) d\varepsilon$, where $\varepsilon = \mathbf{F}/\mathbf{F}_0$. The probability of decay of a given state is $(1-\omega)$. Sevast'yanenko [474] calculated these integrals and used them to determine the populations of bound states:

$$N_n = N_a(g_n^*/\Sigma^*)\exp(-E_n/T)\,,$$
$$g_n^* = \omega_n g_n\,, \qquad \Sigma^* = \sum_n \omega_n g_n \exp(-E_n/T)\,.$$

Figure 8.9 illustrates the comparison of the calculation results [474] with those of the experiments in argon [483]. From the drawing, one can see how important the microfields are. The clearly defined spectral lines lose their intensity. It is shown below that the intensity of the continuous spectrum adjoining the spectral lines increases accordingly.

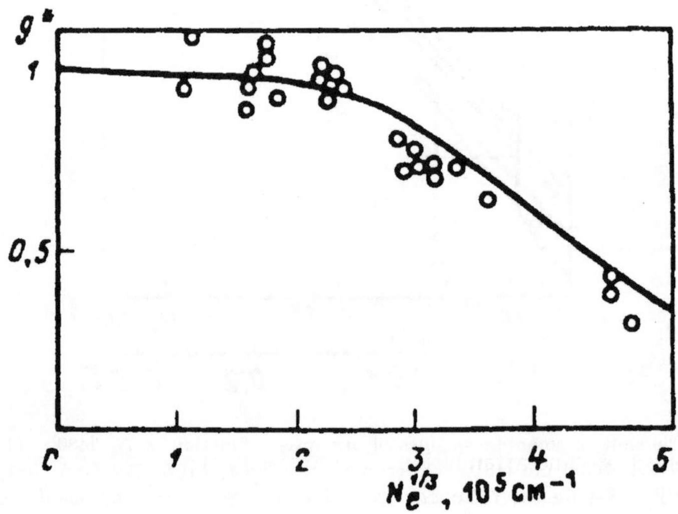

Fig. 8.9. Variation of the effective partition function of the higher level of $5d-4p$ transition in argon ($\lambda = 560$ nm): points, experiment [483]; curve, theory [474].

8.5. The Principle of Spectroscopic Stability

The principle of spectroscopic stability consists of the conservation of the density of oscillator strengths df/dE of discrete and continuous spectra during the superposition of external perturbations (such as interparticle interactions). Physically, this implies that the interaction perturbs the radiators, but does not annihilate them and change their total number. According to this principle, it is assumed that the effect of interaction on the threshold spectra is reduced to the higher terms of the spectral series transform to a continuous spectrum. This is in accordance with the unperturbed density of their oscillator strengths. Therefore, the photoionization cross-section is extrapolated to frequencies less than the ideal threshold frequency according to the behavior of df/dE. For higher frequencies, the photoionization cross-section is assumed to be unchanged. The validity of this procedure is supported by a large body of available experimental data on weakly non-ideal plasma.

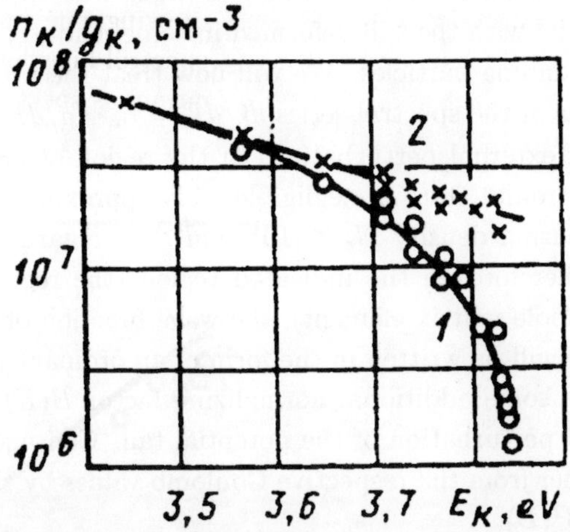

Fig. 8.10. The measured population of atomic levels in a cesium plasma, $N_e = 9.7 \cdot 10^{13}$ cm^{-3}, $T = 0.274$ eV [484]: *1*, without correction; *2*, with correction for the line broadening and slit width.

The effect of "line attenuation" due to the static destruction of the radiating levels by plasma microfields (Fig. 8.9) does not contradict the principle of spectroscopic stability. A continuous background emerges instead of the levels being destroyed. The amplification of the background compensates for the attenuation of the lines. Figure 8.10 gives two distributions of excited atoms over levels measured in a pulsed discharge in cesium vapors. Curve *1* demonstrates the manifestation of line attenuation which is analogous to that demonstrated in Fig. 8.9. Curve *2* corresponds to spectroscopic stability. It is constructed with correction to the results of data processing. This is necessary because of the strong broadening of the lines and the finiteness of the spectral width of monochromator slits as compared to the half-width of the line being recorded.

Kobzev *et al.* [485] proposed that with an increase in density, the spectroscopic stability would be lost and the density of oscillator strengths would cease to persist. The latter must also show up in observations. At the same time, it was proven in [486–488] that this principle could not be violated in a one-article potential approximation. Deviations from this principle must be associated with the effects of many-particle interaction. Any explicit replacement of the many-particle description by some effective interaction in a one-particle potential approximation leads to spectroscopic stability.

We will follow D'yachkov [487] and cite the proof of spectroscopic stability within the model with a central effective potential. The effective potential is a Coulomb potential with the tail deformed in some manner, which allows for the effect of surrounding particles. We will now treat the averaged density of oscillator strengths of the spectral series $df/dE = f_n \cdot dn/dE$.

If the effect of external perturbations in the region of localization of the wave function of ground state is negligible, this approximation appears reasonable for the plasma density $N_e \leq 10^{19}$ cm^{-3} as regards the Lyman and Balmer series. Therefore, in the indicated region (the region of substantial integration in a dipole matrix element), the wave function of the higher state with the energy E will be written in the form of an ordinary regular Coulomb function, but with some additional normalizing factor $B(E)$, which depends on the behavior of perturbation of the potential tail. Consequently, the oscillator strengths differ from the respective Coulomb values by the square of this factor: $f_n = f_{n\text{C}} B^2(E)$.

As a result of perturbation, the density of states dn/dE will also vary. Therefore,

$$df/dE = (df_\text{C}/dE) B^2 (dn/dE)(dn_\text{C}/dE)^{-1} .$$

It may be readily demonstrated that in a quasi-classical approximation, this yields

$$df/dE = df_\text{C} dE . \tag{8.21}$$

Indeed, the normalized quasi-classical wave function of bound state has the form

$$\psi = a p^{-1/2} \cos\left((h/2\pi) \int_{r_1}^{r_2} p dr' - \beta \right) , \qquad a = (2m\Omega/\pi)^{1/2} ,$$

where p is the quasi-classical momentum, Ω is the frequency of classical periodic motion between the turning points r_1 and r_2, and β is some phase. Consequently, $B^2 = (a/a_\text{C})2 = \Omega/\Omega_\text{C}$. The density of states is the inverse distance between levels ΔE^{-1}. However, the Bohr-Sommerfeld quantum rule may readily yield $\Delta E = \hbar\Omega$. Therefore, $dn/dE = (\hbar\Omega)^{-1}$ and, consequently, $B^2 (dn/dE)(dn_\text{C}/dE)^{-1} = 1$. As a result, we derive relation (8.20), which is indicative of the spectroscopic stability of the spectral series.

The most simple way of extending the derived result to cover a continuous spectrum is by placing an atom in a spherical box with a large radius. Relation (8.20) will then automatically extend to cover the region above the photoionization threshold.

As a consequence of the unitary value of the transformation of the final and initial states [489], the principle of spectroscopic stability was proven by D'yachkov [487] for the microfield model, that is, for an atom in a constant, uniform (on the atom scale) external field.

D'yachkov *et al.* [488] have calculated the threshold spectra of hydrogen within the microfield model without dividing them into discrete and continuous components, which is the usual practice, and without assuming spectroscopic stability. A spectrum is represented in the form of a single continuum with the Stark resonances forming lines in the latter. Therefore, the spectroscopic stability in the microfield model is derived by calculation rather than by entry into the latter. Figure 8.11 provides a comparison of experiment with theory.

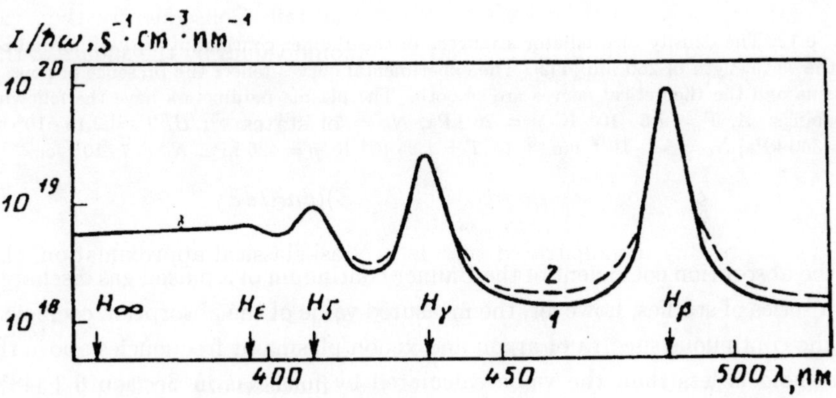

Fig. 8.11. The radiation spectrum of a hydrogen plasma, $N_e = 9.3 \cdot 10^{16}$ cm^{-3}, $T = 1.41 \cdot 10^4$ K [488]: *1*, experiment [490]; *2*, theory [488].

8.6. Continuous Spectra of Strongly Non-ideal Plasma

For small non-ideality, it affects a continuous spectrum within the ideology of spectroscopic instability. It has been revealed that no other manifestations of non-ideality are observed in hydrogen at $N_e \leq 7 \cdot 10^{17}$ cm^{-3} [491, 492] (Fig. 8.12). The Coulomb density of oscillator strengths agrees well with the "experimental" value. The latter was derived from the results of measurement

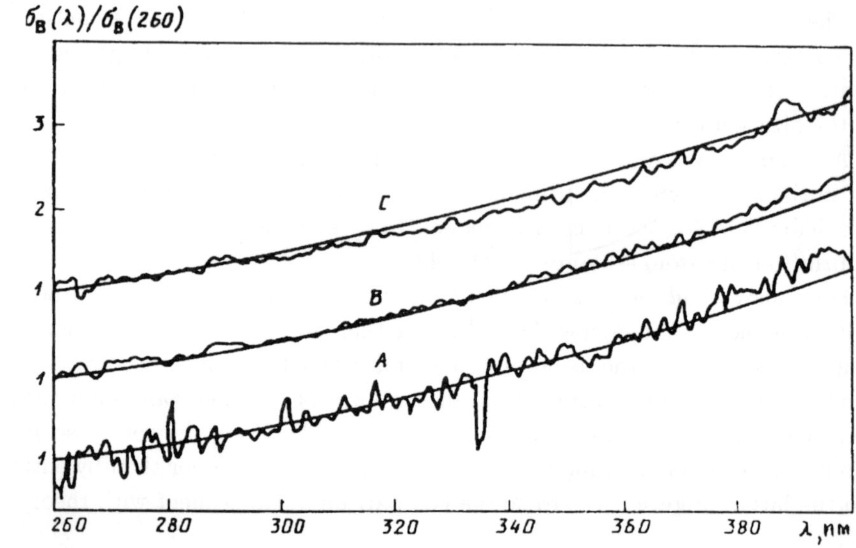

Fig. 8.12. The density of oscillator strengths of the Balmer continuum, $\sigma_B(\lambda)$, normalized to the wavelength of 260 nm [492]. The experimental curves reflect the presence of noise in plasma and the theoretical curves are smooth. The plasma parameters have the following meanings: A, $T = 1.5 \cdot 10^4$ K, $p = 26$ kPa, $N_e = 5 \cdot 10^{16}$ cm^{-3}; B, $T = 2.15 \cdot 10^4$ K, $p = 340$ kPa, $N_e = 5.4 \cdot 10^{17}$ cm^{-3}; C, $T = 1.9 \cdot 10^4$ K, $p = 426$ kPa, $N_e = 7 \cdot 10^{17}$ cm^{-3}.

of the absorption coefficient of the Balmer continuum of a pulsed gas discharge. In a series of studies, however, the measured value of the absorption coefficient in the continuous spectra of argon and xenon plasma at frequencies above the threshold is less than the value calculated by formulas in Section 8.2 [485]. Apparently, these effects were first recorded in [493, 494] at $N_e \geq 10^{18}$ cm^{-3} in xenon.

In argon [151], a wide and continuously varying range of parameters was realized from $N_e \leq 10^{18}$ cm^{-3} and $\Gamma \leq 0.3$ to extremely high parameters $N_e \leq 2 \cdot 10^{20}$ cm^{-3} and $p \approx 0.5$ GPa, where $\Gamma = (1-6)$, in Fig. 8.13. The absorption in the free-free transitions of electrons in fields of ions is described by the Kramers' formula (curve *1*). The inclusion of screening leads to a reduction in absorption (curve *3*). In order to derive the overall absorption coefficient (curve *2*), one must take into account, along with free-free transitions, the photoionization of bound states. One can see that the experimental data in the low-density region are in agreement with the results of numerous

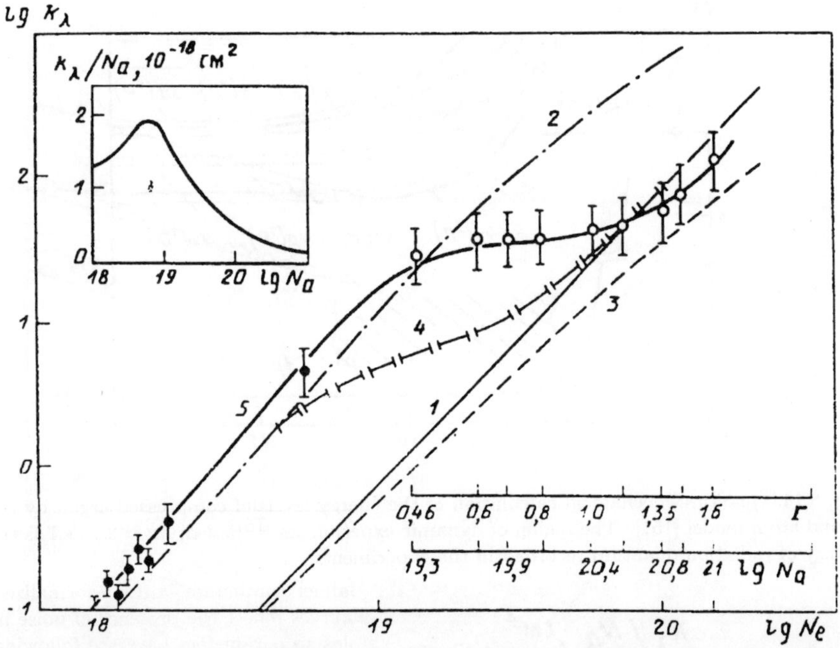

Fig. 8.13. The absorption coefficient of a shock-compressed argon plasma, $T = 20,000$ K [151]. Experiment: ∘, [151]; •, data by other researchers (see [495]). Theory: *1*, free-free transitions; *2*, photoionization included; *3*, free-free transitions in view of plasma screening; *4*, calculation by the bound atom model; *5*, the curve joins the experimental points reduced to the radiation frequency $h\nu = 1.8$ eV and temperature $T = 2 \cdot 10^4$ K.

measurements obtained previously [495]. As the plasma is compressed, the specific absorptivity (k/N_a) decreases monotonically.

The recorded effect was interpreted by Bespalov *et al.* [151] as a result of deformation of the atomic energy spectrum. The interaction brings about a successive transition of excited states to the continuous spectrum as the plasma density increases. A strong deformation of the energy spectrum may be accompanied by changes in the matrix elements of optical transitions.

In order to describe this effect, Bespalov *et al.* [151] used the bound atom model (see Section 5.5). The behavior of deformation of the energy spectrum is illustrated in Fig. 8.14, where one finds the range of energy of the quanta recorded in the experiments, $h\nu = 2.14$ eV, which may cause photoionization from the respective energy states. Electron wave functions were found for these conditions and the photoionization cross-sections of bound atom were

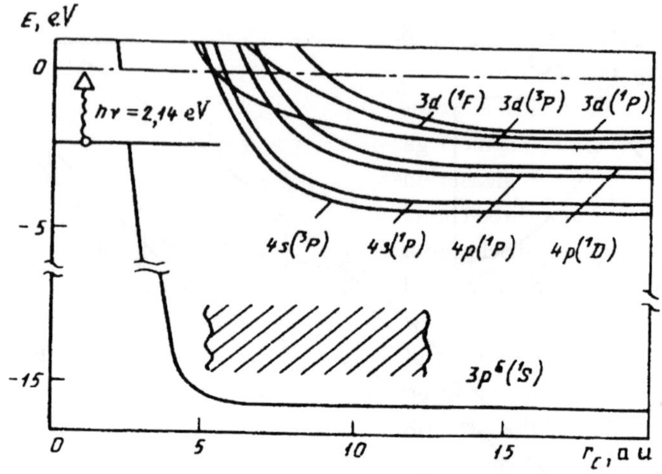

Fig. 8.14. Quantum-mechanical calculation of the energy levels of compressed argon by the bound atom model [151]. The region of dynamic experiments is shaded; $h\nu = 2.14$ eV is the energy of radiation quantum recorded in the experiments.

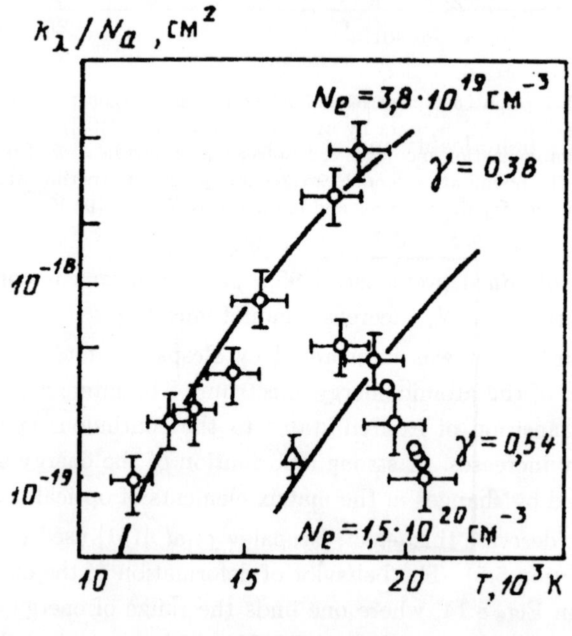

Fig. 8.15. The absorption coefficient in a continuous spectrum, per single atom. Measurement results: •, $\lambda = 444$ nm, xenon [497]; Δ, $\lambda = 580$ nm, argon [496]; $N_e = 4.8 \cdot 10^{18}$ cm^{-3}, $\gamma = 0.21$; ∘, $\lambda = 580$ nm, argon [151]. Calculation: curves.

calculated. One can see (Fig. 8.13) that the theory reproduces the effect recorded in the experiments.

The situation, which is formed with regard to the measurements of the continuous spectra of an inert gas plasma, is discussed in [468]. Figure 8.15 gives the results of the measurements of the absorption coefficient (per single atom) performed in a three-chamber pneumatic shock tube in xenon [497] and in a plasma of pulsed discharge in argon [496]. The results of the shock wave measurements are also shown [151]. They are compared with the results of the calculation for the ideal plasma which, in xenon at $N_e = 3.8 \cdot 10^{19}$ cm^{-3}, agree well with the measurement results. As was obvious from previous discussion, such an agreement was not observed in [151].

The coefficient of the reflection of laser radiation, which has a wavelength of 1.06 μm from a non-ideal shock-compressed xenon plasma with the density $0.5 \le \rho \le 4$ g/cm^{-3} and temperature $T \ge 3 \cdot 10^4$ K, was measured by Fortov *et al.* [499].

Ng *et al.* [201] measured the reflection of laser radiation from an expanding aluminum plasma, which was formed after the arrival of a powerful laser-induced shock wave to a free surface. As in [499], high values of the reflection coefficient were recorded.

The measurement results of [499], given in Fig. 8.16, point to the plasma transition as the non-ideality increases from a weakly reflecting state (ideal

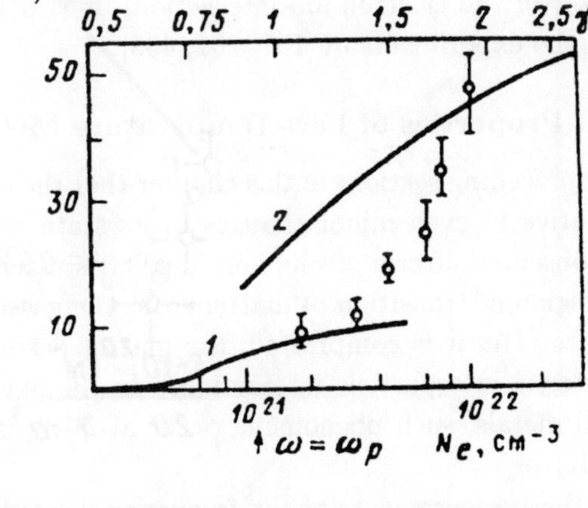

Fig. 8.16. The coefficient of reflection from a non-ideal xenon plasma. Experiment [499]: points. Theory [500]: *1*, extrapolation from the $\gamma \ll 1$ region; *2*, extrapolation from the $\gamma \gg 1$ region.

plasma) to one with high reflectivity (non-ideal plasma). The transition is realized in the range of $\gamma \approx 1$ and $\omega/\omega_p \approx 1$, and cannot be described within any single theoretical model. The curves in Fig. 8.16 are extrapolations to the transition region of the expressions that are valid for an ideal plasma (curve *1* is the extrapolation from the region where $\gamma \ll 1$ and $\omega_p \ll \omega$) and for a strongly non-ideal plasma (curve *2* is the extrapolation from the region where $\gamma \ll 1$ and $\omega_p \gg \omega$). The reflection coefficient was calculated by the Fresnel's formula, $R = |(\sqrt{\varepsilon} - 1)/(\sqrt{\varepsilon} + 1)|$, $\varepsilon = \varepsilon' + i4\pi\sigma/\omega$, where ω is the probing light frequency. The coefficient of high-frequency electrical conductivity $\sigma(\omega)$ was related to the known bremsstrahlung intensity, and the latter was used to derive the dielectric permeability, $\varepsilon'(\omega)$.

The theory of bremsstrahlung and recombination radiation of non-ideal plasma [501–505] allows for the increasing effect made on radiation by the increase in the interion correlation and screening as the non-ideality parameter increases. The radiation is suppressed at low frequency, where it becomes less intensive than its value calculated by the Kramers' formula. This suppression of the radiation is due to the fact that the portions of electron trajectories, on which low frequencies are radiated, are not realized in non-ideal plasma. Two effects are responsible for this: screening and correlation. Artem'ev and Iakubov [503] have cited the characteristic frequency below which the continuum is suppressed,

$$\omega \leq \omega^* \gamma_{ei}^{3/2} , \tag{8.22}$$

where $\omega^* = (6\pi e^2 N_e/m)^{1/2}$ is a frequency close to the plasma frequency and γ_{ei} is the parameter of the electron-ion interaction. It is believed that these effects show up in the experiments of [151, 201, 499].

8.7. The Optical Properties of Low-Temperature Metal Plasma

It follows from the preceding sections of this chapter that the optical properties of matter are sensitive to even minor changes in its state, with qualitatively new effects being observed after a strong non-ideality is reached. This is even more true of the region of transition of matter from the gaseous plasma state to liquid-metal state. (As it is compressed, the matter passes through states characterized by strong interparticle interaction.) In the plasma of vapors of mercury and alkali metals, such phenomena occur at fairly low temperatures (of close to 2,000 K).

The results of the measurements of the frequency dependencies of the coefficients of optical absorption, as well as the reflection for mercury in the "dielectric-metal" transition region, were presented in Chapter 2. (in Fig. 2.16, where the measurement results demonstrate the variation of the optical radiation reflection coefficient). The dependencies characteristic of metal have been obtained at $\rho \geq 12$ g/cm^3. At low frequencies, the reflection coefficient R reaches approximately 90%. It also decreases with density; this is accompanied by a variation of the frequency dependence. At a density of $\rho \sim 9$ g/cm^3, the supercritical mercury metallizes and a maximum value is observed on the $R(\omega)$ curve. This is characteristic of matter with an energy gap. At low and high frequencies, the reflection coefficient drops [59].

Now, the frequency dependence of the absorption coefficient for mercury, $k(\omega)$, at subcritical densities (Fig. 2.15) bears no resemblance to that characteristic of metal. First of all, one's attention is arrested by the sharp edge of wide absorption band shifting towards the red with an increase in density. At low frequencies ($\hbar\omega \leq 1$ eV), with an increase in density and a decrease in temperature, the $k(\omega)$ relationship acquires a plateau. We now discuss these two main peculiarities of $k(\omega)$.

An optical transition between two atomic levels in rarefied gas causes absorption in the spectral line. The spectral line in gas is quite narrow. The line width and shift, as well as the frequency dependence of the absorption coefficient $k(\omega)$, are defined by the interaction processes perturbing the high and low levels.

One can assume that the high-frequency component of the dense mercury spectrum is the edge of a strongly broadened band which emerges upon the transition of mercury atoms from the ground $6s$-state to exciton states. The exciton states occur as a result of the evolution of the first excited states ($6p$ and $7s$) of mercury atoms upon their interaction with the environment, that is, with fluctuations in the density of atoms in the ground state [506]. First, these interactions serve to reduce ΔE_{opt}, the minimum excitation energy, to the so-called optical gap. For an isolated atom, this gap is $E_{\text{exc}} = 4.8$ eV. Secondly, the absorption band broadens strongly with an increase in density. This is because as a result of the interaction with radiation, an electron passes to a greater orbit and finds itself in a cluster containing, with a fairly wide scatter, some or other number of atoms. Therefore, the density of bound electron states is defined by the interaction with its own ion and with the atoms of the medium. Following the reasoning in Section 6.3, one can conclude that in a first approximation,

$$\rho(\hbar\omega) \sim \exp[-(\hbar\omega - \Delta E_{\text{opt}})^2/\Delta E_t^2] \tag{8.23}$$

where ΔE_t is the half-width of absorption band. It is obvious that the $E_{\rm opt}$ and E_t values, which are divided by the average number of atoms with which the bound electron interacts, should be of the order of 0.1 eV (such is the scale of the energy of an electron's interaction with a mercury atom). Assuming that $k(\omega) \sim \rho(\hbar\omega)$, Bhatt and Rice [506] and Popielawski *et al.* [507] performed an evaluation by fitting Eq. (8.22) to the observed $k(\omega)$ relationship, Fig. 2.16. The $\Delta E_{\rm opt}$ and ΔE_t values per single atom are estimated at 0.37 eV and 0.18 eV, respectively.

The experimental data enabled Uchtmann and Hensel [63, 508] to plot the density dependence of $\Delta E_{\rm opt}$ (Fig. 2.13). At low densities, it can be readily extrapolated to the value of mercury atom excitation energy $E_{\rm exc}$. It is very important that the $E_{\rm opt}$ gap decreases as the density increases, and that it closes at $\rho \approx \rho_c$, that is, at the same density as the transport gap ΔE. Therefore, the high-frequency component of spectrum has found a convincing interpretation.

We now discuss the low-frequency plateau of the absorption coefficient. For a long time, no explanation could be found for the emergence of a plateau although the behavior of the density and temperature dependencies indicated clearly that this was a new optical process taking place in a disordered, dense metallic medium. The results of additional measurements, shown in Fig. 8.17, led one to conclude [509] that the plateau had emerged as a result of radiation absorption by charged clusters. Roughly, the plateau in Fig. 8.18 is the new

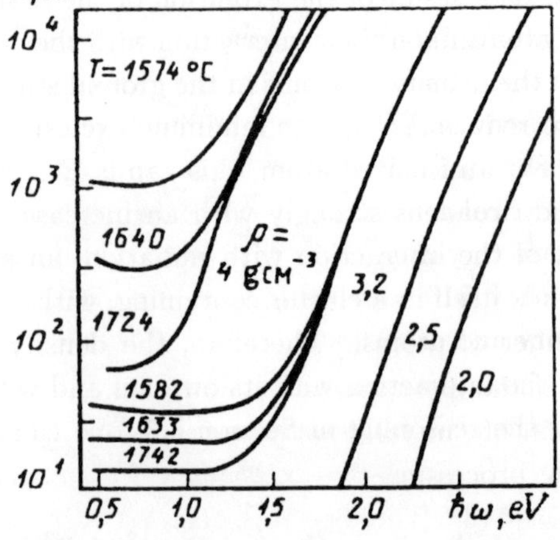

Fig. 8.17. Optical absorption coefficient for mercury [509].

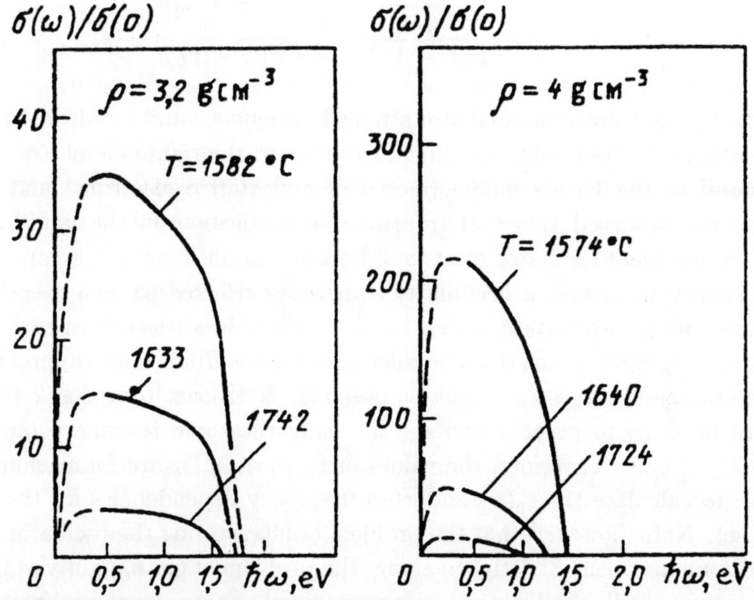

Fig. 8.18. The $\sigma(\omega)/\sigma(0)$ relationship for gaseous dense mercury at different temperatures and densities [509].

absorption band. It follows from Fig. 8.17 that $k(\omega)$ increases with an increase in density and a decrease in temperature on the plateau. The same is true of the cluster concentration (cf., Section 4.3).

Naturally, this gave an absorption band in the neighborhood of the eigenfrequency of an electron bound in a cluster. The conclusions by Uchtmann *et al.* were based directly on the results by Lagar'kov and Sarychev [510]. The latter had calculated $\sigma(\omega)$ using the same techniques employed by them to study the static electrical conductivity $\sigma(0)$ (Chapter 6). The $\sigma(\omega)/\sigma(0)$ relationships obtained are very close to those presented in Fig. 8.18. The results by Lagar'kov and Sarychev [510] lead one to conclude that a qualitative agreement exists between theory and experiment.

We now consider the dielectric permeability, $\varepsilon(\omega)$, of a non-ideal mercury plasma in the region of "metal-dielectric" transition. The $\varepsilon(\omega)$ relationship can be obtained by processing the experimental transmission and reflection coefficients [505, 511–513]. Verleur [511] demonstrated the possibilities of a relatively simple approach based on an approximation of the optical properties of matter by a set of damped classical oscillators:

$$\varepsilon(\omega) = \varepsilon_r + i\varepsilon_i = 1 - \frac{\nu_{po}^2}{\omega(\omega + i\Gamma_0)} - \sum_{\ell \geq 1} \frac{f_\ell \nu_\ell^2}{\omega(\omega + i\Gamma_0) - \nu_\ell^2} \,. \tag{8.24}$$

Here, f_ℓ, ν_ℓ, Γ_ℓ denote the oscillator strength, frequency and bandwidth of the ℓ-th oscillator, respectively. The first two terms in the right side of Eq. (8.23) correspond to the Drude model (model of almost free electrons) and must describe the observed values of transmission coefficients in the liquid-metal region. Subsequent oscillator terms will become significant when a gap occurs in the energy spectrum, a peculiarity sometimes referred to as a pseudogap. The values of the parameters ν_{po}, ν_ℓ, Γ_ℓ, f_ℓ are the values selected for each point in the $\{\rho, T\}$ diagram, such that the calculated curves $R(\omega)$ should approximate the experimental curves as closely as possible. It follows from Fig. 2.16 that this can be done to great advantage by using the three resonance terms in Eq. (8.23), $\ell \leq 3$. Therefore, the values of ν_{po}, ν_ℓ, Γ_ℓ, f_ℓ are found, and it is possible to calculate the $\varepsilon_r(\omega)$ and $\sigma(\omega)$ frequency dependencies for the same $\{\rho, T\}$ set. Note, however, that the problem of interpreting the oscillators thus derived remains open. Strictly speaking, the problem of unambiguity of fitting remains open. Still, the fitting accuracy attained appears very convincing.

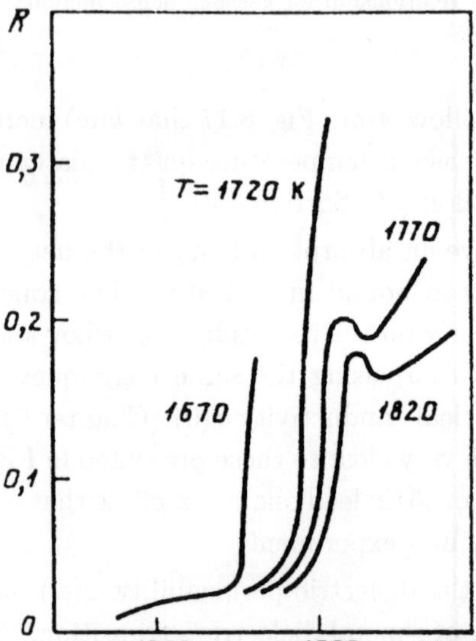

Fig. 8.19. The frequency dependence of the electrical conductivity of mercury obtained from the experimental values of $R(\omega)$ for the same $p - \rho - T$ as in Fig. 2.16.

Figure 8.19 shows the calculated $\sigma(\omega)$ curves. At $\rho \geq 12$ g/cm^3, they correspond rather well to almost free electrons (Drude model). However, even at 11 g/cm^{-3}, qualitative changes are observed which point to a partial loss of metallic properties. At $\rho < 8$ g/cm^{-3}, the quantity ν_{po}^2 disappears altogether and $\sigma(\omega)$ is fully described by the resonance terms. This means that the "metal–non-metal" transition is already accomplished. The values of $\sigma(\omega)$ are directly related to those of $k(\omega)$. Indeed, it turns out that the $\sigma(\omega)$ values on the isochor of 5 g/cm^{-3} presented in Fig. 8.19 agree well with those obtained from $k(\omega)$ in Fig. 2.15. Note further that the static conductivity $\sigma(0) = \nu_{po}^2/(4\pi\nu_0)$ agrees well with the results of static measurements (shown by arrows in Fig. 8.19). In this manner, the results of the $R(\omega)$, $k(\omega)$, $\sigma(\omega)$ measurements are co-ordinated with each other.

It is possible that the most interesting results were obtained at low densities in a dense plasma of mercury vapors [63, 512, 514] (Fig. 8.20). The reflection coefficient increases anomalously at the density values above $3 \cdot g/\text{cm}^3$. The

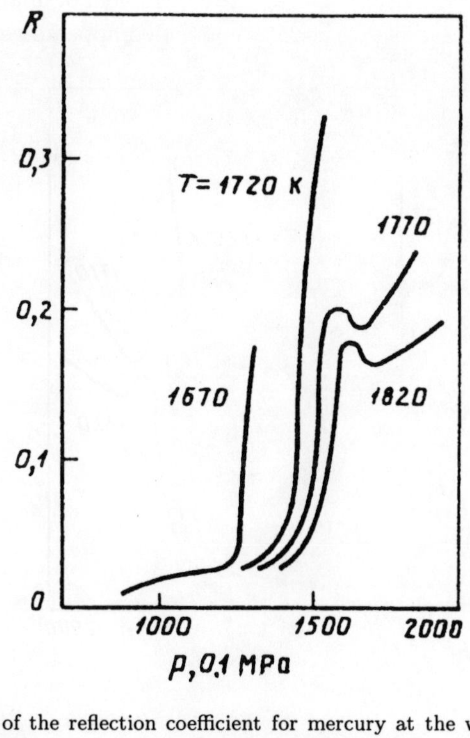

Fig. 8.20. Isotherms of the reflection coefficient for mercury at the wavelength of 650 nm [514].

value of ε_r in this (still far from the critical) region increases accordingly and the $\varepsilon_r(p)$ relationship passes through a maximum. This relationship does not conform to the Clausius-Mosotti formula, which points to a smooth and minor rise of ε_r to 1.6. Instead, in a narrow density range, ε_r passes through a maximum to increase approximately fourfold at the density values where the $k(\omega)$ plateau occurs. The latter leads one to conclude that the dielectric anomaly is likewise related to the transition of charged particles to clusterized states. Moreover, there are good reasons to believe that the clusters are metallized near the saturation line, an assumption made earlier from general considerations. If this is so, the matter is a disordered, two-phase medium where the liquid-metal phase is represented by charge-stabilized, finely dispersed droplets.

Nine

Non-Ideal Plasma with Condensed Disperse Phase (CDP)

9.1. The Classification of States of Plasma with CDP

The emergence of particles of condensed disperse phase (CDP) in plasma leads, in a wide range of conditions, to a significant variation of the electrophysical properties of the plasma. By emitting or absorbing electrons, the condensed particles may vary the concentration of electron in plasma. In doing so, the macroscopic particles may be charged to very high values and as a result of strong interparticle interaction, the system becomes non-ideal.

The interest in the physics of low-temperature plasma with CDP has been sustained owing to a number of new applications. These include the electrophysics of the products of rocket fuel combustion, solid fuel-fired magnetohydrodynamic generators, the technology of plasma sputtering and plasma treatment of materials, and the physics of artificial plasma formations in the atmosphere and in outer space. The plasma with CDP is a surface plasma which occurs under the conditions of high energy contributions to matter and the electrode plasma of some electric discharges.

CDP particles may be introduced into a plasma by design or they may occur as a result of the following: volume condensation upon the efflux of plasma from nozzles; upon the expansion of plasma in the channel of an MHD generator, upon the expansion of aggregates of matter into a void; and in the products of ablation of solid surfaces subjected to the effect of high energy fluxes.

Of the studies devoted to the physics of CDP plasma, special note should be made of [26, 515–518]. As recently as 10 years ago, one almost always referred to the plasma of solid fuel combustion products when speaking about the CDP plasma. Such a plasma is specific in that, thanks to the high values of particle charges, the entire range of states from the Debye plasma to the strongly non-ideal system of charge particles are realized in the plasma under standard conditions ($T = 2,000$–$3,000$ K, $N_e = 10^8$–10^{14} cm^{-3}).

The electron concentration in plasma is defined by ionization of gas and thermal emission. In this section, we shall consider the limiting case, that is, a system of particles and electrons emitted by the latter. Such a system, which characterizes the specific nature of plasma with CDP most clearly, is realized if the atoms or molecules of gas, in which the CDP particles are found, are not ionizable.

The system being considered is characterized by three parameters of the Coulomb interaction: interparticle $\gamma_{pp} = Z^2 N_p^{1/3} e^2/T$, inter-electron $\gamma_{ee} = e^2 N_e^{1/3}/T$ and particle/electron interaction $\gamma_{pe} = Ze^2/\bar{r}T$, where $\bar{r}$ is the mean distance between electron and particle, Z is the mean particle charge, and N_p and N_e are the particle and electron concentrations, respectively. Irrespective of non-ideality, the system may be quasi-homogeneous so that $N_e = ZN_p$, $\bar{r} \approx N_p^{-1/3}$, and if $Z \gg 1$, the inequalities

$$\gamma_{ee} \ll \gamma_{pe} \ll \gamma_{pp} \tag{9.1}$$

are satisfied. Therefore, if $\gamma_{pp} \ll 1$, the whole system is close to ideal: the plasma with CDP is a Debye plasma. If is more naturally characterized by the parameter

$$\begin{aligned} \Gamma_{pp} &= Z^2 e^2/r_{\mathrm{D}} T = \sqrt{4\pi}\gamma_{pp}^{3/2} \\ r_{\mathrm{D}}^{-1} &= [4\pi e^2 Z(Z+1) N_p/T]^{1/2}, \end{aligned} \tag{9.2}$$

where r_{D} is the Debye screening distance. In Fig. 9.1, a curve connects the points at which $\Gamma_{pp} = 1$. Below this curve, the ideal-gas approximation with Debye corrections is valid.

If $\Gamma_{pp} \geq 1$, the system is non-ideal. This is due, primarily, to interparticle interactions and, secondly, to particle/electron interactions. These interactions will be allowed for. Because $\gamma_{ee} = \gamma_{pp} Z^{-5/3}$, the electron gas is non-ideal and non-degenerate, $\gamma_{ee} \ll 1$.

If $\gamma_{pp} \gg 1$, the strong interparticle correlation will lead to the establishment of a short-range order between the particles. Under these conditions, the Wigner–Seitz cell model is valid. It is possible that the boundary separating the unordered and ordered systems of particles lies in the neighborhood of $\gamma_{pp} = 4$ (curve *2* of Fig. 9.1). This number corresponds to the occurrence of a short-range order in a single-component model of an electron-ion plasma (cf., Section 5.1).

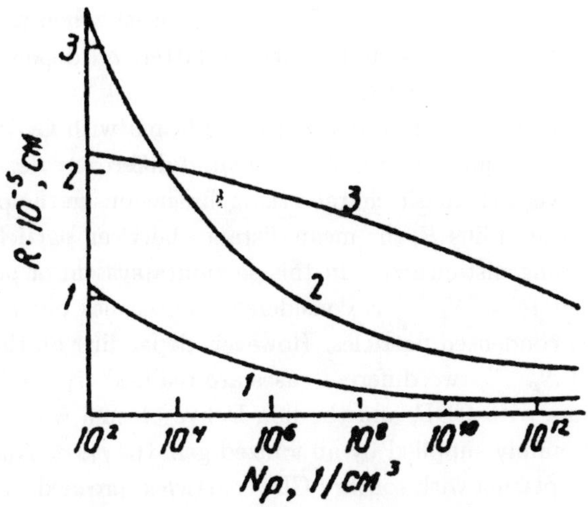

Fig. 9.1. Regions of different states of plasma with CDP [519], $T = 2,500$ K, $W = 2.3$ eV: Curves *1* and *2* correspond to $\Gamma_{pp} = 1$ and $\gamma_{pp} = 4$, respectively; curve *3* corresponds to $æ_s R = (2L)^{1/2}$.

The possibility of strong inhomogeneity in the electron distribution in the neighborhood of particles should be regarded as another peculiar property of a non-ideal plasma with CDP. In Fig. 9.1, the regions of weak and strong inhomogeneities are separated by curve *3*. Above that curve, the charge is strongly screened in the strong inhomogeneity region. This leads to its renormalization, $Z^* < Z$. The renormalization renders the electron system quasi-homogeneous, where curve *3* corresponds to the condition $R = (2L^{1/2})r_{\rm Ds}$ [cf., Eq. (9.59)] and L is some logarithm and $r_{\rm Ds}$. is the Debye's electron radius defined by the electron concentration near the particle surface. Therefore, curve *3* separates the regions of particles with small and great radii.

9.2. Ionization Equilibrium in Weakly Non-ideal Plasma with CDP

The magnitude of the electron concentration N_e in a plasma consisting of ionized gas and CDP is governed by two processes. The first resides in the ionization of gas molecules and recombination, and the second is related to thermal emission and adherence of electrons to the particle surface. The presence of CDP may lead either to an increase in N_e, if the particles assume a positive charge, or to its decrease, if the charge is negative. It is evident that if the particle concentration is sufficiently low, N_e will only depend on the ionization processes in gas. In the other limiting case, when the plasma consists of particles and electrons emitted by the latter, N_e depends only on the emission properties of the particles.

The conditions which can be realized in a plasma with CDP vary greatly. Thus, the ionization equilibrium must be treated differently depending on the relationships between the basic characteristic dimensions in the plasma. These include the particle radius R, the mean distance between particles, $N_p^{1/3}$, and the Debye screening distance r_{D}. In this section a system of particles which takes place when $R \ll N_p^{-1/3}$ is considered. This rules out any immediate contact between condensed particles. However, depending on the relationship between r_{D} and $N_p^{-1/3}$, two different cases are realized: $r_{\mathrm{D}} \gg N_p^{-1/3}$, that is the Debye plasma with CDP, and the non-Debye plasma when $r_{\mathrm{D}} \ll N_p^{-1/3}$. If electrons are mainly supplied by an ionized gas, the $r_{\mathrm{D}} \ll N_p^{-1/3}$ condition corresponds to a plasma with solitary CDP particles screened by the electrons and ions of the plasma. If the gas is practically non-ionizable, the $r_{\mathrm{D}} \ll N_p^{-1/3}$ condition leads to a highly non-ideal system of particles referred to above and discussed in Section 9.3.

We now consider a plasma with CDP where the particle radius R, screening distance r_{D}. and the mean distance between particles are related to each other by the following inequalities:

$$R \ll N_p^{-1/3} \ll r_{\mathrm{D}} \,. \tag{9.3}$$

This means that the CDP particles can be regarded as point particles, which are similar to multiply charged heavy ions. Furthermore, the number of particles in the Debye sphere is great and they participate in screening, $r_{\mathrm{D}}. = [T/4\pi e^2(N_e + N_i + Z^2 N_p)]^{1/2}$. Consequently, a plasma with CDP can be regarded as a mixture of ideal gases in chemical equilibrium.

The statistical equilibrium in a system of CDP particles and electrons emitted by these particles, along with the charge distribution of particles, was treated in [520–526]:

$$P_{z-1} \rightleftarrows P_z + e \,. \tag{9.4}$$

Here, P_z is a particle carrying a charge Z. The particles are assumed to be monodisperse. The equation of ionization equilibrium for Eq. (9.4) takes the form of

$$\frac{N_Z Ne}{N_{Z-1}} = \frac{2\Sigma_Z}{\Sigma_{Z-1}} \left(\frac{mT}{2\pi\hbar^2}\right)^{3/2} \exp(-\Phi_Z/T) \,. \tag{9.5}$$

Equation (9.5) is analogous to Saha's equation for the equilibrium between ions with charges Z and $(Z-1)$. N_Z and N_{Z-1} denote the concentrations of particles with the charges Z and $Z-1$. The Σ_Z/Σ_{Z-1} ratio is the ratio of their internal statistical sums. The quantity Φ_Z is equal to the work done to transfer an electron inside a particle having a charge Z to infinity. It includes the work function, the work done to overcome the electrostatic barrier, and a correction to the work function for the finite radius of particle. Then [520, 521],

$$\Phi_Z = W + \left(Z - \frac{1}{2}\right) e^2/R \,, \tag{9.6}$$

where W is the work function of electron for an uncharged plane sample. Upon introducing the Richardson concentration of electrons,

$$N_{es} = 2(mT/2\pi\hbar^2)^{3/2} \exp(-WT) \,, \tag{9.7}$$

we shall rewrite Eq. (9.5) as

$$N_Z = N_{Z-1}(N_{es}/N_e) \exp[-(\Phi_Z - W)/T] \,. \tag{9.8}$$

In Eq. (9.8), the ratio Σ_Z/Σ_{Z-1} was assumed to be equal to unity. This appears justified in the case of a microscopic metal particle, where the number of electrons is so great that the addition of yet another electron does not affect its electronic statistical sum appreciably. Upon the repeated use of Eq. (9.8), we obtain

$$N_Z = N_0 \left(\frac{N_{es}}{N_e}\right)^Z \prod_{j=1}^{Z} \exp\left(-\frac{\Phi_j - W}{T}\right) = N_0 \left(\frac{N_{es}}{N_e}\right)^Z \exp\left(-\frac{\alpha Z^2}{2}\right) , \tag{9.9}$$

where N_0 is the concentration of particles which remained uncharged. Upon introducing the quantities

$$Z_0 = \frac{RT}{e^2} \ln \frac{N_{es}}{N_e} \,, \qquad \alpha = \frac{e^2}{RT} \,. \tag{9.10}$$

Equation (9.9) is rewritten as

$$N_Z = N_0 \exp\left(\frac{\alpha}{2} Z_0^2\right) \exp\left[-\frac{\alpha}{2}(Z - Z_0)^2\right] \,. \tag{9.11}$$

Equation (9.11) gives the charge distribution of particles. This distribution has the form of a Gaussian exponential curve with a center at the $Z = Z_0$ point and a characteristic half-width of $\Delta Z = (2RT/e^2)^{1/2}$. In most cases that are of practical interest, the width of distribution (9.11) proves very small.

We now turn to the problem of determining the electron concentration. We shall write Saha's equation and the equation of quasi-neutrality as

$$\frac{N_e N_i}{N} = \frac{2\Sigma^+}{\Sigma}\left(\frac{mT}{2\pi\hbar^2}\right)^{3/2} \exp\left(-\frac{I}{T}\right) = K_1(T)\,; \tag{9.12}$$

$$N_e = N_i + ZN_p\,. \tag{9.13}$$

Here, N is the concentration of atoms, Σ^+/Σ is the ratio of the internal statistical sums of ion and atom, and I is the atomic ionization potential. If $N_e \gg ZN_p$, the particles which penetrate a plasma will not affect the electron concentration in the latter. Thus, if we assume, for the sake of simplicity, that $2\Sigma^+/\Sigma = 1$:

$$N_e = N_e^0 = \sqrt{N}(mT/2\pi\hbar^2)^{3/4}\exp(-I/2T) = \sqrt{NK_1}\,. \tag{9.14}$$

Upon substituting the expressions for N_{es} and N_e^0 in Eq. (9.10), we obtain

$$Z = -\frac{RT}{e^2}\left\{\frac{W - I/2}{T} + \ln\left[N^{1/2}\left(\frac{mT}{2\pi\hbar^2}\right)^{3/4}\right]\right\} \approx -\frac{R}{e^2}\left(W - \frac{1}{2}\right)\,. \tag{9.15}$$

The latter approximate equality is valid if $(W - I/2)/T \gg 1$. Hence it follows that, first, Z depends very slightly on temperature and, second, the charge sign depends on the sign of the difference $(W - I/2)$. Therefore, a drop in cesium in its vapor is charged positively while a drop of aluminum in alkali metal vapors is charged negatively.

If $W = 5$ and $I = 4$ eV, the charge of a drop with a radius of 10^{-2} μm is equal to $Z = -17$, the charge distribution of drops being fairly narrow, $\Delta Z < 2$. In doing so, $r_{\mathrm{D}} = 2.5\cdot10^{-5}$ cm if $T = 2.5\cdot10^3$ K and $N_e^0 = 10^{14}$ cm^{-3}, and, by virtue of the inequality (9.3), the particle concentration must be very high: $N_p \gg 10^{13}$ cm^{-3}. However, the $N_e \gg ZN_p$ condition is violated. Consequently, one should calculate N_e by solving the transcendental equation

$$N_e = \frac{RT}{e^2}N_p \ln\frac{N_{es}}{N_e} + (N_e^0)^2[N_e + K_1(T)]^{-1}\,, \tag{9.16}$$

where $K_1(T)$ is the constant of the ionization equilibrium described by the formula (9.12).

We now treat a plasma consisting of particles and electrons emitted by these particles in a difficult ionizable (buffer) gas. One can state in advance that the maximum permissible electron concentration cannot exceed the equilibrium electron concentration over a plane surface, N_{es}:

$$ZN_p = N_{es}\exp(-Ze^2/RT)\,. \tag{9.17}$$

which defines Z as a function of N_p. The work function W also presets the value of N_{es}.

Zhukhovitskii *et al.* [519] generalized the resultant expressions to mixtures of particles characterized by different sizes and work functions. When the mean distance between particles exceeds all other characteristic distances in plasma, the particles may be regarded as solitary. This is discussed in detail in [26, 158].

9.3. Calculation Results

The results of the calculations by Zhukhovitsky *et al.* [519] are presented in Fig. 9.2. The mean electron concentration, which depends on the total concentration of potassium atoms and ions, is shown. The values of the parameters are assumed as follows: $T = 2,500$ K, $W = 5$ eV and $I = 4.34$ eV. Figure 9.2a is plotted for particles with a radius $R = 10^{-6}$ cm at various particle concentrations. Figure 9.2b is plotted for particles with a greater radius, $R = 10^{-5}$ cm. The electron concentration N_e^0, calculated from Saha's formula, is shown for comparison.

Depending on the concentration of potassium atoms, N_K, three regions may be singled out. At low N_K, the plasma presents a system of positively charged particles and electrons emitted by the latter. The properties of the plasma are defined by CDP particles while the presence of potassium only results in minor corrections to N_e. To calculate the N_e in this region, Eq. (9.16) was

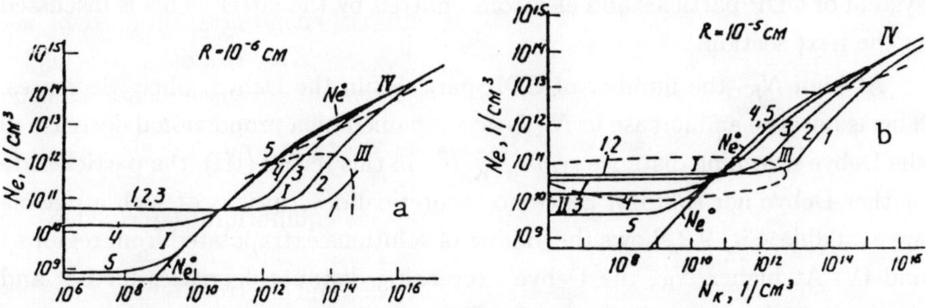

Fig. 9.2. N_e as a function of potassium concentration N_K in a plasma with particles of radii $R = 10^{-6}$ cm (a) and 10^{-5} cm (b) [519]. Particle concentrations N_P: *1*, 10^{14} cm^{-3}; *2*, 10^{13} cm^{-3}; *3*, 10^{12} cm^{-3}; *4*, 10^{10} cm^{-3}; *5*, 10^{8} cm^{-3}, all (a); *1*, 10^{12} cm^{-3}; *2*, 10^{11} cm^{-3}; *3*, 10^{10} cm^{-3}; *4*, 10^{9} cm^{-3}; *5*, 10^{8} cm^{-3}, all (b). The values of N_e^0 correspond to Saha's formula: $T = 2,500$ K, $W = 5$ eV.

used. Upon an increase in the particle concentration, N_e stabilizes at the value $N_{es} = 5 \cdot 10^{10}$ cm^{-3}, which corresponds to the Richardson formula (9.7). This is due to a decrease in the particle charge and its surface potential and, consequently, to a drop in the electrostatic addition to the work function.

At high potassium concentrations, the presence of particles has little effect on N_e. The values of N_e are close to N_e^0 while its deviations from N_e^0 are described by the formulas contained in the preceding section since the particles may be regarded as solitary.

At intermediate $N_{\rm K}$, a reversal of the particle charge occurs. This happens when the $N_e(N_{\rm K})$ curves cross the $N_e^0(N_{\rm K})$ curve. At higher $N_{\rm K}$, $Z < 0$ and $N_e < N_e^0$. Conversely, at lower $N_{\rm K}$, $Z > 0$ and the presence of particles increases the electron concentration.

Figure 9.2 can be subdivided into regions based on a different principle. Region I corresponds to the Debye plasma, with the number of particles in the Debye sphere being $(4\pi/3)N_p r_{\rm D}^3 > 1$. This condition agrees with the plasma ideality condition $\Gamma_{pp} < 1$ within a numerical factor. In its turn, the ideality condition means that only the Debye correlations are significant while the correlations over small distances may be ignored. Hence it follows that the space distribution of electrons may be regarded as quasi-homogeneous. In the case of sufficiently great radii R, the plasma turns non-Debye at both low and high $N_{\rm K}$. At low $N_{\rm K}$, the number of CDP particles in the Debye sphere decreases with N_p. This is because the Debye screening distance (9.2) decreases with an increase in the charge Z. Region II presents a non-ideal system of CDP particles and electrons emitted by the latter. This is discussed in the next section.

At high $N_{\rm K}$, the number of CDP particles in the Debye sphere decreases. This is because an increase in $N_{\rm K}$ is accompanied by a proportional decrease in the Debye screening distance $r_{\rm D}$ to $N_{\rm K}^{-1/4}$. In this region (III), the particles are neither Debye nor solitary. Since no theoretical descriptions of such a plasma are available, Fig. 9.2 shows the joining of solutions extrapolated from regions I and IV. At higher $N_{\rm K}$, the Debye's screening distance decreases further and the particles become solitary (region IV).

A comparison between (a) and (b) in Fig. 9.2 shows that an increase in the particle radius from 10^{-6} cm to 10^{-5} cm results in a sharp narrowing of regions I and IV, which is dicussed above. Regions II and III, in which the plasma becomes non-ideal, expand.

9.4. Ionization Equilibrium Under Conditions of Strong Interparticle Interaction

We now consider a system of CDP particles and electrons emitted by the latter. When $\gamma_{pp} \gg 1$, the Wigner-Seitz cell method is used. This method was first used by Gibson [527] to describe a plasma with CDP. In this and subsequent papers, the problem was solved numerically. The analytical solution and the conditions of validity of the method were obtained by Zhukhovitsky *et al.* [519].

The cell method is well-known in the physics of strongly interacting particles. The matter is broken down into an aggregate of spherical cells, each containing a single particle (in a CDP system). On the whole, each cell is neutral.

The cell radius $\bar{r}$ is found from the condition $4\pi N_p \bar{r}^3/3 = 1$. The value of potential $\varphi(\bar{r})$ at $r = \bar{r}$ is assumed to be zero. Inside the cell, the distribution $N_e(r)$ and the Poisson-Boltzmann equation then take the form of

$$N_e(r) = N_{e0} \exp[e\varphi(r)/T] = N_{e0} \exp[\Phi(r)]\,; \tag{9.18}$$

$$\Delta\Phi = (4\pi e^2 N_{e0}/T) \exp[\Phi(r)] = æ_0^2 \exp[\Phi(r)]\,, \tag{9.19}$$

where N_{e0} is the yet unknown electron concentration on the cell surface. It follows from the condition of electroneutrality of the cell that

$$\Phi'(\bar{r}) = 0\,. \tag{9.20}$$

As will be shown below, the $N_e(r)$ distribution inside the cell is rather homogeneous and one can approximately write $ZN_p = N_{e0}$. The condition $\gamma_{ee} \ll 1$ is then simultaneously valid with $\gamma_{pp} \gg 1$ if $Z \gg 1$.

In a dense system of CDP particles ($N_p R^3 \to 1$), the eigenvolume of particles is comparable to or exceeds the volume taken up by inert gas (or air). The problem of electron number density and particle charge in such a system has been solved in [26, 518]. Assuming that the particle radius is small enough for the condition of weak screening to be valid, that is,

$$\frac{2}{R}\left|\frac{d\Phi}{dr}\right|_{r=R} \gg \frac{æ_0^2}{2}\exp(\Phi_s) = æ_s^2/2\,, \tag{9.21}$$

the solution of the homogeneous equation of (9.19) at $r \to R$ is written as

$$\Phi(r) = A/r + B\,, \tag{9.22}$$

where A and B are constants. On the other hand, at $r \to \bar{r}, \exp\Phi \to 1$. Therefore, $\Phi(r)$ can be presented as [527]

$$\Phi(r) = (1/3)\bar{r}^2 æ_0^2[\bar{r}/r - 3/2 + (r/\bar{r})^2/2]\,. \tag{9.23}$$

Upon comparing Eqs. (9.22) and (9.23), one can readily see that at $R \ll \bar{r}$, Eq. (9.23) is valid in the entire range. Thus,

$$\Phi_s = \frac{1}{3}\bar{r}^2 æ_0^2 \left[\frac{\bar{r}}{R} - \frac{3}{2} + \frac{1}{2}\left(\frac{R}{\bar{r}}\right)^2\right] \approx \bar{r}^3 æ_0^2/(3R)\,. \qquad (9.24)$$

From Eqs. (9.24) and (9.19), an equation for $æ_0$ (or N_e^0) is derived:

$$\ln(æ_s/æ_0) = \bar{r}^3 æ_0^2/(6R)\,.$$

Here, $æ_s = (4\pi e^2 N_{es}/T)^{1/2}$. Upon substituting $æ_s$, $æ_0$, $\bar{r}$ in Eq. (9.21), we obtain

$$N_{e0} = (N_p TR/e^2)\ln(N_{es}/N_{e0})\,. \qquad (9.25)$$

Therefore, in a weak screening mode, N_{e0} is given by the ideal-gas approximation even in the case of strong interparticle correlations. In doing so, the electron concentration is constant almost everywhere except in the surface regions. The system may then be regarded as approximately a system of positive charges against a compensating negative background.

Upon substituting Eq. (9.23) in Eq. (9.21), the weak screening condition takes the form of

$$(Ræ_s)^2 \ll 2L\,, \qquad (9.26)$$

where $L = \ln(N_{es}/N_{e0})$.

The cell model criterion of validity can be obtained using the same method that is instrumental in obtaining the validity criterion of the Wigner electron crystal [528]. Our system differs only by the reversal of the charge and background signs. We now consider the small amplitude vibrations of particle near the cell center. Insofar as the electron concentration in a rarefied system is constant almost everywhere, a particle being displaced for a distance Δr from the position of equilibrium coinciding with the cell center is acted upon by a force

$$F(\Delta r) = (4\pi/3)Ze^2 N_{e0}\Delta r \approx (4\pi/3)Z^2 e^2 N_p \Delta r\,.$$

The particle performs vibrations in a spherical potential well:

$$U(\Delta r) = (2\pi/3)Z^2 e^2 N_p (\Delta r)^2\,.$$

On demanding that the amplitude of such thermal vibrations be very much less than the cell radius, the condition

$$\gamma_{pp}^{1/2} = (Z^2 e^2/\bar{r}T)^{1/2} \gg 1 \qquad (9.27)$$

is obtained.

9.5. Great-Radius Particles. Charge Renormalization

In the strong screening mode, when $R\text{æ}_{s\cdot} \gg (2L)^{1/2}$, the formulas obtained in Section 9.4 are inapplicable. However, the problem may be reduced to that already discussed if a charge renormalization is performed [519]. The renormalization takes place due to the emergence of a thin electron layer which strongly screens the particle charge. The renormalized charge Z^* turns out to be much lower than the total charge Z. Therefore, it is in the weak screening mode. To find Z^*, the behavior of $\Phi(r)$ and $N_e(r)$ near the surface and at a great distance from it is analyzed. $\Phi(r)$ and $N_e(r)$, which assume high values of Φ_s and N_{es} on the surface, decrease sharply with an increase in r. In doing so, the local value of the important parameter $[r\text{æ}(r)]^2$ decreases. This continues until the parameter assumes a value of $2L_1 = 2\ln[N_e(r_1)/N_{e0}]$, after which there occurs a transition to the weak screening mode described in Section 9.4.

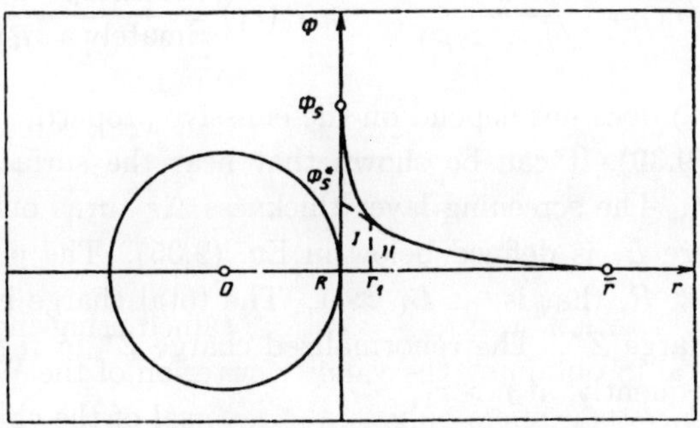

Fig. 9.3. Qualitative pattern of potential distribution in the neighborhood of particle in the strong screening mode.

The potential distribution around a great-radius particle is shown in Fig. 9.3. A strong screening region I is found inside a sphere r_1 and a weak screening region II, outside the sphere. First, consider the inner region. Let $(r_1 - R)/R \ll 1$, that is, in solving Eq. (9.19), one can use a one-dimensional approximation if

$$(2/r)|d\Phi/dr| \ll \text{æ}^2(r)\exp[\Phi(r)]\,. \tag{9.28}$$

In region I Eq. (9.19) may be rewritten as

$$\begin{aligned} d^2\Phi/dr^2 &= \text{æ}^2(r_1)\exp[\Phi(r)]\,; \\ N_e(r) &= N_e(r_1)\exp[\Phi(r)]\,. \end{aligned} \tag{9.29}$$

In the case of strong screening, it is evident that $\Phi_s \gg \Phi(r_1)$ and $\Phi'(R) \gg \Phi'(r_1)$. The solution of Eq. (9.29) will then be written as

$$\Phi(r) = \ln[1 + tg^2((r - r_1)æ(r_1)/\sqrt{2})]\,. \tag{9.30}$$

Therefore,

$$\Phi_s = \ln[1 + tg^2((R - r_1)æ(r_1)/\sqrt{2})]\,. \tag{9.31}$$

We shall find the value of $æ(r_1)$ by substituting Eq. (9.31) into Eq. (9.29):

$$æ_s^2/æ^2(r_1) = 1 + tg^2[(R - r_1)æ(r_1)/\sqrt{2}]\,. \tag{9.32}$$

The electron distribution near the particle surface is strongly inhomogeneous, $N_{es} \gg N_e(r_1)$ or $æ_s \gg æ(r_1)$. In accordance with Eq. (9.31), this is only possible when

$$æ(r_1) = \frac{\pi}{\sqrt{2}(r_1 - R)}\,, \qquad N_e(r_1) = \frac{\pi T}{8e^2(r_1 - R)^2}\,. \tag{9.33}$$

Note that $N_e(r_1)$ does not depend on the emissive properties of the particle.

Using Eq. (9.30), it can be shown that near the surface, $N_e(r)[1 + æ_s (r - R)/2^{1/2}]^{-2}$. The screening layer thickness Δr turns out to be equal to $R/(L_1)^{1/2}$, where L_1 is defined below in Eq. (9.35). The present discussion is valid at $\Delta r \ll R$, that is, at $L_1 \gg 1$. The total charge in region I is the renormalized charge Z^*. The renormalized charge Z^* in region II is weakly screened. Consequently, at $r > r_1$,

$$\Phi(r) \approx \Phi^*(r) = Z^* e^2/rT$$

while the condition of quasi-neutrality takes the form of $N_{e0} = Z^* N_p$. Taking into account the relation $r_1^2 æ^2(r_1) = 2L_1$, we obtain

$$Z^* = \frac{r_1 T}{e^2} \ln \frac{N_e(r)}{N_{e0}} = \frac{RT}{e^2} L_1\,. \tag{9.34}$$

The electron concentration on the cell boundary, which is close to the mean concentration within the cell, is

$$N_{e0} = (N_p RT/e^2)L_1\,, \quad L_1 = \ln[N_e(r_1)/N_{e0}] = -\ln(2\pi N_p R^3)\,. \tag{9.35}$$

Equation (9.35) does not include the quantity $æ_s$. Therefore, the concentration N_{e0} does not depend on the emissive properties of the particle matter. The dependence of Φ_s^* on $æR$ for small values of $æR$ coincides with $\Phi_s(æR)$.

However, in the case of $æ_2 R > \sqrt{2L_1}$, the values of Φ_s^* become constant and independent of the work function. Accordingly, effective charge Z^*, may be substantially lower than the true charge defined by the formula

$$Z = 4\pi\sqrt{2}R^2 N_{es}/æ_s\,.$$

The results of the calculations according to Eqs. (9.25) and (9.35) are in good agreement with the results of numerical calculations [527, 529].

The results of the calculations of the electron concentration in systems of CDP particles with different radii are presented in Fig. 9.4. The calculations [519] have been performed in two approximations: for an ideal system (curves *I*) and for a highly non-ideal system using charge renormalization (curves *II*). It is assumed that $W = 2.7$ eV, $T = 2,500$ K. The values of the non-ideality parameters $\gamma_{pp}^{1/2}$ and Γ_{pp} are indicated near the calculated curves.

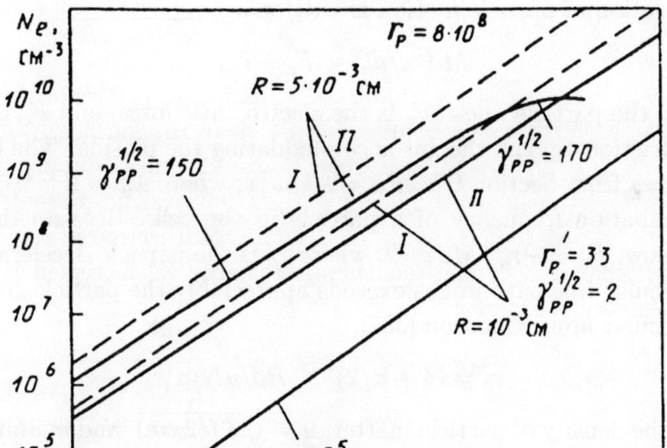

Fig. 9.4. Electron concentration in an ideal (curves *I*) system and non-ideal (curves *II*) system of CDP particles of different radii R at $T = 2,500$ K and $W = 2.7$ eV [519]. Indicated near the curves are the values of γ_{pp} and Γ_{pp}.

For particles with a radius of 10^{-3} cm, Eq. (9.23) is valid since $\gamma_{pp}^{1/2} \gg 1$, $R^2 æ_s^2 \gg 2L_1$. For particles having a radius of 10^{-5} cm, neither approximation is applicable although at low N_p, Eq. (9.25) is preferable and at high N_p,

Eq. (9.23). Their use yields fairly similar results. A great difference (up to a factor of 10) between the values of N_e given by Eqs. (9.35) and (9.23) is only observed in the case of large particles of $R = 5 \cdot 10^{-3}$ cm.

9.6. Relaxation of CDP Particle Distribution

The results of the treatment of the dynamics of particle motion [530] are used to determine the time taken for an equilibrium to be established in a strongly non-ideal plasma. It is estimated below. The behavior of a strongly non-ideal plasma is also investigated in a turbulent flow of gas, where it turns out that the short-range order may exist when the level of turbulent pulsations is not too high. The process of structure development occurs, for example, during the injection of particles into a gas flow, when the particles are first arranged chaotically relative to one another. We will estimate the time taken by a particle deflected from the cell center to change to the equilibrium position.

The equation of particle motion is written as

$$Md^2x/dt^2 = F_e + F_{v1}\,, \tag{9.36}$$

where M is the particle mass, F_e is the electrostatic force, and F_v is the force of viscous friction against the gas accommodating the plasma. The expression for F_e follows from Section 9.5: $F_e = -M\omega_0^2 x$, where $\omega_0^2 = Z^2e^2/\bar{r}^3 M$ is the natural oscillation frequency of a particle in the cell. Because the particle velocity is low, $F_v = -\tau_p^{-1} M dx/dt$, where τ_p is the particle deceleration time. If the molecular free path in gas exceeds appreciably the particle size, then in a free molecular approximation [387],

$$\tau_p = (4 + \pi/2)^{-1} \cdot R\rho/uNm\,,$$

where ρ is the density of particle matter, $u = \sqrt{T/(2\pi m)}$, and m and N denote the mass and concentration of gas molecules, respectively. Thus,

$$d^2x/dt^2 + \tau_p^{-1} dx/dt + \omega_0^2 x = 0\,. \tag{9.37}$$

We will evaluate the quantities τ_p and ω_0 under the typical conditions of $T = 2{,}500$ K, $N = 2.9 \cdot 10^{18}$ cm^{-3}, $m = 5 \cdot 10^{-23}$ g, $R = 10^{-4}$ cm, $\bar{r} = 10^{-3}$ cm and $W = 2.5$ eV. For $\rho = 2.5$ g/cm^3, we derive $\tau_p = 9 \cdot 10^{-6}$ s and $\omega_0 = 5 \cdot 10^3$ s^{-1}. Therefore, a gas at atmospheric pressure is a very viscous medium for CDP particles.

The solution of (9.37) for $\tau_p \ll \omega_0^{-1}$ is written as $x = x_0 \exp(-t/\tau_{\mathrm{R}})$, where x_0 is the initial deviation and τ_{R} is the relaxation time,

$$\tau_{\mathrm{R}} = (\tau_p \omega_0^2)^{-1} = (4\pi/3)(4 + \pi/2) R^2 uNm\bar{r}^3/Z^2e^2\,. \tag{9.38}$$

It is possible to talk of the ordering of CDP particles only if the lifetime of the plasma exceeds $\tau_{\rm R}$ considerably. In our example, $\tau_{\rm R} = 510^{-3}$ s.

We will now treat the stability of an ordered structure under the conditions of a turbulent flow of plasma with CDP [530]. In a turbulent flow, the pulsations of the gas velocity and the resultant pulsations of the force of viscous friction acting on the particles tend to render the space distribution of particles chaotic. The forces of electrostatic origin counteract this.

If the mass fraction of particles is small, they have no considerable effect on the gas motion. In turbulent flow, this implies that the characteristics of turbulence (velocity pulsation u', scale of fluctuations $l_{\rm E}$, and characteristic lifetime $\tau_0 \approx l_{\rm E}/u'$) do not change during the injection of particles.

In order to maintain the cellular structure, it is necessary that enough time be allowed for the processes of relaxation to take place inside cells. It is characterized by $\tau_{\rm R}$, that is, there should be

$$\tau_{\rm R} \ll \tau_0 \,. \tag{9.39}$$

For the largest scale of turbulence in the flow core, the semi-empirical theory of turbulence gives $\tau_0 \approx (d/u)\mathrm{Re}^{-1/8}$, where d is the tube diameter and u is the averaged cross-section velocity of flow. For the smallest scale (scale of energy dissipation), $\omega_0 \approx (d/u)\mathrm{Re}^{-3/8}$. Because $\mathrm{Re} \gg 1$, the most destructive effect on the cellular structure is made by small-scale pulsations. The foregoing expressions enable one to define the range of parameters, in which the inequality (9.39) is valid.

9.7. Electron Mobility

The coefficient of electrical conductivity is defined by the number density of electrons and their mobility [(6.1),(6.3)]. While increasing the total collision frequency (6.12), the electron-particle collisions may reduce the electron mobility,

$$\sum_i \nu_i = \nu + \nu_i/\gamma_{\rm E} + \nu_p \,.$$

Here, ν_p is the electron-particle collision frequency $\nu_p = \nu_{p1} + \nu_{p2}$, where ν_{p1} and ν_{p2} denote the frequency of elastic and inelastic electron-particle collisions, respectively. Expression (6.6) is suitable for the frequency ν_{p1}. The expression for the Coulomb logarithm $\ln \Lambda_p$ for an ideal plasma with CDP is found in [517]. It is defined by three lengths: $r_{\rm D}$, Ze^2/T and R. The frequency ν_{p2} characterizes the collisions that take place when the electron reaches the particle surface. In view of the electrostatic barrier,

$$\nu_{p2} = N_p \pi R^2 v \cdot \exp(Ze^2/RT)\,, \quad Z < 0\,;$$

$$\nu_{p2} = N_p \pi R^2 v \cdot (1 + Ze^2/RT)\,, \quad Z > 0\,.$$

If $Z < 0$, the electron must overcome the barrier $(-Ze^2/R)$ in order to get to the particle surface. This reduces ν_{p2}. If $Z > 0$, ν_{p2} naturally proves to be higher.

Upon comparing the values of the collision frequency, one can write the inequalities pointing to the conditions in which collisions with particles reduce the mobility. These are, for example,

$$\begin{aligned} \nu_{p1}/\nu_i &= Z^2 N_p \ln \Lambda_p / N_i \ln \Lambda \geq 1\,, \\ \nu_{p2}/\nu &= \pi R^2 \Phi N_p / \cdot q_{ea} N \geq 1\,, \end{aligned} \tag{9.40}$$

where q_{ea} is the electron-atom scattering cross-section (for buffer gas or alkali admixture). Only small fine particles, whose concentration is high, may produce an appreciable effect on mobility.

Interesting effects may take place in a plasma in a strong electric field which heats electrons and disturbs the system's equilibrium. A system of CDP particles and electrons emitted by those particles in a constant electric field **F** was treated by Samuilov and Gorbatov [531]. The field heats the electrons strongly if its intensity exceeds the critical value F_c:

$$eF_c \approx TNq_{ae}\sqrt{\delta(1+\gamma)}\,.$$

This expression differs from the standard one by the factor $\sqrt{(1+\gamma)}$. The heating of electrons is made difficult by heat transfer with particles.

An interesting result is a decrease in σ with an increase in F. Indeed, the electron-particle collision frequency increases with T_e, while the rate of emission remains unchanged. The positive charge of particles decreases and this leads to a decrease in N_e : $N_e = N_{es}(T/T_e)^{1/2}$.

9.8. Radiation of Plasma with CDP

The continuous spectrum of radiation of a plasma with CDP may be due to the bremsstrahlung of electrons from CDP particles rather than from the atoms of a buffer gas. It is not to be assumed, however, that this radiation is formed as a result of pair collisions between electrons and particles interacting under the Coulomb's law. One also cannot use the known formulas (see Section 8.2) to calculate the continuous spectrum. Note three important peculiarities. First, the finiteness of the particle radius R limits the frequency range considerably.

Second, in view of the fact that a particle is screened over the mean interparticle distance, the electron motion in plasma under high values of γ_{pp} cannot be treated as a series of scattering from individual cells separated in time. Third, bound, rather than free, electrons make the main contribution to the spectrum being emitted. Artem'ev and Iakubov [532] have discussed these singularities and derived formulas that enable one to calculate the spectral intensity of radiation.

All electrons are divided into four groups:

(i) Free electrons moving by infinite orbits without crossing the particle surface.
(ii) Electrons moving by infinite orbits and crossing the particle surface.
(iii) Orbital electrons moving by finite orbits without crossing the particle surface (the absorption of radiation by these electrons was investigated by Zhukhovitskii [533]).
(iv) Suborbital electrons moving by finite orbits and crossing the particle surface.

The interaction of a free electron with a particle and its environment is described by the cell potential (9.23). In view of our interest in frequencies that are much higher than the plasma frequency $\omega \gg \omega_p = \sqrt{4\pi e^2 N_e/m}$, and of the fact that $\gamma_{pe} \approx 1$, we must treat the portions of electron trajectory that are close to the particle. In this case, the cell potential is almost the Coulomb one. The standard technique is also applicable in calculating the bremsstrahlung of electrons from the Coulomb centers developed by Kramers [534]. However, the non-pointness of the scattering center places considerable limitation on the parameters of trajectories:

$$b_0 > R\,,$$

where b_0 is the distance of the minimum convergence of an electron with a particle. In view of this condition, the expression is written for S_ω, the impact parameter-average spectral energy emitted by an electron during a single interaction event.

The spectral radiation power is equal to $\langle S_\omega \rangle / \langle \tau \rangle$, where $\langle \tau \rangle$ is the average time during which the electron passes one cell, $\langle \tau \rangle \approx (4/3)\Gamma_{pe}^{-1/2}\bar{r}/v$. Under these conditions, we average the spectral radiation power over the Maxwellian distribution of electrons to derive the power of plasma radiation per unit volume. Instead of using the absolute form, it is convenient to represent the result as a ratio to the assumed radiation power for scattering from point Coulomb centers derived by Kramers:

$$W_\omega^{Kr} = (32/3\sqrt{3})Z^2e^6N_pN_e/m^2v_T^2c^3\,.$$

This relation may be referred to as the "Gaunt factor". For electrons in the first group, that is, free electrons that do not cross the particle surface, Artem'ev and Iakubov [532] derived the following expressions:

$$g_1 = 1 - 2^{-2/3}3^{5/6}\pi^{-1}\Gamma^2(2/3)[1-(2/3)\Phi](\omega/\omega^*)^{2/3}, \omega^*\Phi^{-3/2} \ll \omega \ll \omega^*\,;$$
$$g_1 = 3^{1/2}2^{-1}[1-2^{3/2}\omega/(3\omega^*\Phi)]^{-1}\exp(-2^{5/2}\omega/3\omega^*), \omega^* < \omega \ll \Phi\omega^*\,. \tag{9.41}$$

Here, $\omega^* = [Ze^2/mR^3]^{1/2}$ is the characteristic frequency of the problem under consideration and $\Phi = Ze^2/(RT)$ is the relative particle potential, which is high under the given conditions $\Phi \gg 1$. The Gaunt factor g_1 proves to be less than unity and decreases rapidly at frequencies exceeding ω^*.

The bremsstrahlung of free electrons crossing the particle surface is not marked because the fraction of such electrons is small.

An orbital electron moves by an elliptical orbit characterized by the eccentricity ε and semi-axis a, and possesses a bond energy equal to $-Ze^2/(2a)$. Because such motion is periodic, the electron radiates at frequencies that are multiples of $\omega_0 = (Ze^2/(ma)^3)^{1/2}$. The total spectral power radiated by an electron is

$$I_\omega = \sum_{n=1}^{\infty} I_n(a,\varepsilon)\delta(\omega - n\omega_0)\,,$$

where the component $I_n(a,\varepsilon)$ is emitted by the n-th harmonic.

The averaging of I_ω over the Boltzmann distribution of electrons, over the parameters ε and a on condition that the electron does not cross the particle surface $a(1-\varepsilon) > R$, gives the Gaunt factor.

The asymptotic expression for g_3 has the form

$$g_3 = \frac{3\sqrt{6}}{32}\Phi\frac{\omega^*}{\omega}\exp\left(-\frac{4\sqrt{2}}{3}\frac{\omega}{\omega^*}\right), \quad \omega \gg \omega^*\,. \tag{9.42}$$

The quantity g_3 proved to be much greater than g_1 (Fig. 9.5). This is in spite of the fact that the number of these electrons is small. The thing is that they must be positioned in the vicinity of the particle and radiate in the neighborhood of the frequency ω^*. Zhukhovitskii [533] was the first to head the importance of these electrons. He has also pointed out the important role played by the characteristic frequency ω^* from the standpoint of the diagnostics of plasma with CDP.

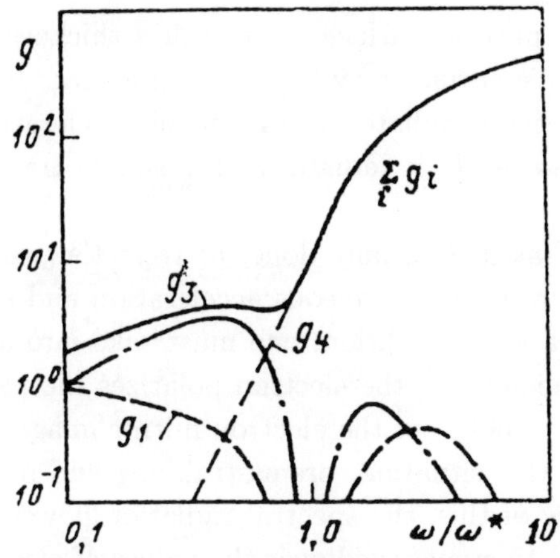

Fig. 9.5. The Gaunt factor of total radiation of plasma, Σg_i, and the contribution made by the individual components g_1, g_3 and g_4, $\Phi = 10$ [532].

It is known that true bremsstrahlung is sometimes accompanied by intense polarization radiation [535]. The effect of the own field of electron on a particle brings about a polarization of the latter. The dipole moment of a metal particle is

$$D_p = R^3 e\mathbf{r}r^{-3} = -e\mathbf{r}(\omega^*)^{-2} ,$$

where $\mathbf{r}$ is the radius vector of an electron relative to the particle center. As a result, the particle radiates and the energy of the total bremsstrahlung and polarization radiation is given by the expression

$$E_\omega^P + E_\omega^T = \left[1 - \left(\frac{\omega}{\omega^*}\right)^2\right] E_\omega ,$$

where E_ω is the spectral density of the true bremsstrahlung of electrons of the first and third groups. The polarization radiation effect leads to a marked suppression of radiation at frequencies $\omega \approx \omega^*$. Figure 9.5 gives $g_1(\omega)$ and $g_3(\omega)$ in view of the polarization component.

The polarization radiation may be interpreted as radiation of the electrostatic image of an electron moving inside a particle. In doing so, the acceleration of the dipole moment of the electron is directed opposite to that of the electron image. With $\omega \to \omega^*$, the orbits of the electron and image draw closer together and their radiation is mutually cancelled out.

We will further treat the radiation of suborbital electrons. If $\Phi \gg 1$, the main fraction of bound electrons crossing the particle surface will be concentrated in the particle surface layer with a thickness of $l \approx R/\Phi \ll R$. Therefore, the particle surface may be regarded as flat. In the free portion of its trajectory, an electron radiates as a result of deceleration. At the moment of emission and absorption, it radiates as a result of transition to a different medium.

It is convenient, as it is usually done, to treat the transition radiation as bremsstrahlung upon an instantaneous acceleration and stoppage of electron on the particle surface. In addition, one must take into account that during its flight towards the surface, the electron polarizes the particle, which shows up effectively in the "flight" of the electron mirror image towards the surface of the particles. In this situation, bremsstrahlung and polarization radiation are inphase. In view of this, the spectral radiation power of a single electron is written down and then averaged over the values of electron escape velocity v. In the end, the appropriate Gaunt factor g_4 was derived in [532]:

$$g_4 = \frac{4\sqrt{3}}{\pi}\exp(\Phi)\Phi^{-2}\left(\frac{\varepsilon}{\varepsilon+1}\right)^2, \quad \Phi^{1/2}\omega^* \ll \omega \ll \omega^*(Z/\Phi)^{1/2}, \qquad (9.43)$$

where ε is the dielectric permeability of the particle matter and Z is the particle charge. Figure 9.5 gives g_4 at $\Phi = 10$. The high value of g_4 at high frequency is due to the large number of electrons in the close neighborhood of the particle.

In studying the radiation at a higher frequency, the details of the description of the electron motion in the immediate proximity of the surface become important. With the frequency $\omega > \omega^*(Z/\Phi)^{1/2}$, the main part is already played by the portions of trajectory in the field of image forces $F = -e^2/[4(R-x)^2]$. The Coulomb field of the particle may also be ignored.

Figure 9.5 gives the Gaunt factor for total plasma radiation. All the components discussed above contribute to this factor. Figure 9.5 is constructed for conditions with the relative potential of the particle being $\Phi = 10$. These conditions are typical of a plasma with CDP. Such, for example, is the plasma with the particle number density $N_p = 10^9$ cm^{-3} and particle radius $R = 4\cdot10^{-5}$ cm. The particle charge $Z = RTe^{-2}\ln(2\pi R^3 N_p)^{-1}$ at $T = 2.5\cdot10^3$ K is close to 600, with the electron number density being $N_e = 6\cdot10^{11}$ cm^{-3}. The radiation being investigated corresponds to the frequency interval 10^{12}–10^{14} s^{-1}.

Therefore, the main contribution to the radiation of non-ideal plasma with CDP is made by bound electrons. This radiation cannot be calculated using expressions that are common to plasma. The bremsstrahlung from atoms is not very pronounced if the pressure does not exceed 10^5 Pa. At $\omega \leq 10\omega^*$, the thermal radiation of particles compared to CDP-plasma radiation may be ignored. However, as the frequency increases, the intensity of the thermal radiation of particles increases very rapidly and defines the continuous spectrum (see [536]).

9.9. Some Results of Experimental Investigations

The size distribution of particles in the plasma of solid-fuel combustion products is, as a rule, bimodal. The submicron component ($R \approx 10^{-2} - 10^{-1} \mu$m) emerges in a flow of combustion products as a result of volume condensation of volatile components. The coarsely component ($R \approx 1 - 10 \mu$m) is formed by particles of high-melting components which do not evaporate in the combustion zone, but are instead carried out from this zone. Downstream in the flow, the size of the particles in the submicron component increases as a result of coalescence and coagulation. The distribution curve of the submicron component also approaches that of the micron. Figure 9.6 gives the size distribution of particles measured in a gas duct simulating the channel of a coal-fired MHD generator [537]. The sampling of particles was made with the aid of an impactor.

An analogous picture is observed when a powerful pulse of laser radiation acts on a metal surface. Large droplets of metal are ejected from the surface, either because of its instability or as a result of being washed out by vapor flows. The finely divided component emerges under conditions of volume condensation.

The space distribution of CDP particles can be judged from the results of the electron microscopy of samples taken by a probe introduced for some time into a plasma flow. Such micro-photographs of samples were obtained

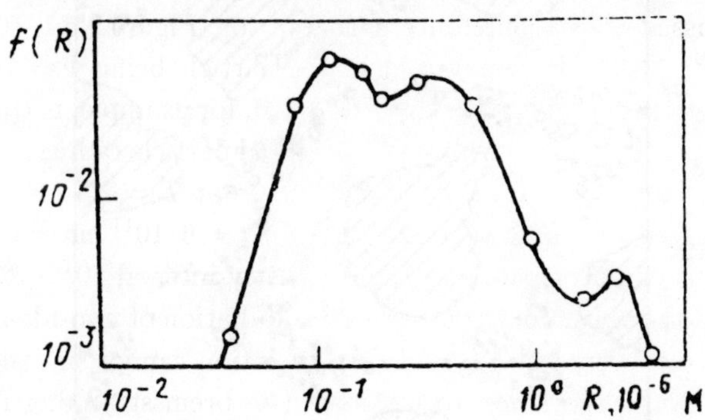

Fig. 9.6. The size distribution of CDP particles measured in [537]: $f(R)$ is the mass fraction.

by Dragan *et al.* [538] in the flame of synthetic aluminized fuels. A typical sample is shown in Fig. 9.7. The characteristic size of the large particles of Al_2O_3, formed during the combustion of fuel, is close to 1 μm. In a flame containing a small natural impurity of alkali atoms ($N \approx 10^8$ cm^{-3}), the contribution by the latter to ionization is small and the plasma of particles and electrons emitted by those particles is realized. Under the conditions of this experiment, $\sqrt{\gamma_p} \approx 50$, that is, the system is strongly non-ideal. The arrangement of particles in the photograph is indicative of the presence of an ordered structure of the polydisperse system of particles. There is reason to believe that this reflects the presence of ordering in the plasma.

Zimin and Popov [539] have measured the coefficient of electrical conductivity of combustion products using methane-air mixtures seeded with fine powders of materials which have a low work function. The two-wave method was used to measure the coefficient of electrical conductivity. The concentration of powder was varied from 0.2% to 18% by mass. The treatment of the measurement results, illustrated in Fig. 9.8, enables one to determine the work function and average radius of the particles [540]. Figure 9.8 is indicative of a good agreement between the results of measurements of the coefficient of electrical conductivity and the calculation results.

Goncharov *et al.* [541] have analyzed the results of measurements of the coefficients of absorption, scattering and transmission of the plasma of an erosion flame. The latter emerges as a result of a laser pulse of millisecond duration with an irradiation of the order of 1 MW/cm^2 (Fig. 9.9).

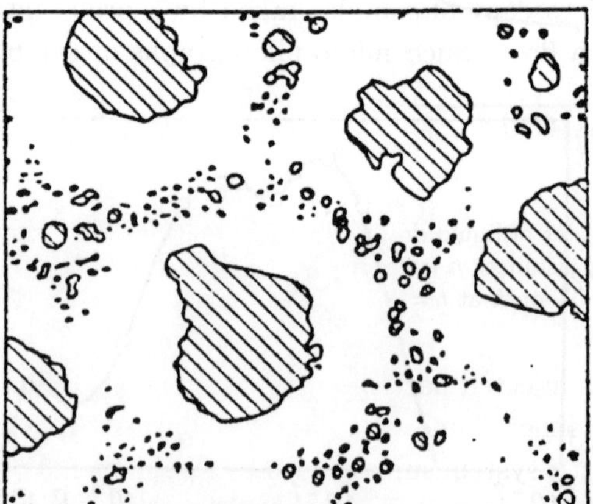

Fig. 9.7. Micro-photograph of a sample in a flame of aluminized fuel [538].

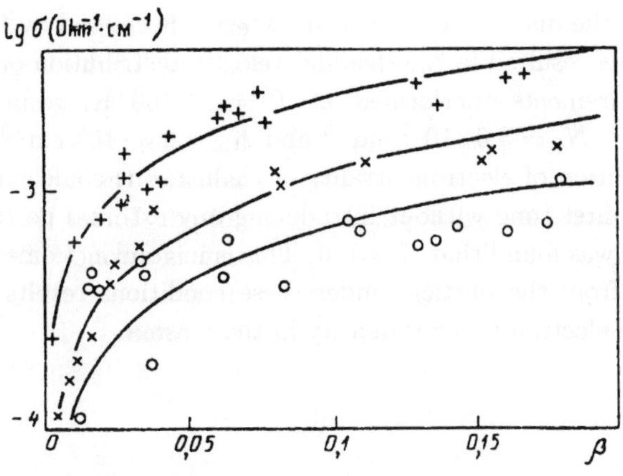

Fig. 9.8. The coefficient of electrical conductivity as a function of the mass fraction of calcium oxide powder [540]. The measurement results [539]: ○, $T = 2,200$ K; ×, 2,400 K; +, 2,530 K.

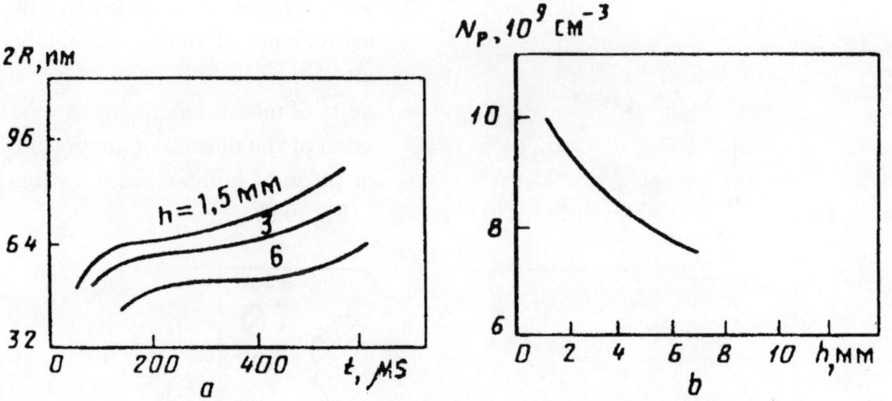

Fig. 9.9. (a) The diameter of liquid droplets $2R$ as a function of the neodymium laser pulse duration t at different distances h from the lead target surface. (b) The dependence of the droplet concentration N_p on h at $t = 300$ μs [542].

Petrov [542] has injected particles of cerium oxide (average diameter $2R \approx 0.25$ μm) into the inner flame of a Maecker burner, in which propane was burnt. The diagnostic system supported the measurements of the temperature of plasma and particles, as well as the number density of electrons, atoms and ions. The CDP particles were sorted out by a pulse analyzer with respect to the amplitudes and durations of the pulses (caused by these particles) of attenuation of the diagnostic laser beam when it is crossed by flying particles separately. This resulted in the size and velocity distribution of particles.

The measurements, performed at $T = 2,100$ K, resulted in $N_i = 1.0 \cdot 10^{10}$ cm^{-3}, $N_e = 3.6 \cdot 10^{10}$ cm^{-3} and $N_p = 2.3 \cdot 10^{8}$ cm^{-3}. As a result, using the condition of electroneutrality of plasma, one could find the particle charge for the first time without introducing any external perturbations into the plasma. It was found that $Z = 160$. This is in good agreement with (9.34). The emission from the particle under these conditions results in a two-fold increase in the electron number density in the plasma.

Ten

The Dynamics and Stability of Non-Ideal Plasma

The unusual properties of non-ideal plasma have a fundamental effect on the following: the plasma behavior in the external electric and magnetic fields; the dynamics of plasma flow in channels and during external flow past objects; and on the stability of the very state of non-ideal plasma. These aspects of the physics of non-ideal plasma are especially important in applications and they must be taken into consideration when designing new experimental and technical devices.

10.1. The Thermal Instability of Non-ideal Plasma with Current. Stratification

Under supercritical pressure, the matter experiences, as a result of isobaric heating, a continuous transition from the liquid metal to gaseous plasma state. At first the electrical conductivity decreases slowly and then very rapidly in the $T \approx T_c$ range. It only starts to increase at high temperatures. The negative values of the $d\sigma/dT$ derivative lead to the development of thermal instability [85].

If the resistance of the electronic circuit, in which the plasma gap is connected, is high, the electric current i can be regarded as being well-stabilized. Small temperature fluctuations ($\delta T > 0$) will then lead to a decrease in σ and a subsequent increase in Joule heat, i^2/σ, that is, to the heating of matter. The process of heating also continues to develop. Therefore, the homogeneous state of plasma with electric current proves unstable. As a result of instability, the matter changes to a new state which is spatially inhomogeneous [543].

It is shown below that inhomogeneities which are oriented across the plasma column develop most rapidly. They tend to break the matter up into layers transverse to the current. In the physics of gas discharge, such layers are referred to as strata. If the thermal instability leads to such strata, the latter can be referred to as thermal strata. This pattern was observed by Dikhter and Zeigarnik [78, 84] after the isobaric heating of cesium with current (cf., Fig. 2.30).

10.1.1. *Isobaric Joule heating*

Consider a column of current-heated matter based on the assumption that it is homogeneous. Its homogeneous state is the background against which instability is developed. The energy balance takes the form of

$$\rho c_p dT/dt = J^2/\sigma - W\,, \tag{10.1}$$

where c_p is the specific heat at constant pressure, J is the current density and W is the energy losses.

When the external circuit includes a capacitor bank with a capacitance C and a ballast resistor R, the current occurs upon a discharge of the bank through the latter and through a gap closed by the material being investigated. The strength of current i decays exponentially with time, $i(t) = i_0 \exp(-t/\tau)$, $\tau = RC$. The current density $J = i/s$, where s is the cross-section of the discharge column. In the experiments [79, 80], the discharge was restricted by the capillary walls, thus defining the value of s. In the experiments by Dikhter and Zeigarnik [78, 84], the pressure was maintained by inert gas (cf., Chapter 2) and the matter expanded upon heating. This led to an increase in the cross-section of the plasma column.

In this case, the cross-section s can be found by using the equation of state and equation of balance for mass:

$$p = p(\rho, T)\,, \qquad M = \rho s L\,, \tag{10.2}$$

where M is the mass of metal and L is the length of the discharge column, that is, the inter-electrode gap. Here, in accordance with the estimates given by Dikhter and Zeigarnik [78], we ignore the diffusion mixing of the matter of metal and inert gas. If the metal has changed to the state of ideal plasma, Eq. (10.2) gives $s = (M/m_a)T(1 + x\bar{Z})/pL$, where m_a is the atomic mass of metal, x is the degree of ionization and $\bar{Z}$ is the mean ion charge.

In this formulation, the problem has two characteristic times: the time of current decay, τ, and the time of heating of the material, $t_h = \rho c_p T \sigma J^{-2}$. If $\tau \gg t_h$, quasi-stationary treatment is valid. In this case, the state of matter (its temperature) has enough time to adjust to the value of relatively slowly varying current. The energy balance (10.1) reduces to

$$i^2/(s^2\sigma) - W = 0\,. \tag{10.3}$$

This approximation was realized in the experiments by [79, 80]. The radiation losses of energy prevailed. If t_h is great, the left-hand side of Eq. (10.1) is also great and the Joule heat leads to an increase in temperature over time. In the limit of ideal heat insulation ($W = 0$), Eq. (10.1) assumes the form

$$\rho c_p dT/dt = i^2/(s^2\sigma)\,. \tag{10.4}$$

The temperature increases sharply in the initial stage and then continues to rise smoothly.

Given the $\sigma(T)$ and $W(T)$ relationships, one can use Ohm's law, $i = \sigma s F$, and Eq. (10.3) to plot the current-voltage characteristic $F(i)$, where F is the electric field intensity, for a stationary homogeneous discharge.

10.1.1.1. *Thermal instability*

For using the instability, we consider a set consisting of equations of balance of energy and mass, Ohm's law and field equations:

$$\rho c_p dT/dt = \sigma E^2 - W\,,\ i = \sigma s F\,,\ M = \int_V \rho d\mathbf{r}\,; \tag{10.5}$$

$$\mathrm{div}\mathbf{J} = 0\,, \quad \mathrm{rot}\mathbf{F} = 0\,. \tag{10.6}$$

For an analysis of instability, the set (10.5)–(10.6) is linearized in a standard manner relative to small perturbations. Since the current is assumed to be stabilized, homogeneous fluctuations are impossible. Now, consider the inhomogeneous fluctuations of the form $\delta T(r,t) = \delta T(t)\exp(i\mathbf{kr})$, where $\mathbf{r}$ is the radius vector and $\mathbf{k}$ is the wave vector. Following a series of transformations, we obtain

$$\delta\mathbf{F} = -\frac{1}{\sigma s}\frac{d(\sigma s)}{dT}(\mathbf{Fk})\frac{\mathbf{k}}{\mathbf{k^2}}\delta T\,;$$

$$\frac{d\ln\delta T}{dt} = \frac{\sigma F^2}{c_p}\left\{-\frac{2k_y^2}{k^2}\frac{d\ln(\sigma s^2)}{dT} + \frac{d\ln\sigma}{dT} - \frac{d\ln c_p}{dT}\right\}$$
$$+\frac{W}{c_p}\left\{\frac{d\ln c_p}{dT} - \frac{d\ln W}{dT}\right\} \equiv \gamma(t)\,. \tag{10.7}$$

In Eq. (10.7), the y-axis is directed along the field $\mathbf{F}$.

Equation (10.7) is obtained for small wavelengths $k_y L \ll 1$, $k_x r \ll 1$, where $r \sim s^{1/2}$ is the transverse dimension of the plasma column. The need for these inequalities is caused by the integral form of the balance of mass of matter in Eq. (10.5), in which the integration is performed over the plasma volume $V = sL$. In the approximation (10.7), this volume is assumed to be very large. There is one more restriction pertaining to the size of k which, however, can be readily removed if the heat conductivity is included. It is not included in Eq. (10.5). Therefore, the wavelengths must be less than the characteristic size of the heat conductivity band, that is, $k^{-1} \ll (æ t_h \rho c_p)^{1/2}$, where $t_h = (c_p \rho T)\sigma / J^2$ is the heating time and æ is the coefficient of heat conductivity.

The time dependence of the increment γ occurs due to the non-stationary nature of the background. The fluctuations in temperature against such a background develops according to the law

$$\delta T(t) = \delta T(t_0) \exp\left[\int_{t_0}^{t} \gamma(t')dt'\right] .$$

The system is unstable if the increment $\gamma(t) > 0$ and, in addition, if the perturbations develop faster than the background at roughly, $1/\gamma(t) < \tau$, where τ is the development time of the background.

If $d\ln\sigma/dT < 0$, it follows from Eq. (10.7) that the maximum increment is characteristic of the fluctuations which develop with respect to the current $k_y = k$. In this case,

$$\gamma(t) = -\frac{1}{t_h}\left[\frac{d\ln(\sigma s^2 \rho c_p)}{d\ln T} + \frac{t_h}{t_c}\frac{d\ln(W/\rho c_p)}{d\ln T}\right] . \tag{10.8}$$

In addition to the heating time t_h, the cooling time $t_c = (\rho c_p T)/W$ is introduced. Under the conditions of full heat insulation of plasma, the increment takes the most simple form,

$$\gamma(t) = \frac{-1}{t_h}\frac{d\ln(\sigma s^2 \rho c_p)}{d\ln T} , \tag{10.9}$$

when $W = 0$. In the opposite, quasi-stationary case:

$$\gamma(t) = \frac{-1}{t_h}\frac{d\ln(\sigma s^2 W)}{da\ln T} . \tag{10.10}$$

The quantity $\gamma(t)$ is greater than zero if the decrease in electrical conductivity with a rise in temperature is sharper than the increase in $s^2 \rho c_p$, or even in $s^2 W$. The region of plasma metallization satisfies this condition fully. Even if the background development time τ is of the order of the heating time t_h, the instability of the development time may be small due to the great value of the parameter $|d \ln \sigma \ln T|$, that is a sharp decrease in σ with an increase in temperature.

After simplifying the formula (10.9) for the liquid state region, the heat capacity c_p varies very slowly at $T \approx T_c$. Therefore, one can write

$$\gamma(t) = -\frac{1}{t_h} \frac{d \ln(\sigma/\rho)}{d \ln T} \tag{10.11}$$

instead of Eq. (10.9). This value is positive because its variation upon heating is stronger than that of density. Moreover, the closer we are to T_c, the greater the instability increment becomes.

Such instability leads to the emergence of layers, that is inhomogeneities of discharge develop in the direction across the current. They can be referred to as thermal strata because they occur in the presence of local equilibrium in plasma. They have little in common with ionization strata. The latter are due to the singularities of the kinetics of excitation, ionization and recombination. They also occur in the plasma of gas discharges far away from the state of local thermodynamic equilibrium.

The linear analysis prevents a correct opinion of the non-linear stage from being formed. However, if the strata are indeed the result of the development of thermal instability, the matter in hot layers is then in the plasma state while in the cold layers, it apparently has a density which is close to critical.

10.1.2. *Comparison With Experiment*

For a comparison with experiment, evaluations for the conditions of experiment [78] involving the heating of cesium wire in an argon atmosphere are made under the following conditions: $p = 500$ atm, the initial current density $J = 1.5 \cdot 10^6$ A/cm^2, the initial wire diameter of $4.4 \cdot 10^{-2}$ cm, the wire length $L = 1.2$ cm, and current time $\tau = 720$ μs. During the experiment, 4–8 μs after the heating starts, regular dark and light transverse layers were observed. In the hundreds of microseconds that follow, the light layers expanded gradually. It can be assumed that these strata are the result of the development of thermal instability.

First, we evaluate the time $t(T_c)$ during which the matter is heated, say, to the critical temperature. It follows from Eq. (10.4) that

$$t(T_c) = \frac{1}{i_0^2} \int_{T_0}^{T_c} \rho c_p \sigma s^2 dT .$$

The density decreases by a factor of two and σ by a factor of approximately five. Meanwhile, the heat capacity of liquid $c_p \approx 6/m_a$. As a result, $t(T_c)$ amounts to several microseconds. Close to this value is the heating time t_h, which is evaluated by the moment of taking the first frame of streak picture. The results of the processing of experimental data, obtained by Dikhter and Zeigarnik [78, 84], indicate that the cooling (losses of W) can be ignored. One can then use Eq. (10.9) for the increment. The values calculated for l/γ are several times less than t_h. Consequently, the instability has enough time to develop and reach the non-linear mode.

It is possible that the observed strata are the result of the development of this instability. A correct analysis of the non-linear stage could provide the solution to a very complicated mathematical problem. It must be non-stationary and two-dimensional. In doing so, it is natural that one cannot ignore the longitudinal thermal conductivity. On the contrary, it plays a major role. Based on the dimensional considerations, we shall try to evaluate the width of the layers and the rate of their movement: $\Delta l \sim (æ t_h/c_p)^{1/2}$ and $u \sim (æ/t_h c_p)^{1/2}$, where æ is the coefficient of thermal conductivity.

In the third frame of the streak picture ($t = 20$–30 μs), the temperature of the hot region is close to $2.0 \cdot 10^4$ K. This means that the thermal conductivity is mainly radiative, $æ = (16/3)\sigma_{\rm SB} l_{\rm R} T^3$, where $\sigma_{\rm SB}$ is the Stefan-Boltzmann constant and $l_{\rm R}$ is Rosseland's length. The latter is given by Kobzev and Kurilenkov [544], $l_{\rm R} \approx 5 \cdot 10^{-3}$ cm. Taking into consideration that $c_p = 5/m_c$ in a fully lonized plasma, we perform calculations: $t_h = 400\mu$s, $\Delta l = 0.2$ cm and $u = 4$ m/s. The velocity proves to be very low and Dikhter and Zeigarnik [78] also observed no movement in the layers. The layer thickness Δl is estimated to be close to the observed value.

10.2. Non-uniform Steady-state Conditions of Joule Heating of Plasma

In some of the experiments discussed in Chapter 2, a non-ideal plasma was developed upon the isobaric Joule heating of matter in capillaries. The results of measurements performed under the conditions of stationarity can be used to derive information on the properties of matter (such as electrical conductivity and thermal conductivity). However, in view of the complex temperature dependencies of the coefficients of electrical and thermal conductivity, one should bear in mind that such states of matter can be strongly non-uniform. Rutkevich and Sinkevich [545] have analysed non-uniform stationary states which identify the conditions of stability and the emergence of contraction or domains, as well as the effect of inhomogeneities on effective conductivity.

Consider a plasma column with a length L connected to an external circuit having an e.m.f. Φ and resistance R. The original set of equations are

$$\operatorname{div}(\text{æ}\operatorname{grad} T) + \sigma F^2 = 0\,, \quad \operatorname{div} \mathbf{J} = 0\,, \quad U = \Phi - iR\,, \tag{10.12}$$

where U is the potential difference across the plasma column. The $\sigma(T)$ and æ(T) relationships, as well as the wall and electrode temperatures T_w, are pre-assigned. The problem is to determine the characteristic of the current-voltage.

Rutkevich and Sinkevich [545] have solved two problems: the variation of temperature along the column (longitudinal problem) and, transversely, to the column (transverse problem). In the former, the heat is extracted *via* electrodes (the heat flux, ægrad T, is parallel to the current i) while the walls are ideally heat-insulated. In the transverse problem, the heat flux is perpendicular to the current, that is, heat is only extracted through the walls. For concreteness, consider a longitudinal problem. Direct the y-axis along the column and place the origin of co-ordinates in the middle of the column (the top and bottom electrode co-ordinates $y = \pm L/2$). As is customary in heat transfer problems, make a transition to a new "temperature" θ uniquely related to T:

$$\theta(T) = \text{æ}_{\text{W}}^{-1} \int_{T_{\text{W}}}^{T} \text{æ}(T')dT'\,, \quad \text{æ}_{\text{W}} = \; tae(T_{\text{W}})\,.$$

The $\theta(y)$ distribution is then found from the solution to the following non-linear boundary problem:

$$\frac{d^2\theta}{d^2y} + \frac{1}{m} f(\theta) = 0\,, \quad \left.\frac{d\theta}{dy}\right|_{y=0} = 0\,, \quad \theta\Big|_{y=\pm L/2} = 0\,, \tag{10.13}$$

where $m^{-1} = J^2\text{æ}_{\text{W}}/\sigma_{\text{W}}$, $\sigma_{\text{W}} = \sigma(T_{\text{W}})$ and $f(\theta) = \sigma_{\text{W}}/\sigma(\theta)$.

Equation (10.13) is integrated by quadratures:

$$|y| = \sqrt{m/2}\frac{L}{2}\int_0^{\theta} \frac{d\theta}{\sqrt{V_0 - V(\theta)}}\,,$$
$$V(\theta) = \int_0^{\theta} f(\eta)d\eta\,, \quad V_0 = V[\theta(y=0)]\,. \tag{10.14}$$

Equations (10.14) uniquely define the solution because $\theta_0 = \theta(y = 0)$ is not an independent parameter. Indeed, the evident relationship

$$\sqrt{m/2}\int_0^{\theta_0,} d\theta[V_0 - V(\theta)]^{-1/2} = 1 \tag{10.15}$$

is obtained.

Integrals in the form of Eqs. (10.14) and (10.15) are typical of the dynamics of one-dimensional motion, in which an equation emerges that is analogous to Eq. (10.13):

$$d^2s/dt^2 + m^{-1}f(s) = 0\,, \quad f(s) = dV/ds\,.$$

Here, s, t, $V(s)$, m and $f(s)$ denote the co-ordinate, time, potential, mass and force, respectively. This analogy makes simpler the analysis of non-linear functionals of potential. The analysis is performed in a conventional manner [543]. We now consider the results obtained by Rutkevich and Sinkevich [545].

The qualitative structure of current-voltage characteristics has been studied. It is shown that for the case of non-monotonic dependence of electrical conductivity on temperature, which is characteristic of the $\sigma(T)$ isobar of non-ideal cesium plasma, the current-voltage characteristics can have portions of non–single-valued current and voltage. In the case of the longitudinal problem, the current-voltage characteristic is N-shaped (Fig. 10.1a) with a portion of negative differential resistance, $R_d = dU/di < 0$. In the transverse problem, the current-voltage characteristic is S-shaped (Fig. 10.1b). In the portions of $R_d < 0$, thermal instability can develop. The emergence of N-type portions is due to a sharp decrease in electrical conductivity and electronic thermal conductivity upon heating (in the $T \approx T_c$ region). The development of S-type portions is due to the rise of the $\sigma(T)$ isobar in the high-temperature region.

Rutkevich and Sinkevich [545] have also considered the effect of finite thermal resistance of the walls on the structure of current-voltage characteristics. The rise in thermal resistance may bring about the emergence of new portions with $R_d < 0$. In doing so, the qualitative features of the current-voltage characteristics with different orientations of heat flux converge.

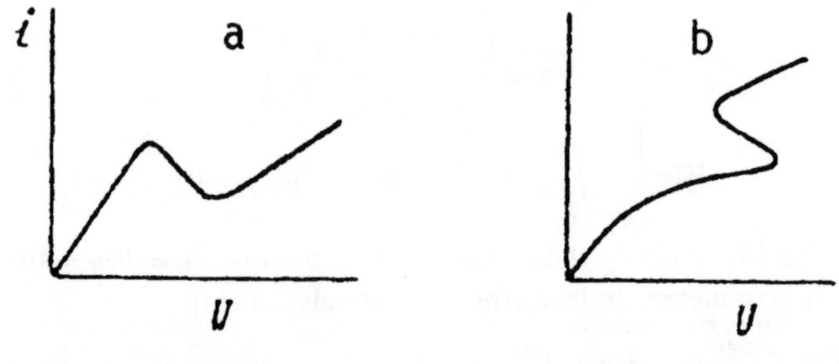

Fig. 10.1. Current-voltage characteristics of plasma column: *a*, N-shaped; *b*, S-shaped.

Stationary states which correspond to the pre-assigned values of the external circuit parameters (Φ, R) are shown by the points of intersection of current-voltage characteristic and the load curve $U = \Phi - iR$. The non-single-valuedness of current-voltage characteristic may result in the number of such points exceeding unity. Only stable solutions are realized physically. The spatial temperature distribution becomes non-uniform, that is, two- or three-phase, depending on the behavior of $\sigma(T)$ and æ(T).

The inhomogeneity of the plasma column may bring about an appreciable difference between the measured electrical conductivity $\sigma_{\text{eff}}(T) = \langle J\rangle/\langle F\rangle$ (where $\langle J\rangle$ and $\langle F\rangle$ are the average values of J and F) and the true electrical conductivity $\sigma(T)$. Rutkevich and Sinkevich [545] have studied the relationship between $\sigma(T)$ and $\sigma_{\text{eff}}(T_0)$, where T_0 is the maximum temperature in the plasma volume. Figure 10.2 illustrates the results of the calculation of $\sigma_{\text{eff}}/\sigma$ for the longitudinal $(\sigma_{\parallel}/\sigma)$ and transverse $(\sigma_{\perp}/s)$ problems. This depends on the parameter $\theta_0/\theta_{\text{W}}$ uniquely defining T_0/T_{W} and growing with T_0/T_{W}. In doing so, the true $\sigma(T)$ relationship characteristic of non-ideal metal plasma was modeled by a two-step function. This is also shown in Fig. 10.2.

It turns out that the difference between $\sigma_{\text{eff}}(T)$ and $\sigma(T)$ may be qualitative. When the $\sigma(T)$ minimum is sufficiently narrow, the $\sigma_{\text{eff}}(T_0)$ relationship, after passing the minimum, has a maximum at T_m. Therefore, at $T_0 > T_m$,

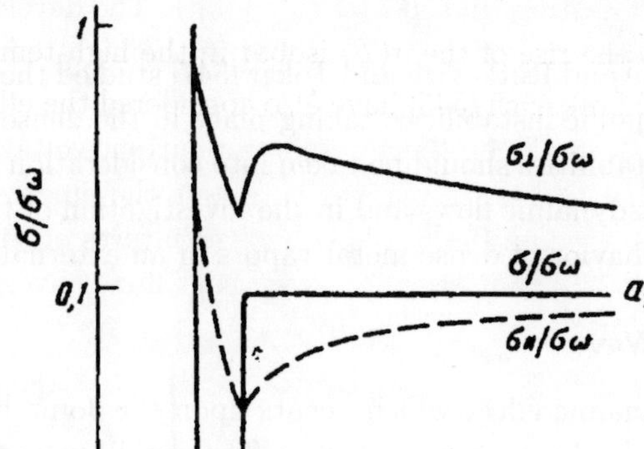

Fig. 10.2. Effective conductivity under the conditions of longitudinal $(\sigma_{\parallel}/\sigma)$, transverse $(\sigma_{\perp}/\sigma)$, heat transfer, and true conductivity (σ) as a function of reduced temperature $\theta_0/\theta_{\text{W}}$ [545].

the effective conductivity, as distinct from the true one, decreases with an increase of in T_0. The difference between σ_{eff} and σ is due to the fact that the spatial distribution of conductivity is single-phase at $\theta < \theta_1$, two-phase at $\theta_1 < \theta_0 < \theta_2$, and three-phase at $\theta_0 > \theta_2$. When θ_0 passes through θ_1, a zone of lower conductivity, ($\sigma = a_1\sigma_{\text{W}}$), emerges in the center. In doing so, $\sigma_{\text{eff}}(\theta_0)$ starts decreasing continuously rather than by jumps. Hence for a sufficiently small $(\theta_2 - \theta_1)$, there will be $\sigma_{\text{eff}}(\theta_2) > a_2\sigma_{\text{W}}$. Upon the passage of θ_0 through θ_2, the conductivity in the center increases to the $a_2\sigma_{\text{W}}$ level and $\sigma_{\text{eff}}(\theta_0)$ starts to grow. However, a monotonic growth of σ_{eff}, on the condition that $\sigma_{\text{eff}}(\theta_2) > a_2\sigma_{\text{W}}$, is impossible. This is because at very high temperatures, the value of electrical conductivity must reach the $a_2\ \sigma_{\text{W}}$ level.

This effect must be included when interpreting the experimental data obtained from σ_{eff} measurements under the conditions of strong Joule heating. According to the studies by Dikhter and Zeigarnik [43, 78] discussed in the preceding section, the electrical conductivity of cesium measured on isobars at $T = (10\text{–}20) \cdot 10^3$ K decreased with an increase in temperature. Such behavior of $\sigma_{\text{eff}}(T)$ appears incomprehensible and, on the other hand, is inconsistent with the results of experiments performed in this field in other laboratories (Chapter 7). In this temperature range, one must expect an increasing $\sigma(T)$ relationship. It can be assumed that the values of $\sigma_{\text{eff}}(T)$ measured by Dikhter and Zeigarnik [43, 78] were affected by the inhomogeneity of the plasma column.

Rutkevich [546] and Rutkevich and Tokar [547] studied the possible types of magnetohydrodynamic instabilities taking place in the dense plasma of metal vapors. These instabilities should be taken into consideration both in the analysis of magnetogasdynamic flows and in the investigation of the effect of Joule heating on the behavior of dense metal vapors in an external magnetic field.

10.3. Heating Wave

An interesting dynamic effect which occurs upon the Joule heating of matter in a capillary is the heating wave [548]. The capillary is filled with liquid metal beforehand. Since the electrical conductivity of the latter, σ_{M}, is high, the heat release is low (J^2/σ_{M}) and the metal is heated slowly. However, if a hot discharge occurs in some part of the current volume, its front will start to propagate over cold matter. The plasma conductivity σ_{P} is much lower than σ_{M}, that is, J^2/σ_{P} is high. On the contrary, the plasma thermal conductivity æ$_{\text{P}}$ is very high. Due to the thermal conductivity, cold matter is heated and the discharge phase interface (heating wave front) moves along the capillary at a constant rate. The hot plasma then propagates gradually over the entire discharge. The matter passes through the front and is heated on the supercritical isobar so that it changes to the state of non-ideal plasma from the liquid-metal state before transforming itself into a strongly ionized plasma.

The wave propagation mechanism is analogous to the slow flame propagation mechanism. As distinct from heat transfer problems, heat release is localized in the wave front and moves therewith. This mode of hot discharge propagation over cold matter is well-known in the physics of low-temperature plasma. It is realized as ionization wave in gas discharge [549] under the conditions of high-frequency discharge in a flow of gas and laser spark propagation [456, 550]. In all these cases, hot plasma absorbs the energy of the electromagnetic field, for which the cold non-ionized matter is transparent.

A heating wave in cesium was apparently observed by Barol'skii *et al.* [551] at $p = 500$ atm, $J = 10^4$ A/cm^2. Heat release in liquid metal is not high, $\sigma_{\rm M} \approx$ 400 Ohm$^{-1} \cdot$ cm^{-1}), and cesium was heated to within a critical temperature for more than 10 ms. The discharge developed is strongly inhomogeneous. To overcome this difficulty, a diaphragm was introduced in the discharge gap. This led to the development of an initial increase in resistance and the ignition of a discharge at one of the ends of a quartz capillary. In the region of ignited hot discharge, the temperature increases rapidly ($T_{\rm P} \approx 10^4$ K). As a result, the discharge in the initial stage acquires a two-phase structure. The photographs of the initial stage of the discharge clearly indicate the presence of these two regions. The wave velocity was of the order of 10 m/s.

10.3.1. *Formulation of the Problem*

Let a current flow through a liquid metal-filled capillary with a radius R. The current density J is such that the Joule heating of cold metal is insignificant. When a region of the capillary is filled with a hot plasma, the heat released in it is transferred by thermal conductivity (and by radiation transfer) in all directions, including those along the capillary. The hot plasma layer which directly adjoins the cold metal is referred to as "front". A constant temperature T_k is maintained in the hot plasma behind the front. This is due to the balance between Joule heating and radial heat losses. As a result of longitudinal thermal conductivity, the front moves over cold matter.

We now consider the steady-state mode in the system of co-ordinates where the front is resting. Cold metal with a density of ρ_0 and a temperature of T_0 flows into the front at velocity u_0. The matter is heated in the front so that it expands in a longitudinal direction. Therefore, the velocity of hot plasma which exits from the front, u_k, is much higher than u_0. According to the equation of continuity, $u_0\rho_0 = u_k\rho_k$, where ρ_k is the plasma density at $T_k \gg T_0$. All velocities are subsonic and the expansion occurs at constant pressure p, which is equal to the pressure in cold matter.

One must include the following in the energy balance: longitudinal heat transfer through the front; heat transfer by moving matter; Joule heating; and the radial losses of heat by conduction and radiation. We shall assume that the radiant thermal conductivity approximation is valid. By directing the x-axis along u_0, we can write the energy balance equation:

$$\frac{\partial}{\partial x}\left[æ(x)\frac{\partial T}{\partial x}\right] - u_0\rho_0 c_0\frac{\partial T}{\partial x} + \frac{J^2}{\sigma(T)} + \frac{1}{r}\frac{\partial}{\partial r}\left[ræ(T)\frac{\partial T}{\partial r}\right] = 0\,. \tag{10.16}$$

Here, $æ(T)$ is the overall coefficient of thermal conductivity, c_p is the heat capacity per unit mass which varies slightly with temperature, and $\sigma(T)$ is the electrical conductivity. The two-dimensionality of the problem greatly complicates the solution. We shall therefore consider a simplified one-dimensional problem. By averaging Eq. (10.16) over the cross-section of the capillary, and in accordance with Raizer [550], we express the energy losses through the capillary side surface, $F_-(T) = (2/R)[æ(T)\partial T/\partial r]_{r=R}$, as

$$F_-(T) = A\int_0^T æ(T)dT/R^2\,, \tag{10.17}$$

where $A = 2$ is the numerical coefficient defined by the radial temperature profile. As a result, we obtain the equation

$$\frac{d}{dx}\left[æ(T)\frac{dT}{dx}\right] - \rho_0 u_o c_o\frac{dT}{dx} + \frac{J^2}{\sigma(T)} - F_-(T) = 0\,. \tag{10.18}$$

Let us now formulate the boundary conditions. The $T(x)$ (front) solution, which is of interest to us, has the form in Fig. 10.3a. The main drop in termperature occurs in the region of the size L_f, which defines the front width. Beyond this region, $dT/dx \to 0$ while the natural boundary condition resides in the absence of heat flux away from the front:

$$æ(T)\frac{dT}{dx}\bigg|_{x\to-\infty} = æ(T)\frac{dT}{dx}_{x\to+\infty} = 0\,.$$

At $x \to -\infty$ in cold metal, $T = T_0$; at $x \to +\infty$ in hot plasma, $T = T_k$. The solution of Eq. (10.18) with the boundary conditions is possible at a certain value of the parameter u_0. The latter which is the velocity of front movement relative to liquid metal.

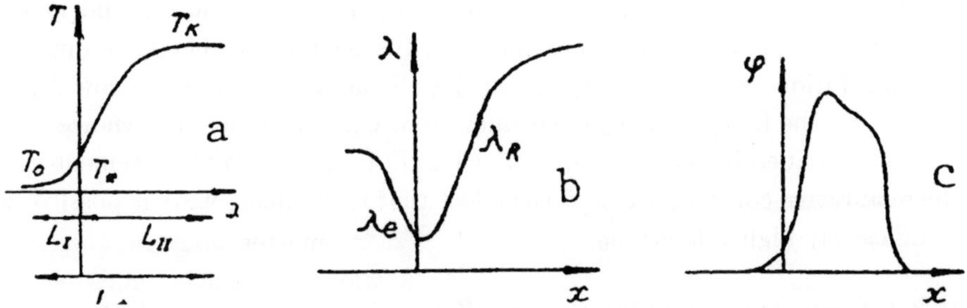

Fig. 10.3. Profiles of the parameters of matter in the heating wave front: *a*, temperature; *b*, thermal conductivity æ; *c*, source function ϕ [548].

We now introduce the quantity z, which is the ratio between the thermal conductivity and transfer heat fluxes:

$$z(T) = \text{æ}(T)(dT/dx)(u_0\rho_0 c_p T)^{-1}\,. \tag{10.19}$$

Equation (10.18) is, relative to z, a first-order equation with the boundary conditions $z(T_0) = 0$, $z(T_k) = 0$:

$$\begin{aligned} zdz/d\ln T &= -(z^2 - z + V\varphi(T))\,, \\ V &= \frac{\pi^2}{3e^2}J^2\frac{1}{(u_0\rho_0 c_p)^2}\,, \\ \varphi(T) &= \frac{3e^2}{\pi^2}\frac{\text{æ}(T)}{T\sigma(T)}\left[1 - \frac{F_-(T)\sigma(T)}{J^2}\right]\,. \end{aligned} \tag{10.20}$$

The wave solution is possible if the function of the heat source, $W = J^2/\sigma - F_-(T)$, reaches zero at the initial and final temperatures T_0 and T_k, respectively. Accordingly, $\varphi(T_k)$ and $\varphi(T_0)$ are likewise equal to zero.

A stationary moving interface between cold and hot phases is possible if the Joule release in the cold region is fully absent. It is only in this case, as shown by Zel'dovich [552], that a mathematically correct wave solution exists. In the theory of the thermal conductivity of the motion of flame, this was ensured by the exponential temperature dependence on the rates of chemical reaction. In our problem, $\varphi(T_0)$ is more than zero. This corresponds to the heating of cold metal by current until the front is passed. Nevertheless, a mode with a moving quasi-stationary interface is possible if the time during which the wave passes the capillary is very much less than that of the Joule heating of cold matter. To this end, the condition $V_{\varphi}(T) \ll 1$ must be valid at low temperatures ($T < T_*$). If this is the case, the technical difficulty of deriving such a solution can be obviated by cutting off the $\varphi(T)$ function at low T (for all $T < T_*$, the function $\varphi(T)$ is assumed to be equal to zero). The choice of T_* will be explained below. The condition $V\varphi(T < T_*) \ll 1$ is in agreement with the qualitative considerations to the effect that the heating wave is possible at a sufficiently high velocity u_0.

10.3.2. *Characteristics of the Heating Wave*

The thermal conductivity coefficient æ(T) experiences considerable qualitative changes with an increase in temperature. In the case of liquid metal, this is the electronic thermal conductivity æ$_e$, which is related to the electrical conductivity by the Wiedemann-Franz law:

$$æ_e = (\pi^2/3e^2)\sigma T\,. \tag{10.21}$$

Therefore, in the T_c region, one must observe, along with the decrease in σ, a decrease in electronic electrical conductivity.

Under certain conditions, radiative heat transfer is included in the radiant thermal conductivity approximation. The coefficient of the radiant thermal conductivity is

$$æ_{\mathrm{R}}(T) = (16/3)\sigma_{\mathrm{SB}} l_{\mathrm{R}} T^3\,, \tag{10.22}$$

where $\sigma_{\mathrm{SB}} = 4.1\cdot 10^{11}\ \mathrm{cm}^{-2}\cdot s^{-1}\cdot K^{-3}$ is the Stefan-Boltzmann constant and l_{R} is the Rosseland's length equal to the mean path length of radiation. Studies [544] on the optical properties of non-ideal cesium plasma indicate that the density effects can lead to a weak dependence of the Rosseland's mean path length of radiation, l_{R}, on pressure and temperature in the ranges of 300–1,000 atm and $4\cdot 10^3$–$1.5\cdot 10^4$ K. The value of l_{R} in this region of parameters is estimated at $5\cdot 10^{-3}$ cm. Since l_{R} is very much less than the capillary radius ($R = 0.1$ cm), the radiant thermal conductivity approximation is valid.

The æ(T) = æ$_e(T)$ + æ$_{\mathrm{R}}(T)$ plot is given in Fig. 10.4. At temperatures below 3,000 K, the inequality æ$_{\mathrm{R}} \ll$ æ$_e$ takes place and heat transfer is realized by electronic thermal conductivity. However, even at $T \geq 5,000$ K, æ$_{\mathrm{R}}$ exceeds æ$_e$ considerably and the thermal conductivity is radiative. The mechanism of heat transfer varies and æ(T) is no longer related to $\sigma(T)$.

We now turn our attention to the $\varphi(T)$ function in Eq. (10.20). With due regard for Eq. (10.21), the expression for $\varphi(T)$ can be rewritten as

$$\varphi = [(æ_e + æ_{\mathrm{R}})/æ_e][1 - F - \sigma/J^2]\,.$$

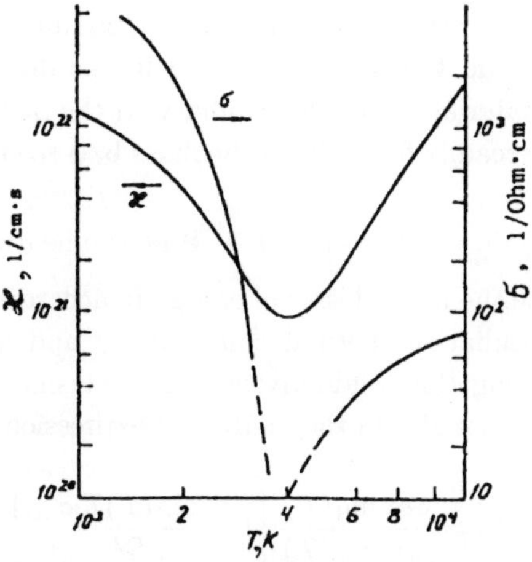

Fig. 10.4. The temperature dependence of the electrical (σ) and thermal (æ) conductivity of cesium plasma along the $p = 500$ atm isobar [548].

We will now discuss the front structure and evaluate its width. The profiles of the thermal conductivity coefficient æ(T) and source function $\varphi(T)$ in the front are shown in Fig. 10.3b. The peculiarities of the temperature dependence of the coefficients $\sigma(T)$ and æ(T) give rise to two characteristic regions. In region I, which adjoins the cold matter, heat transfer is accomplished by the electronic thermal conductivity ($T < T^*$).

In region II, the thermal conductivity increases as a result of radiant heat transfer. It is in this region that an intensive heat release takes place. The assumptions made are valid if the width of this region, L_{xx}, exceeds l_R. By integrating the expression æ$(T)dT/dx = z(T)u_0\rho_0 c_p T$ in the (T_*, T_k) temperature range, we obtain

$$L = (u_0\rho_0 c_p)^{-1}\int_{T_*}^{T_k} æ(T)T^{-1}z^{-1}(T)dT$$

for L. In view of $z(T) < 1$, one can evaluate L from below:

$$L > (u_0\rho_0 c_p)^{-1}\int_{T_*}^{T_k} æ(T)T^{-1}dT\,.$$

10.3.3. *Front Velocity*

The quantity u_0 is the velocity of travel of the heating wave front over cold material. The velocity observed will depend on the experimental conditions.

If the discharge is restricted by the walls, it is only possible for the test substance heated behind the front to expand in the direction of front travel and force the cold material out of the capillary. If the discharge tube is closed at one end, it can be readily found from the mass balance that the rate of front travel will be

$$u = 2\rho_0 u_0/\rho_k \gg u_0 \, .$$

We shall calculate the radial losses of energy in accordance with Eq. (10.17) including only the radiative thermal conductivity and assuming that l_R = const. Upon calculating the conductivity of hot plasma ($T > T^*$) from the Spitzer's formula σ_SP, we obtain the analytical expression for the source function $\varphi(T)$:

$$\varphi(T) = \frac{3e^2}{\pi^2} \frac{æ_\mathrm{R}(T)}{\sigma_\mathrm{Sp}(T)} \left[1 - \frac{æ_\mathrm{R}(T)T}{2R^2} \frac{\sigma_\mathrm{Sp}}{J^2}\right] .$$

A strong temperature dependence is contained in braces which disappear at $T \to T_k$. This enables one to use, for the evaluation of the average level of φ, an approximation:

$$\varphi = \frac{3e^2}{\pi^2} \frac{æ_\mathrm{R}(T_k)}{\sigma_\mathrm{Sp}(T_k)} \approx \left(\frac{æ_\mathrm{R}}{æ_e}\right)_{T_k} \gg 1 \, .$$

It follows from the solution (10.23) to the eigenvalue problem that, at sufficiently high T_k values, $V_\varphi \approx 0.5$. From this, we derive the expression for the velocity u_0:

$$u_0 = (\rho_0 c_p)^{-1} \left[\frac{J^2}{\sigma_\mathrm{Sp}(T_k)} \frac{2æ_\mathrm{R}(T_k)}{T_k}\right]^{1/2} = \frac{æ_\mathrm{R}(T_k)}{\rho_0 c_p R} \, .$$

Finally, in the same approximations, Eq. (10.24) gives $L_f \geq 0.4R$. Therefore, L_f turns out to be of the order of the discharge radius and the condition of radiant thermal conductivity across the front coincides with the condition $R \gg l_\mathrm{R}$.

The effect observed in a capillary discharge by Barol'skii *et al.* [551] may be associated with the notions of the heating wave passage. The following are the results of calculation for concrete conditions: $J = 10^4$ A·cm^{-2}, $p = 500$ atm, $R = 0.17$ cm; the temperatures $T_* = 3 \cdot 10^3$ and $T_k = 1.6 \cdot 10^4$ K, $\varphi = 8$. Hence $u_0 = 26$ cm/s. The rate of front travel along the capillary $u = 2u_0\rho_0/\rho_k = 1.9 \cdot 10^3$ cm/s. The time of heating wave propagation over a 2 cm long capillary is of the order of 1 ms. This is less than the time of the Joule heating of liquid metal. The observed values of the wave velocity range from 3 to 17 m/s.

10.4. The Recombination Coefficient in a Plasma With Multiply Charged Ions

Under high density, the recombination is due primarily to three-particle electron–electron-ion collisions. In accordance with Thomson's theory [6], a free electron gives up its energy during time τ and, if it finds itself in the vicinity of an ion, is bound by the latter. Therefore, the rate of triple recombination in an ideal plasma is written as

$$dN_e/dt = -N_e \mathrm{w}/\tau \,.$$

Here, w is the probability of an electron finding an ion in the neighborhood of its radius $r_{\mathrm{T}} = Ze^2/T$. At this distance, the electron-ion interaction becomes strong. In an ideal plasma, the probability w is small:

$$w = 1 - \exp[-(4/3)\pi r_{\mathrm{T}}^3 N_i] \approx (4/3)\pi r_{\mathrm{T}}^3 N_i \,.$$

The time τ in the elementary theory is the time required for an electron to meet another electron:

$$\tau = (N_e q \nu)^{-1}\,, \quad q \approx \pi(e^2/T)^2 \ln\Lambda\,, \quad \nu = \sqrt{T/m}\,.$$

Here, q is the inter-electron collision cross-section, and $\ln\Lambda$ is the Coulomb logarithm. As a result, we derive

$$\frac{dn_e}{dt} = -\alpha \frac{e^{10} Z^3 \ln\Lambda}{\sqrt{m}\, T^{9/2}} N_e^2 N_i \,. \tag{10.23}$$

In reality, the kinetics of energy loss by electron are much more complicated. An electron recombines at a much higher rate if it loses its energy in small batches rather than all of it in a single collision. An electron in the field of ion diffuses under the effect of collisions in the energy space and makes a transition from the positive to negative energy region. This problem was solved by Gurevich and Pitaevskii [553]. The literal part of (10.23) remained the same, but the constant $\alpha = 4\pi\sqrt{2\pi}/3$ and logarithm $\ln\Lambda$ were calculated.

In a non-ideal plasma, formula (10.23) is invalid for two reasons [554]. If $\gamma_{ZZ} \gg 1$, the system of ions forms a cellular structure (see Chapter 5) and, because $\gamma_{Ze} \geq 1$, the cell radius $\bar{r} = [3/4\pi N_i]^{1/3}$ proves to be greater than the Thomson radius r_{T}. This implies that an electron always interacts strongly with an ion. Therefore, the probability $w = 1 - \exp(-\gamma_{Ze}^3)$ is always close to unity. Furthermore, an electron, which is at the distance of $N_i^{-1/3}$ from an ion on the average, possesses the mean potential energy $U = -Ze^2 N_i^{1/3}$. Because the total energy of free electron $E \approx T$, the mean kinetic energy

$K \approx Ze^2N_i^{1/3} + T$. If, in the limiting case, $\gamma_{Ze} \gg 1$, then $K = Ze^2N_i^{1/3} \gg T$. In an ideal plasma, the reverse, happens $K \approx T$.

In order to be bound by an ion, an electron must transfer itself to the region of negative energy $E < 0$ and loses energy of the order of T. It is important that this energy is lost gradually. This is because the cross-section of collision with a loss of energy ΔK is $q \approx e^2(K\Delta K)^{-1}$ [555]. We will estimate the time @ as the time of electron diffusion to the energy distance $T: \ \tau = T^2/D$. Naturally, the diffusion coefficient D is proportional to N_e. We will write the expression for D literally from dimensional considerations by using the parameters N_e, e^2, and m, which are available in the problem. Hence, $D \approx N_e e^4\sqrt{K/m}$. Finally, we will derive

$$dN_e/dt = -\alpha_1 N_e^2 e^4 \sqrt{K} w \ln\Lambda/(T^2\sqrt{m}) \,. \qquad (10.24)$$

for the recombination rate.

In the limiting case of strong electron-ion interaction:

$$dN_e/dt = -\alpha_1 N_e^2 e^4 \sqrt{Ze^2N_i^{1/3}} \ln\Lambda/(T^2\sqrt{m}) \,. \qquad (10.25)$$

Upon comparing (10.25) and (10.23), we see how the dependence of the recombination coefficient on N_e, Z and T has changed. The process of recombination in this case appears to be a binary, rather than three-particle, process. Numerically, the recombination rate decreased by a factor of approximately $\gamma_{Ze}^{5/2}$. The coefficient in (10.25) was calculated by Biberman *et al.* [554] using the constant sink approximation to solve the diffusion equation in the energy space. The constant $\alpha_1 = 7.4$ was derived within the cell model.

We will write the expression for the Coulomb logarithm as $\ln\Lambda = \ln(\rho_{\max}/\rho_{\min})$, where $\rho_{\max}$ and $\rho_{\min}$ denote the maximum and minimum impact parameters, respectively. It is evident that $\rho_{\max} = (3/4\pi N_i)^{1/3})$. In a classical plasma, $\rho_{\min} = e^2/K$. Thus, $\ln\Lambda = \ln Z$. If

$$\lambda_e/(e^2K^{-1}) = \sqrt{\gamma_{Ze}T/\mathrm{Ry}} > 1 \,,$$

then

$$\rho_{\min} = \lambda_e \,, \qquad \ln\Lambda = \ln(Z\sqrt{\gamma_{Ze}T/\mathrm{Ry}}) \,.$$

In Chapter 1, the conditions are discussed, under which a non-ideal plasma with multiply charged ions arises: $Z \gg 1$. Apparently, the collision recombination dominates when the density is still very high. Therefore, the remaining recombination channels are closed.

As for the ionization coefficients, the density corrections to them occur even with moderate non-ideality. These corrections are due to the shift in the atomic (ionic) levels and a decrease in the boundary of the continuum. These quantities cause changes in the energy thresholds of the excitation and ionization processes. They also enter the exponents by appearing in the expressions for the rates of these processes [556, 557].

10.5. Hydrodynamic "Explosion" Under Conditions of Low-Temperature Plasma Flow

The processes of dissociation and ionization lead to sharp non-monotonic temperature dependencies of the coefficients of viscosity and thermal conductivity (Fig. 10.5). This fact may cause a qualitative variation in viscous plasma flow in the channels of MHD generators, in the fuel elements of gas-phase nuclear reactors, in plasmochemical facilities, and other technical apparatuses. Indeed, the viscous dissipation of energy under conditions of limited heat transfer leads to a substantial increase in temperature, thereby causing a decrease in viscosity and an "explosion" acceleration of flow [558, 559].

Consider the conditions of existence and the stability of steady-state one-dimensional flow of plasma. This is assuming that both thermal conductivity and viscosity define the most important dissipation processes taking place in such a medium. When an equilibrium plasma moves along the z-axis between plates, whose temperature T_0 does not vary under the effect of constant pressure gradient, the equations of motion, energy and boundary conditions for such flow take the form of

$$\frac{\partial}{\partial x}\left(\eta\frac{\partial p}{\partial x}\right)-\frac{\partial p}{\partial z}=0\,;\ \frac{\partial}{\partial x}\left(\lambda\frac{\partial T}{\partial x}\right)+\eta\left(\frac{\partial V}{\partial x}\right)^2=0\,;\qquad V'(0)=T'(0)=0\,;\qquad T(x_0)=T_0\,, \tag{10.26}$$

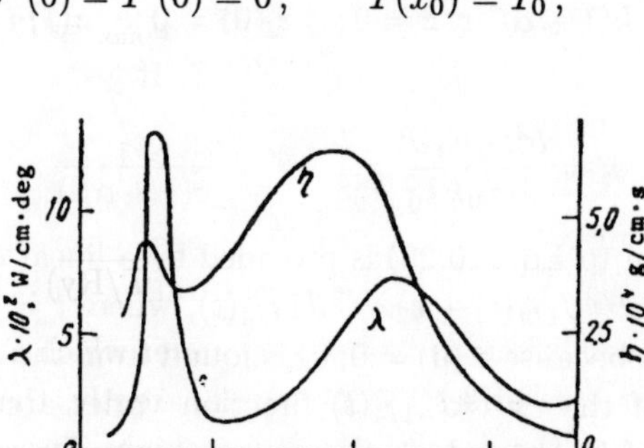

Fig. 10.5. Viscosity and thermal conductivity æ of helium plasma at $p = 1$ bar.

where η are the coefficients of shear viscosity, $\lambda = \lambda' + \lambda_R$ is the sum of the coefficients of "molecular" and "reactive" thermal conductvity, and x_0 is the distance between the plates. In considering the case of intensive dissociation or ionization of plasma, one can assume

$$\eta = \eta_0 \exp\left(\frac{E_1}{kT}\right) ; \ \lambda = \lambda_0 \exp\left(\frac{E_0}{kT}\right) , \ E_1 > 0 ; \qquad (10.27)$$
$$\lambda_0^2 , \ \eta_0 = \text{const} .$$

In the dimensionless variables,

$$T = T_0 = \frac{\theta k T_0^2}{E_1} ; \quad V = \frac{b x_0^2}{\eta_0} \exp\left(-\frac{E_1}{kT_0}\right) w ;$$
$$r = \frac{x}{x_0} ; \quad \eta = \frac{E_2 - E_1}{kT_0} ; \quad \beta = \frac{kT_0}{E_1} .$$

Equations (10.26) will be written as

$$1 + \frac{\partial}{\partial r}\left[\frac{\partial w}{\partial r} \exp\left(-\frac{\theta}{1+\beta\theta}\right)\right] = 0 ;$$
$$\exp\left(-\frac{\theta}{1+\beta\theta}\right)\left(\frac{\partial w}{\partial r}\right)^2 e^{-\alpha k} + \frac{\partial}{\partial r} \exp\left(\frac{\theta - 2}{1+\beta\theta}\right) = 0 ; \qquad (10.28)$$
$$w'(0) = \theta'(0) = 0 ; \quad w(1) = \theta(1) = 0 .$$

at $|x\beta\theta| \ll 1$, they are reduced to

$$z^4 + k(1+\alpha\beta) r^2 z = 0 ; \quad z'(0) = 0 ; \quad z(1) = 1 , \qquad (10.29)$$

where

$$k = \frac{b^2 x_0 E_1 e^{\alpha}}{\lambda_0 \eta_0 k T_0^2} ; \qquad \theta' = \frac{z'}{z(1+\alpha\beta)} .$$

The general solution to Eq. (10.29) is provided by a linear combination of the Bessel functions $c_1 x^{1/2} J_{1/4}(t) + c_2 x^{1/2} J_{-1/4}(t)$, where $t = [(1+\alpha\beta)k]^{1/2} x^2/2$ with $c_1 = 0$. This is because $z'(0) = 0$; c_2 is found from the condition $z(1) = 1$.

The behavior of the $cx^{1/2} J_{-1/4}(t)$ function varies according to the sign $1+\alpha\beta : 1+\alpha\beta > 0$, if $E_2 > 0$, which corresponds to the exponentially increasing thermal conductivity $1+\alpha\beta < 0$ at $E_2 < 0$. At $1+\alpha\beta > 0$, the $cx^{1/2} J_{-1/4}(t)$ function changes its sign with an increase in argument. Since $|cx^{1/2} J_{-1/4}(t)|$ must be finite for $0 < x < 1$, the condition $J_{-1/4}[(1+\alpha\beta)^{1/2} k^{1/2}/2] = 0$

defines the critical value of the parameter k, which turns out to be equal to $16(1+\alpha\beta)^{-1}$. At $k < k_c$, the solution to Eq. (10.26) exists while at $k > k_c$, the "explosion" mode of the plasma flow is realized.

In the case of $1+\alpha\beta < 0$, $cx^{1/2}J_{-1/4}(t)$ is a monotonic function of x and, therefore, the solution to Eqs. (10.26) exists at any positive value of k. In doing so, the temperature in the flux core proves to be less than the temperature on the plates: $|\beta\theta| \ll 1$ must be valid in the range of existence of the solution because $|\alpha| \sim 1$.

We now investigate the stability of the solution to the set of equations in (10.26) in one-dimensional perturbations. The non-stationary version of Eqs. (10.26) ($\theta'(0,t) = 0$; $\theta(1,t) = 0$; $\theta(r,0) = \theta(r)$) is reduce to

$$e^{-\theta}\frac{x_0^2 pc_p}{\tau\lambda_0}\frac{\partial\theta}{\partial t} = \theta'' + (1+\alpha\beta)(\theta') + kr^2 \tag{10.30}$$

on the condition that $c_P\eta_0/\lambda_0 \ll 1$. Following the substitution of variables and linearization, we obtain the equation in perturbation, $z_1(z = z_0 + z_1)$, z_0-, the stationary solution to Eq. (10.29) with the corresponding boundary, and the initial conditions:

$$(1+\alpha\beta)^{-1}\frac{x_0^2 pc_p}{\tau\lambda_0} z_0^{-1/(1+\alpha\beta)}\frac{\partial z_1}{\partial t} = \frac{z_1''}{(1+\alpha\beta)} + kr^2 z_1\,. \tag{10.31}$$

At $1+\alpha\beta < 0$, the solution to Eq. (10.26) is unstable. This is because the linear differential operator which corresponds to Eq. (10.31) is positive [560, 561]. For stability of the solution to Eqs. (10.26) at $1+\alpha\beta > 0$, it is sufficient (which can be demonstrated) to satisfy the inequality $k < (\pi^2/4)(1+\alpha\beta)^{-1}$.

A study of the equations which correspond to an axisymmetric flow of the plasma reveals that the character of the stationary solutions does not vary qualitatively as compared to the solutions to Eqs. (10.26). In the case of $1+\alpha\beta > 0$, there is $k_c \sim 10^2(1+\alpha\beta)^{-1}$. As in the plane case, the stationary solution at $1+\alpha\beta < 0$ is unstable. The sufficient condition of stability of the stationary solution at $1+\alpha\beta > 0$ is $k < (1+\alpha\beta)^{-1}\pi^2$. Note that the traditional theory of hydrodynamic explosion [559] gives the values of 0.88 and 2.00 for k_c in the plane and axisymmetric cases, respectively.

Previously, the boundary condition of the first type was used for $\theta(1)$ in Eq. (10.28) (or for z in Eq. (10.29)). The choice of the boundary conditions for the second and third types for $\theta(1)$ will probably fail to bring about a qualitative variation in the character of the solution, the conditions of existence of steady-state, and the explosion modes of plasma flow.

Table 10.1. Parameters of plane flow of low-temperature cesium and potassium plasma ($T_0 = 2,000$ K, $p = 1$ at) at $k < k_c$.

Element	η_0	λ_0	E_1	E_2	x_0	b	Re	$c_p\eta_0/\lambda_0$
K	$1.28 \cdot 10^{-4}$	$1.47 \cdot 10^5$	$0.88 \cdot 10^{-12}$	$0.84 \cdot 10^{-12}$	1.0	$0.704 \cdot 10^3$	$2.05 \cdot 10^3$	10^{-2}
Cs	$1.98 \cdot 10^{-5}$	$4.17 \cdot 10^5$	$1.678 \cdot 10^{-12}$	$1.63 \cdot 10^{-12}$	1.0	$3.91 \cdot 10^2$	$2.8 \cdot 10^3$	10^{-3}

Note: The data are given in the units of CGS electrostatic system.

As an example of the realization of the explosion behavior of plasma flow discussed above, Table 10.1 presents the parameters of plane flow of low-temperature cesium and potassium plasma ($T_0 = 2,000$ K, $p = 1$ atm) at $k\square k_c$. Under these conditions, heat transfer is accomplished by the molecular and "reactive" mechanisms of thermal conductivity. Meanwhile, the exponential decreases in viscosity and increase in thermal conductivity are due to the processes of dissociation and ionization taking place in the plasma. The use of the boundary condition of the first type for temperature comes naturally in view of the smallness of T_0. It follows from the data in Table 10.1 that the low-temperature plasma flow is laminar, $|\alpha\beta\theta| \ll 1$ (because $\theta \sim 1$), with the plasma heating being such that T_0 is in the region of exponential temperature dependence of η, λ. It should be noted that under certain conditions, the explosion mode of plasma flow may become turbulent.

Therefore, insofar as the meaning of the assumptions made and the nature of η, λ data are concerned, the foregoing analysis is applicable in an investigation of flows in devices where the plasma is at fairly low temperatures, such as in the channels of magnetohydrodynamic generators. In studying the flows of a more heated and dense plasma ($T \geq 10,000$ K, $p \geq 10$ atm), one should, generally speaking, change the boundary conditions for temperature and include heat transfer by radiation in, for example, a radiant thermal conductivity approximation. The coefficient of the radiant thermal conductivity is characterized by a variation in its value with an increase in temperature, which is even sharper than in the case of λ, thus favoring the emergence of the "explosion" modes of plasma flow.

10.6. Hydrodynamic Effects in Non-ideal Plasma

One of the most interesting peculiarities in the behavior of plasma, under the conditions of strong non-ideality, is the possible separation of the system into two phases: plasma vapor and a quantum plasma liquid with strong electrostatic interaction between charged particles (cf., Section 5.7). The possibility of such a separation is associated with the loss of the thermodynamic stability of the model equations of state for non-ideal plasma. This includes, in a certain manner, the collective and quantum effects of the Coulomb interaction of charged particles. Because of the absence of a quantum-mechanical theory of plasma which can be used with any interaction parameters, as well as of experimental data in many regions of the phase diagram of states, it is difficult to draw a conclusion on the existence (or absence) of the plasma phase transition predicted by the models. Therefore, in analyzing hydrodynamic phenomena in a superdense plasma developed by pulsed local concentration of energy, one cannot rule out the possibility of the passage of the phase separation regions.

In order to construct the phase equilibrium curve for the system being considered, one needs the data on the chemical potential $\mu = (p, T)$ for both phases, which are in thermodynamic equilibrium. This is because the data provided by the existing models are ambiguous. In view of this, no restrictions are pre-assigned on the form of the phase diagram. The purely thermodynamic relationships used in analysis are also not related to a particular model of the equation of state for plasma.

Bearing in mind the concrete experimental apparatuses that can be used for the dynamic generation of plasma (Chapter 3), we shall consider [371] three methods of pulsed stimulation of plasma: shock compression, isentropic compression and the adiabatic expansion of matter pre-compressed by shock wave.

10.6.1. *The Position of Isentropes and Phase Boundaries in the p-V-Plane*

In considering a mixture of two phases in the thermodynamic equilibrium [371, 562, 563], the value of entropy S_c in a two-phase region can be written as

$$S_c = S_1 + (V_c - V_1)\Delta S/\Delta V \,, \tag{10.32}$$

where $\Delta S = S_1 - S_2$ and $\Delta V = V_1 - V_2$ denote the variation of entropy and specific volume upon transition from the first (p_1, V_1) to second (p_2, V_2) phase, respectively. The entropy variation dS_1/dT (total derivatives are calculated along the phase boundary) is related to the slope of phase separation curve dp/dT by the relationship

$$\left(\frac{dS_1}{dT}\right)_p + \left(\frac{\partial S_1}{\partial p}\right)_T \frac{dp}{dT} = \frac{dS_1}{dT} \,. \tag{10.33}$$

By differentiating Eq. (10.32) with respect to p, but with due regard for the Clapeyron-Clausius equation $dp/dT = \Delta S/\Delta V$ and Eq. (10.33), we obtain

$$\frac{\partial V_c}{\partial p} = \frac{dV_1}{dp} - (V_c - V_1)\frac{d}{dp}\left[\ln \frac{dp}{dT}\right] - \frac{dS}{dT}\left(\frac{dT}{dp}\right)^2 + \left(\frac{\partial S_1}{\partial T}\right)_p \left(\frac{dT}{dp}\right)^2 + \left(\frac{\partial S_c}{\partial p}\right)_T \frac{dT}{dp}.$$

It follows that the slopes of isentrope in the two-phase region $(\partial V/\partial p)_S$ (at $V_c \to V_1$) and phase boundary dV_1/dp are

$$\left(\frac{\partial V_c}{\partial p}\right)_S - \frac{\partial V_1}{\partial p} = -\frac{dS_1}{dT}\left(\frac{dT}{dp}\right)^2. \tag{10.34}$$

Using thermodynamic identities [1], the slopes of isentrope and isotherm in the single-phase region are written as

$$\left(\frac{\partial V_1}{\partial p}\right)_S - \left(\frac{\partial V}{\partial p}\right)_T = \frac{T}{c_{p_1}}\left(\frac{\partial V_1}{\partial T}\right)_p. \tag{10.35}$$

Consequently, in the p-V-plane, the isentrope, in accordance ($c_p = c_V - T(\partial V/\partial T)_p^2(\partial p/\partial V)_T > 0$) with the condition of stability of matter, ($c_V > 0 : (\partial p/\partial V)_T < 0$), is always steeper than the isotherm.

The slopes of the isotherm and boundary curve are related to each other by

$$\frac{dV_1}{dp} - \left(\frac{\partial V}{\partial p}\right)_T = \left(\frac{\partial V}{\partial T}\right)_p \frac{dT}{dp}. \tag{10.36}$$

By subtracting Eq. (10.34) from Eq. (10.36) and using the relationship $dS_1/dT = c_{p1}/T - (\partial V/\partial T)_p(dp/dT)$, we obtain

$$\left(\frac{\partial V_c}{\partial p}\right)_S - \left(\frac{\partial V_1}{\partial p}\right)_T = 2\left(\frac{\partial V_1}{\partial T}\right)_p \frac{dT}{\partial p} - \frac{c_{p_1}}{T}\left(\frac{dT}{dp}\right)^2. \tag{10.37}$$

Analogously, by using Eqs. (10.35) and (10.36) and the Maxwell's relation, $(\partial S_1/\partial p)_T = -(\partial V/\partial T)_p$, one can obtain the relationship between the slopes of the first-phase isentrope and boundary curve:

$$\left(\frac{\partial V_1}{\partial p}\right)_S - \frac{dV_1}{dp} = -\left(\frac{\partial V_1}{\partial T}\right)_p \frac{dS_1}{\partial T}\frac{T}{c_{p_1}}\frac{dT}{dp}. \tag{10.38}$$

By subtracting Eq. (10.37) from Eq. (10.35), the relationship for the kink of isentrope upon its transition from the single-phase to two-phase region is derived:

$$\left(\frac{\partial V_1}{\partial p}\right)_S - \left(\frac{\partial V_c}{\partial p}\right)_S = \left[\sqrt{\frac{T}{c_p}}\left(\frac{\partial V_1}{\partial T}\right) - \sqrt{\frac{c_{p1}}{T}}\frac{dT}{dp}\right]^2 > 0\,. \qquad (10.39)$$

The signs of the relationships (10.35) and (10.39) can also be set according to the Le Chatelier's principle (the external stimulation which disturbs the thermodynamic equilibrium of the system is due to a variation of the pressure along the isentrope).

The set of equations (10.34) to (10.39) can be used to analyze, in the p-V-plane, the processes of isentropic compression and expansion of non-ideal plasma. It follows from the set of equations (10.34) to (10.39) that, for a pre-set phase boundary in the p-V-plane, the possibility of a phase transition taking place under the conditions of isentropic processes depend on the signs of the derivatives $(\partial V_1/\partial T)_p$, dT/dp and dS_1/dT.

We will now consider several examples. If, for plasma vapor (phase I), $(\partial V_1/\partial T)_p > 0$ (the case when $(\partial V_1/\partial T)_p$ changes the sign is discussed in Section 10.6.4), the case which appears most likely is that of phase transitions with a reduction in volume, $\Delta V > 0$, and an establishment in the system of some order $\Delta S > 0$ so that the slope of the phase curve $dT/dp = \Delta S/\Delta V > 0$. However, one cannot restrict oneself to this case only. This is because there are experimentally known transitions which have both $dT/dp < 0$ and a temperature maximum on the phase curve [564].

First, consider the transitions with $dS_1/dT > 0$. Irregardless of the sign of dT/dp, it follows from Eq. (10.34) that

$$\left(\frac{\partial V_c}{\partial p}\right)_S < \frac{\partial V_1}{\partial p}\,. \qquad (10.40)$$

If the phase equilibrium curve has a negative slope in the p-V-plane and is located to the right of the initial state in the case of compression, and to the left of the initial state in the case of expansion, the inequality of (10.40) means that it is impossible to change from a single-phase region to a two-phase region by means of an isentropic variation of the pressure. This is only possible when $dV_1/dp > 0$.

Assuming that the initial states of the isentropes belong to the first phase, let us consider the conditions on which they cross the phase boundary. At $dT/dp < 0$, Eq. (10.38) yields

$$\left(\frac{\partial V_1}{\partial p}\right)_S < \frac{dV_1}{dp}\,. \tag{10.41}$$

This leads to the case shown in Fig. 10.6a. Boundaries with positive slope are impossible in this case because they are inconsistent with Eq. (10.41) and with due regard for

$$\left(\frac{\partial p}{\partial V}\right)_S = \left(\frac{\partial p}{\partial V}\right)_T - \frac{T}{c_V}\left(\frac{\partial p}{\partial T}\right)_V^2 < 0\,. \tag{10.42}$$

When $dT/dp > 0$, Eq. (10.38) yields $(dV_1/dp)_s < dV_1/dp$. Hence Fig. 10.6b shows the phase boundaries which be investigated in this case under the conditions of isentropic variation of pressure.

Based on the empirical equations of state for both the solid and liquid phases, Urlin [565] obtained an increase in the solid phase entropy along the melting curve for metals, $dS_1/dT > 0$. Since the entropy of transition is almost constant on the melting curve [564–566], it can be assumed that $dS_2/dT > 0$ for the liquid phase along the melting curve. The foregoing analysis of $dT/dp > 0$ indicates (Fig. 10.6c) that if the matter did not melt upon shock compression, it can change to the liquid phase upon isentropic expansion (release). If the matter behind the front is in the liquid state, it cannot change to the solid phase upon release. These results are consistent with the conclusions reached by [565].

Now, consider the transitions with $dS_1/dT < 0$. In this case (observing, as previously, the condition $(\partial V_1/\partial T)_p > 0$), one must consider only the negative slopes of the boundary of the two-phase region, $dV_1/dp < 0$. Indeed, in accordance with Eq. (10.34),

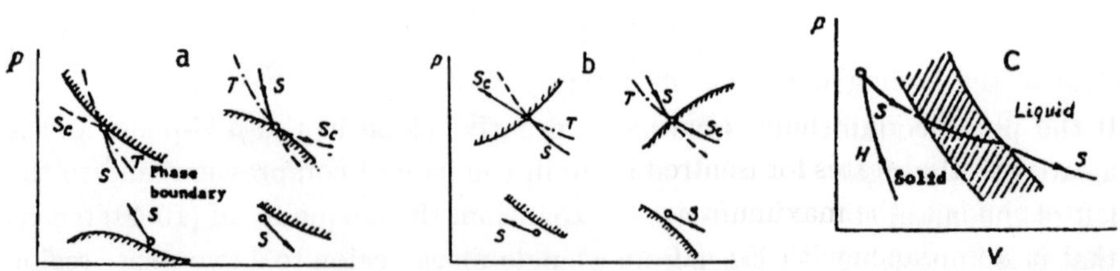

Fig. 10.6. The position of isentropes and phase boundaries in the p-V-plane $(dS/dT) > 0$; the two-phase region is shaded. S, the isentrope of the first phase, with the broken line showing its continuation into the two-phase region; S_c, the isentrope of the two-phase region; T, the isotherm, a, $(dV_1/dp) < 0$, $(dT/dp) < 0$; b, $(dV_1/dp) > 0$, $(dT/dp) > 0$; c, melting in the release wave, $(dT/dp > 0$, $(dS_1/dT) > 0$.

$$\left(\frac{\partial V_c}{\partial p}\right)_S > \frac{\partial V_1}{\partial p}. \tag{10.43}$$

However, in view of Eq. (10.39), $(\partial V_1/\partial p)_S > (\partial V_c/\partial p)_S$ which, along with Eq. (10.42), rules out the possibility of $\partial V_1/\partial p > 0$. The isentropes of the initial phase cross the two-phase region in the case of $dT/dp > 0$. Otherwise, they are flatter in the p-V-plane than the phase separation curve. Such a phase transition cannot be recorded under the conditions of isentropic variation of pressure (Fig. 10.7).

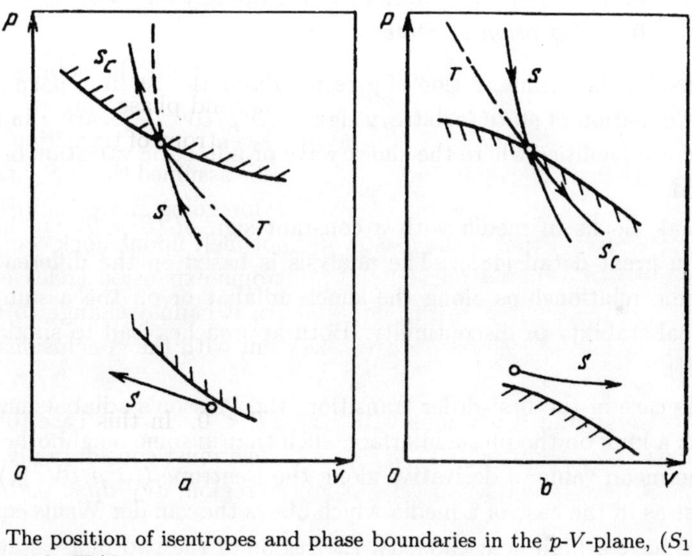

Fig. 10.7. The position of isentropes and phase boundaries in the p-V-plane, $(S_1/dT) < 0$: *a*, $dT/dp > 0$; *b*, $dT/dp < 0$.

For the boiling curve of one-component systems, it is usually $dT/dp > 0$ and for the right-hand boundary curve (evaporation), $dS_1/dT < 0$. By selecting initial states for isentropic compression on the saturation curve (which corresponds to the maximum density of the gas phase at a given temperature), and in accordance with Eq. (10.43), the isentrope in the p-V-plane is steeper than the right-hand boundary curve of boiling. Therefore, the compression isentropes cannot be employed to investigate the critical point neighborhood.

As a rule, for most substances, the conditions $(\partial p/\partial S)_V > 0$ and $(\partial^2 p/\partial V^2)_S > 0$ are valid. The shock adiabat also passes through the p-V-plane above the Poisson's adiabat to emerge from the same point [458].

Therefore, the critical point region cannot be investigated under the conditions of shock compression of states lying on the saturation curve. It follows from the foregoing inequalities that when a shock wave propagates over a material with normal thermodynamic properties, with the initial states of the said wave lying in the p-V-plane on (or above) the saturation curve, the precipitation of the condensate of liquid is impossible behind the shock front. In case the initial states are in the vapor-gas region, the possibility of partial or full condensation of the medium in the process of shock compression will depend on the degree of supersaturation of vapor and the shock wave intensity.

10.6.2. *Disintegration of Discontinuity in a Plasma with an Arbitrary Equation of State*

Now, consider the disintegration of pressure discontinuity in a plasma with an arbitrary equation of state (arbitrary sign of $(\partial^2 p/\partial V^2)_S$, bearing in mind the experimental facilities where the shock wave or adiabatic variation of pressure is realized.

Normal shocks in media with a constant sign of $(\partial^2 p/\partial V^2)_S$ have been studied in great detail [485]. The analysis is based on the differential thermodynamic relationships along the shock adiabat or on the assumptions of mechanical stability of discontinuity. Both approaches lead to similar results (cf., [458]).

In the case of the first-order transition, the Poisson's adiabat may experience such a kink on the phase interface, such that in some neighborhood of the latter the mean value of derivative along the isentrope $\langle(\partial^2 p/\partial V^2)_S\rangle$ is negative. Just as in the case of a media which obeys the van der Waals equation of state and have a high heat capacity, the region of the two-phase states in the p-V-plane may be adjoined by a region in which $(\partial^2 p/\partial V^2)_S < 0$ [567, 568]. Moreover, in a non-ideal plasma, one cannot rule out in advance the possibility of a second-order transition occurring with a kink on the Poisson's adiabat.

If, assuming [569] that a medium, prior to shock compression, possesses normal thermodynamic properties with $(\partial^2 p/\partial V^2)_S > 0$, Fig. 10.8 shows, in the p-V-plane, a plurality of states (wave adiabat [570]) emerging in the medium following the disintegration of pressure discontinuity; the density profiles in the corresponding waves are also shown in the figure.

In the $p_0 < p < p_1$ pressure range, when $(\partial^2 p/\partial V^2)_S > 0$ is valid, the solution is constructed in the usual manner [458] so that the sought curve is the Hugoniot adiabat $0a(p_0 < p_a < p_b < p_2)$.

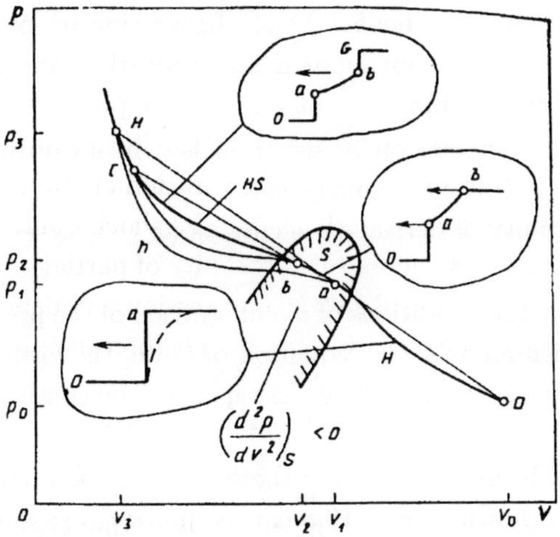

Fig. 10.8. Wave adiabat in a medium with alternating sign $(\partial^2 p/\partial V^2)_S$. The density profiles corresponding to different portions of the wave adiabat are shown alongside it.

At $p_1 < p < p_2$, the shock adiabat $0a$ is complemented by a segment of isentrope S, which corresponds to the value of entropy at point a, while p_2 is the second kink on that isentrope. The matter is compressed to the a state in a direct shock wave propagating in unperturbed medium at a rate of $D_a = V_0[(p_1 - p_0)/(V_0 - V_1)]^{1/2}$. The matter is also compressed to the b state (on the isentrope segment ab) in a centered compression wave, which is separated from the region of translational motion (at point b) by a surface of weak discontinuity.

Since the propagation rates of the "head" and "tail" of this wave (defined by the slopes of tangents to the isentrope at points a and b) are different from each other, the pressure profile of ab will spread out upon motion in space.

In the $p_2 < p < p_3$ region, the flow pattern differs from the preceding case in that the interface between the compression wave and region of translational motion is a surface of strong discontinuity. In this case, the disintegration of arbitrary discontinuity proceeds on the HS curve, which is the locus of intersection of the Hugoniot adiabats h emerging from the states on the isentrope S. The tangents to this isentrope are drawn from the same initial points.

Therefore, the matter in this case first compresses in the shock wave to the a state and then, along the isentrope in the centered compression wave, to the b state. Finally, in the secondary shock wave, it compresses to the final state c. The rate of secondary wave propagation in pre-compressed matter depends on the slope of the Rayleigh line bc. Thus, the wave in motion will lag behind the primary shock wave by moving at a rate defined by the Rayleigh line $0a$. Such a configuration is stable.

At $p_3 < p$, the compression is accomplished in an ordinary shock wave analogous to the first of the foregoing cases (at $p < p_1$). Note that at the point (p_3, V_3), a discontinuity of derivative occurs along the curves HS and H if, at this point, $dS \neq 0$ [570].

Therefore, under the conditions of disintegration of the pressure discontinuity in a medium with an arbitrary equation of state, the formation of complex flow modes, including several shock waves separated by continuous compression waves, is possible.

In case the equilibrium time behind the surface of the discontinuity is great, the density profiles shown in Fig. 10.8 can be markedly distorted (the broken line in Fig. 10.8). The theory of shock waves in media with delayed excitation of internal degrees of freedom is exposed in detail by Zel'dovich [567].

The most well-studied case corresponds to a kink on the wave adiabat interface when the segment of isentrope S in Fig. 10.8 is contracted to a single kink point. In doing so, shock compression to pressures in the $p_1 < p < p_3$ range causes the propagation of two shock waves without the formation of a intermediate centered compression wave. The formation of a two-wave structure has now been experimentally recorded in two cases. They are the shock wave separation into both elastic and plastic waves, which emerge due to the finite value of the shear modulus of solids, and a two-wave structure caused by polymorphic transformations with volume reduction [458]. For a two-wave structure to form, it is necessary for the slope of adiabat at the point of kink to be less in magnitude than the slope of the Rayleigh line at the same point (Bethe-Weyl condition).

In the case of shock wave separation, the region of a two-wave structure can be used to obtain quantitative information on phase transition [571]. The thermodynamic discussion, which is analogous to the derivation of Eq. (10.34), leads (at $V_0 = V_1$) to the quadratic equation in dp/dT:

$$\left(\frac{dp}{dT}\right)^2 + \frac{2(\partial V_1/\partial T)_p}{(\partial V_1/\partial p)_T - (\partial V/\partial p)_{HC}} \frac{dp}{dT} + \frac{c_{p_1}}{T[(\partial V_1/\partial p)_T - (\partial V_1/\partial p)_{HC}]} = 0 . \qquad (10.44)$$

The thermodynamic description of the first quasi-ideal plasma phase is usually adequately accurate; the shock compressibility of the shock mixture is related to the experimentally found velocity of a weak secondary wave, D_2, by

$$\left(\frac{\partial V_1}{\partial p}\right)_{HC} = \frac{V_1^2}{D_2^2}.$$

Therefore, the coefficients in Eq. (10.44) are determined. This enables one to find, experimentally, the important quantity such as the slope of the phase equilibrium curve at the kink of the shock adiabat.

Given the position of the shock adiabat in the two-phase region and an appropriate thermodynamic description of the initial phase, the phase equilibrium curve is found by a numerical integration of the relationship [572]

$$\frac{dp}{dT} = \frac{(p_c - p_1)(V_1 - V_c)}{2T_c(V_{1c} - V_c)} + \left[\int_{T_1}^{T_c} c_V dT + \int_{V_1}^{V_{1c}} T\left(\frac{\partial p}{\partial T}\right)_V dV + \int_{p_1}^{p_c} V dp\right]_1,$$

where the integrals are calculated for the inital phase parameters along the phase equilibrium curve; subscript 1 corresponds to the kink of the adiabat when crossing the phase boundary. For the slopes of the shock adiabats and isentropes in the single-phase region, the following relationship [371] is valid:

$$\left(\frac{dp_1}{dV}\right)_H = \frac{(\partial p_1/\partial V)_S + (\gamma/2V_1)(p_1 - p_0)}{1 - (\gamma/2V_1)(V_0 - V_1)}, \tag{10.45}$$

where p_0, V_0, are the states prior to shock compression and the Grüneisen parameter is

$$\gamma = V\left(\frac{\partial E}{\partial p}\right)_V^{-1} = -\frac{V}{T}\left(\frac{\partial p}{\partial V}\right)_S\left(\frac{\partial T}{\partial p}\right)_S.$$

Using the expression for thermodynamic quantities in the two-phase region, we obtain

$$\left(\frac{dp}{dV}\right)_{HC} = \frac{(\partial p_c/\partial V)_S + (\gamma_c/2V_1)(p_1 - p_0)}{1 - (\gamma_c/2V_1)(V_0 - V_1)} \tag{10.46}$$

for the state on the phase boundary, where the Grüneisen parameter of the mixture $\gamma_c = -\frac{V}{T}\left(\frac{\partial p_c}{\partial V}\right)_S \frac{dT}{dp}$.

It can be seen from Eqs. (10.45) to (10.46) that the kink of the shock adiabat depends on the properties of the first phase and the slope of the phase equilibrium curve, $dT/\partial p$, at the kink point. Similar to the isentropes, the information on the properties of the second phase is superfluous, given the pre-assigned phase equilibrium curve.

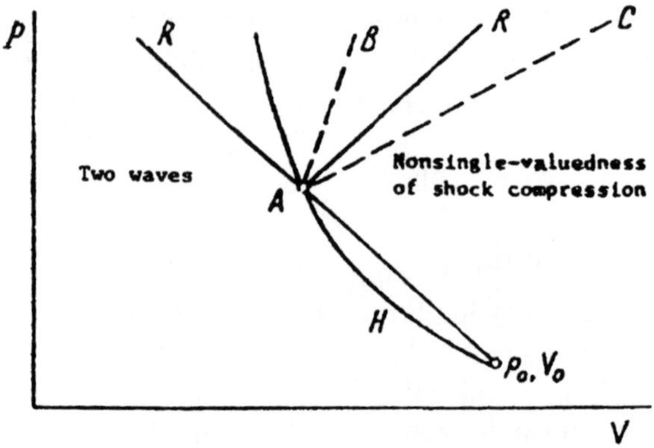

Fig. 10.9. Abnormalities of the shock compression of plasma. H, Hugoniot adiabat; R, Rayleigh line $(p_A - p_0)/(V_A - V_0)$ = const.; R', mirror reflection of Rayleigh line; AC, instability boundary according to Eq. (10.51); AB, the boundary of spontaneous radiation of sound by shock explosion (10.52); the slope of AB may be negative.

The variation of the Grüneisen parameter, upon transition from the single-phase to two-phase region, is

$$\gamma - \gamma_c = -\frac{V}{T}\left[\left(\frac{\partial p}{\partial V}\right)_S \left(\frac{\partial T}{\partial p}\right)_S - \left(\frac{\partial p_c}{\partial V}\right)_S \frac{dT}{dp}\right], \tag{10.47}$$

where

$$\left(\frac{\partial T}{\partial p}\right)_{S1} - \frac{dT}{dp} = \frac{(\partial V_1/\partial p)_S - (dV_1/dp)}{(\partial V_1/\partial T)_p} \tag{10.48}$$

and $(\partial V_1/\partial p)_S$ is related to $\partial V_1/\partial p$ by Eq. (10.38).

An analysis of the foregoing formulas indicates that, depending on the actual conditions, the difference $\Delta = (dp/dV)_{HC} - (dp_1/dV)_H$ can have any sign. If additional conditions are given, the case of $\Delta > 0$ leads to a two-wave structure. At $\Delta < 0$, in a number of cases, the emergence of shocks on the shock adiabat due to phase transition is possible. Under the conditions of a strong kink of the shock adiabat ($\Delta < 0$), when the shock compression curve in the p-V-plane proceeds to the right of the mirror reflection of the Rayleigh line kink (the line from Fig. 10.9),

$$\left|\frac{dp}{dV}\right| < R = \frac{p - p_0}{V_0 - V} \tag{10.49}$$

and the shock compression becomes anomalous. This is reflected by a decrease in mass velocity with an increase in pressure on the Hugoniot adiabat. In this case, the condition of the uniqueness of the solution to the Riemann's problem upon the disintegration of random discontinuity [573] is violated. Hence the shock compression becomes non–single-valued. Therefore, the conditions of abnormality of shock compression are prover to be symmetric [572]; the compression is anomalous if the shock adiabat extends beyond the angle formed by the Rayleigh line and its mirror reflection (Fig. 10.9). Upon exiting from the said angle to the left, the wave splits and on exiting to the right, the shock compression mode becomes non–single-valued. The stability conditions may further narrow down the region of normal shock wave compression of the medium (cf., the third section).

Under the conditions of isentropic expansion of shock-compressed matter, the release isentrope may cross a region of states with anomalous thermodynamic properties, $(\partial^2 p/\partial V^2)_S < 0$, which is due to the plasma phase transition. In this case, in order to construct the expansion wave in some pressure range, one must introduce an expansion shock. However, a compression shock is impossible [458]. A clear experimental proof of the existence of expansion shocks was obtained by Ivanov *et al.* [575], who had studied spallation phenomena upon detonation of charges on the surface of steel plates. Kutateladze *et al.* [576] described the results of precision experiments to register the structure of expansion waves in the highly critical region of Freon, which supported the predictions of Zel'dovich [458] on the emergence of expansion shocks in the critical point neighborhood. States close to near-critical have been realized in the release wave during experiments in the adiabatic expansion of shock-compressed plasma (Section 3.5). In this case the formation of expansion shock waves in expanding non-ideal plasma is also possible.

The unusual character of current decay and the anomalous behavior of resistance in exploding wire experiments have been associated by Bennet *et al.* [577] with the formation of expansion shock waves upon the expansion of metal heated by an electric discharge. The formation of expansion waves in such gas is unlikely, although at high heat capacities ($c_V \geq 17R$) a region with $(\partial^2 p/\partial V^2)_S < 0$ exists. This has been indicated by Zel'dovich [458] previously.

The possibility of the formation of the so-called electroadiabatic expansion shock waves in the discharge chamber of a shock tube was studied theoretically by Paxton [578]. The anomaly in thermodynamic functions is caused by the flow of strong electric current in plasma (leading to an inequality in the electron and ion temperatures).

10.6.3. *Shock Wave Stability Relative to Periodic Front Perturbations*

The stability of plane stationary discontinuity in a medium with an arbitrary equation of state relative to weak, non-uniform periodic front perturbations was investigated by D'yakov [579] and a number of researchers after him [574, 580–587]. The linearization of the hydrodynamic equations describing the motion behind the discontinuity front and the inclusion of boundary conditions at the front helped in the derivation of a characteristic equation. Its subsequent investigation has enabled one to reveal the various types of instability of the shock discontinuities.

The region of absolute instability corresponds to an exponential rise with time in the amplitude of perturbations emanating from the shock front. In doing so, the amplitude of perturbations decreases exponentially as its distance from the surface of discontinuity increases. In the case of absolute stability, the amplitude of perturbations decreases according to the power law [580]. Freeman [581] and Zaidel' [582] obtained the power law of decay for a different formulation of the boundary conditions. The region of absolute instability is defined by the inequalities

$$\left(\frac{\partial V}{\partial p}\right)_H < -\frac{V_0 - V}{p_0 - p}; \tag{10.50}$$

$$\left(\frac{dV}{dp}\right)_H > \frac{V_0 - V}{p - p_0}\left(1 + 2\beta^{-1}\frac{D}{c}\right), \tag{10.51}$$

where V_0, p_0 are the initial states $\beta = V_0/V$; D is the rate of travel of the discontinuity front in the laboratory co-ordinate system; and $c = [-V^2 (\partial p/\partial V)_S]^{1/2}$ is the velocity of sound behind the discontinuity front.

The condition (10.50) leads to the flow pattern which contains several shocks separated by continuous centered waves. This case was discussed in the second section. The condition (10.51) is stronger than (10.49) and it leads to the non–single-valuedness of the shock compression.

The development of perturbations in plane shock waves was considered by Zaidel' [582], who had studied their asymptotic behavior for important cases. The results of the analysis of the effect of low viscosity on the perturbation development were used [587] to treat the experimental data on recording the decay of the shock front perturbations. This enables one to determine the viscosity of a number of materials at extremely high pressures. A sharp variation of the viscosity in such experiments helped to reveal the melting of matter in shock wave [588]. In these experiments, the perturbations occurred at the moment when the shock wave had passed the butt joint of two samples of one and the same material with sinusoidal recesses in one of the butt-joined samples.

The study [589] is devoted to the derivation of the general thermodynamic criteria of absolute shock wave stability.

The possibility of the formation of a peculiar region of "spontaneous sound radiation by discontinuity" on the shock adiabat, in which perturbations assume the form of persistent travelling waves moving in the direction of discontinuity while falling behind the latter with time, appears to be of great practical interest from the standpoint of plasma physics. The additional (as compared to those referred to by D'yakov [579]) conditions for the existence of such a region are written as [579]

$$\frac{V_0 - V}{p - p_0}\frac{1 - \beta^{-2}(D/c)^2 - \beta^{-1}(D/c)^2}{1 - \beta^{-2}(D/c)^2 + \beta^{-1}(D/c)^2} < \left(\frac{dV}{dp}\right)_H < \frac{V_0 - V}{p - p_0}\left(1 + 2\beta^{-1}\frac{D}{c}\right) . \qquad (10.52)$$

For the sound instability criterion to be valid, the return run portion $(\partial p/\partial V)_H > 0$ of the shock adiabat is desirable. This form of shock adiabat in the p-V-plane is typical [458] of a plasma in which a rise in temperature is accompanied by an excitation of the internal degrees of freedom (such as ionization, dissociation, vibrational, rotational and electronic excitation).

Figure 10.10 shows the shock adiabats of a superdense tungsten plasma, H_{TF}, extracted from Fortova *et al.* [590]. It is calculated from the approximation formulas of the quasi-classical Thomas-Fermi theory. The wavy line in Fig. 10.10 marks the lower boundary of the instability region (10.52) while the upper boundary lies beyond the range of validity of the quasi-classical model. This is in view of the spontaneous emergence of electron-positron pairs in such plasma. Note that the inclusion of energy and the pressure of equilibrium radiation, H_{L}, renders shock compression stable in the entire pressure range.

The shock adiabats of the two-phase, "liquid/metallic vapor", mixture calculated from the semi-empirical equation of state [269] for copper are presented in Fig. 10.11. The return run of the shock adiabats and the resulting "sound instability" (10.52) will depend on the phase transformation processes.

The kink of the shock adiabats, as defined by thermal ionization, proves to be especially significant for non-ideal cesium plasma [592]. This is because the values of the first and second ionization potentials differ greatly (3.89 and 25.1 eV). The shock adiabats for cesium plasma calculated within the ring

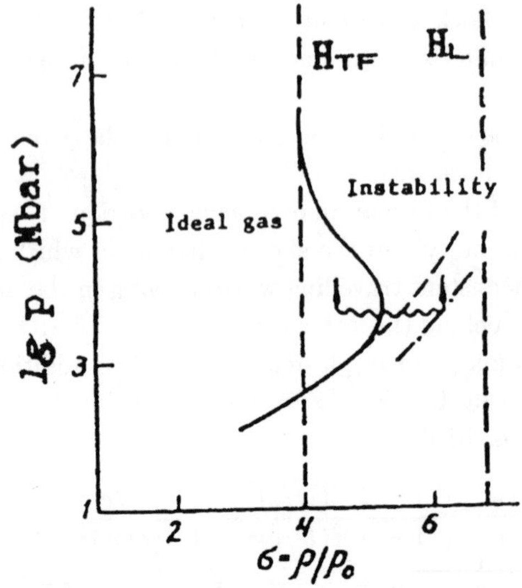

Fig. 10.10. Shock adiabat of tungsten in the superhigh pressure range [590]. The wavy line marks the instability region (10.52).

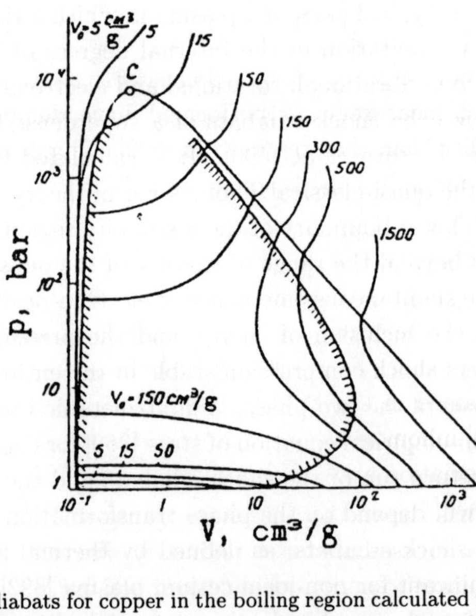

Fig. 10.11. Shock adiabats for copper in the boiling region calculated from the semi-empirical equation of state [269]. The shades mark the instability region (10.52). The solid curve marks the phase boundary.

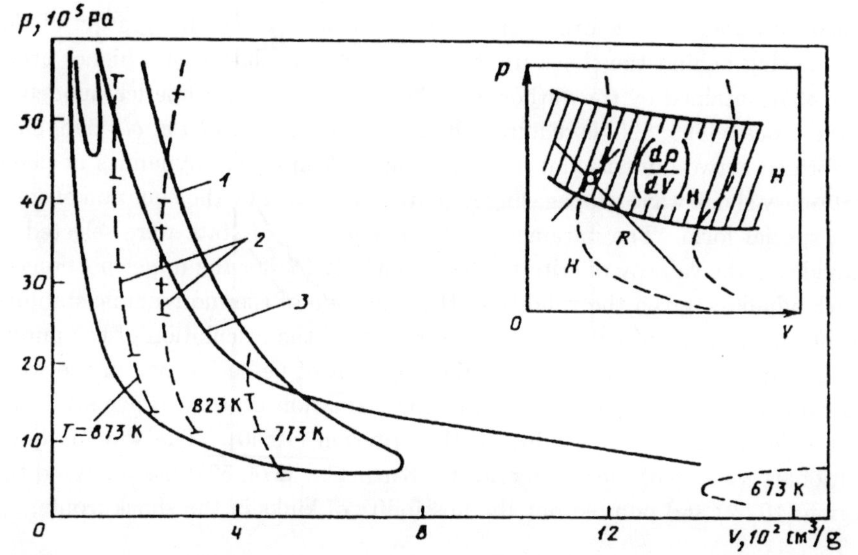

Fig. 10.12. Shock adiabats of cesium plasma. *1*, $\Gamma = e^2/kT\rho_D$ = const curves; *2*, Hugoniot adiabats with the initial temperatures indicated next to them; *3*, instability boundary according to Eq. (10.52). Shown above right is a diagrammatic view of the shock adiabats and the instability region (shaded) at high pressures.

Debye approximation in a grand canonical ensemble are shown in Fig. 10.12. The thermodynamic measurements performed have demonstrated a lower plasma compressibility than that reported in [592]. This somewhat shifts the boundary of sound instability, as shown in Fig. 10.12, to the region of higher pressures.

In the experiments [593, 594] in plane shock wave compression of gaseous Ar, Xe, CO_2, the interferometric technique helped to register the development of instabilities behind the wave front. The effects observed are associated with the return run of shock adiabats although the sound instability criterion (10.52) was not attained in any of the investigated cases. It is possible that the observed [68, 69] inhomogeneities are due to the kinetic phenomena which occur upon the dissociation and ionization of plasma.

Pleshanov [589] has shown that shock waves may lose their stability when the Hugoniot adiabat crosses the phase boundary, thereby causing the precipitation of the condensate to occur behind the shock front.

The physical meaning of the emergence of the possibility of sound radiation by discontinuity is not yet clear. Possibly, it is a peculiar form of instability which may lead to a deformation of the front or to the disintegration of the plasma slug behind the shock front. In this case, a solution of a higher order than that obtained by the authors of [579–586] or a direct numerical modeling of the process [595, 596] is required. Fortov *et al.* [595] and Ni *et al.* [596] have performed a two-dimensional numerical simulation of the dynamics of plane stationary shock wave propagating in matter described by the equation of state of a special form. The parameters of this equation of state were selected to provide for the validity of criteria (10.50) and (10.52) for the upper portions of shock adiabats, given the validity of the condition of thermodynamic stability in the entire range of parameters. The results of the calculation of the purely non-linear stage demonstrated the development of perturbations in the wave front in the situation of (10.52), and the formation of a two-wave structure under the conditions of validity of the criterion (10.50). The wave is quite stationary outside of these two criteria. Kuznetsov [574, 591] has analyzed the region (10.52) and pointed out the possibility of kinks in the shock fronts.

10.6.4. *The Ambiguity of Representation of Thermodynamic States in the p-V-Plane. Shock Waves Under Conditions of Decreasing Plasma Temperature*

In the preceding discussion, it was assumed that $(\partial V/\partial T)_p > 0$. Analogously, the analysis can be performed at $(\partial V/\partial T)_p < 0$. If $(\partial V/\partial T)_p$ changes its sign in the region of parameters being considered, difficulties would arise due to the ambiguity of representation of the thermodynamic states in the p-V-plane. This leads, in the case of thermodynamically equilibrium systems, to the possibility of the intersection of isentropes in the p-V-plane without violating the Planck-Kelvin formulation of the second law of thermodynamics [597]. For a strongly non-ideal plasma, the ambiguity of representation of the states in the p-V-plane is also possible (cf., the intersecting isotherms in Fig. 10.13). This calls for a special consideration of the problem.

The equilibrium states of a single-component system are represented in a three–coordinate space by a two-dimensional parametric surface (equation of state). The various thermodynamic processes in this surface are shown by appropriate curves. On such a surface, the isentropes cannot intersect without violating the second law of thermodynamics (which was also proven with the aid of the Pfaffian differential equations of isentrope [598]). However, when projected onto the p-V-plane, the intersections of isentropes become possible. This is due to the unfortunate choice of the pair of variables. Kestin [598] has formulated sufficient conditions for the choice of single-valued thermodynamic

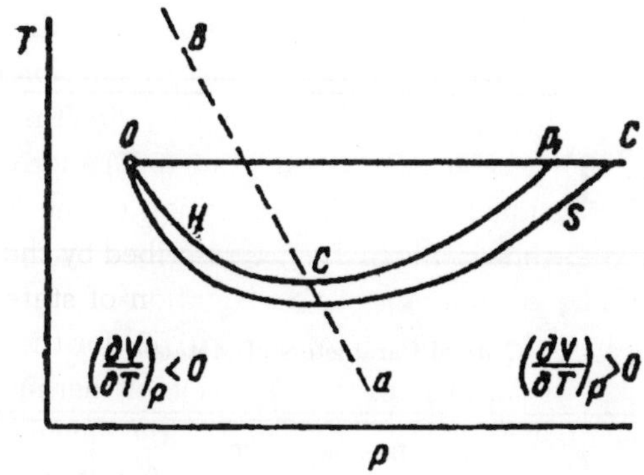

Fig. 10.13. Shock wave under the conditions of decreasing plasma temperature. On *ab*, $(dV/dT)_p = 0$; *H*, shock adiabat; *S*, isentrope.

co-ordinates. These co-ordinates include the (T, p), (T, V), (S, p) and (S, V) pairs. The nonsingle-valued pairs are the (p, V) and (T, S) variables, which are widely used in analyzing thermodynamic processes and in proving the theorems of thermodynamics. It is common knowledge that a consideration of the processes in these variables lead to a number of paradoxes. For instance, Sommerfeld proposed the Carnot cycle for water at a temperature of 4°C. As a result of this cycle, useful work was performed using heat from a single source [599]. However, at atmospheric pressure, water has a density maximum for temperatures of 4°C. This clearly affects the unambiguity of representations of the thermodynamic states in the p-V-plane. Note that in dynamic experiments, the thermodynamic states are registered quite unambiguously. This is because, along with p, V, the internal energy E is also determined (the equilibrium temperature is determined from these variables just as unambiguously [cf. Chapter 5]).

In a media with the alternating sign $(\partial V/\partial T)_p$, shock waves may be formed under the conditions of a decreasing temperature [561]. Figure 10.13 is a diagrammatic representation of the isentrope and Hugoniot adiabat when $(\partial V/\partial T)_p < 0$ in the medium before shock wave. When the states behind the front lie in the $(\partial V/\partial T)_p > 0$ region, the possible modes of shock waves, under the conditions of decreasing temperature, correspond to the portion with p_1. In doing so, the pressure, density and entropy in the shock wave increase. An analysis of the modes for water (use was made of an empirical equation of state) has shown [600] that such a shock wave, is in all respects (except the temperature jump), analogous to a regular shock discontinuity.

Appendix 1

Critical Parameters of Metals.

Element	T_c, K	p_c kbar	ρ_c $g{\cdot}cm^{-3}$	S_c $\frac{cal}{mol{\cdot}K}$	Element	T_c K	p_c kbar	ρ_c $g{\cdot}cm^{-3}$	S_c $\frac{cal}{mol{\cdot}K}$
Zr	16,250	7.52	1.79	40.45	Hg	$1,763 \pm 15$	1.51 ± 0.025	$4.22 + 0.4$	33.790
Zn	3,190	2.63	2.29	31.27	Hg	$1,753 \pm 10$	1.52 ± 0.01	5.7 ± 0.2	33.241
Y	10,800	3.74	1.30	38.92	Na	2, 573	0.275	0.206	31.683
Yb	4,280	1.38	2.36	36.92	K	2, 223	0.152	0.194	33.931
V	12,500	10.78	1.86	37.26	Rb	2, 093	0.159	0.346	36.485
U	11,630	6.11	5.30	43.07	Cs	$2,057 \pm 40$	0.144	0.428 ± 0.02	38.111
W	21,010	15.83	5.87	43.34	Li	3, 223	0.689	0.105	27.666
Ti	11,790	7.63	1.31	37.92	Mn	5, 940	6.28	2.46	35.68
Sn	8,200	3.35	2.05	39.44	Mg	3, 590	1.98	0.56	29.48
Th	14,950	4.88	3.21	43.67	Pb	4, 980	1.84	3.25	39.81
Tu	5,910	2.65	3.22	38.52	La	11, 060	3.35	1.78	40.58
Tl	4,470	1.63	3.16	38.67	Fe	9, 600	8.25	2.03	37.20
Tb	8,060	3.08	2.57	42.34	Ir	15, 380	12.78	6.77	41.91
Te	1,850	0.75	2.21	35.22	In	6, 420	2.43	1.84	37.95
Ta	20,570	13.5	5.04	43.07	Ho	7, 240	2.94	2.84	40.63
Sr	3,860	0.90	0.86	35.23	Hf	18, 270	9.38	3.88	42.59
Ag	7,040	4.50	2.93	36.86	Au	8, 970	6.10	5.68	39.40
Sc	8,350	4.08	0.93	36.82	Ge	9, 170	4.90	1.64	37.77
Sm	5,340	2.10	2.51	40.62	Ga	7, 210	4.31	1.77	35.91
Ru	15,500	13.74	3.79	37.14	Gd	8, 670	3.25	2.50	42.79
Rh	13,510	11.23	3.62	39.66	Er	8, 250	3.34	2.86	40.51
Re	19,600	15.7	6.32	43.42	Eu	4, 680	1.21	1.67	41.36
Ra	3,830	0.77	1.93	38.32	Dy	7, 240	2.91	2.76	41.05
Pr	9,160	2.85	1.86	41.39	Cu	8, 390	7.46	2.39	35.30
Tc	15,930	11.81	3.09	41.00	Co	10, 460	9.23	2.20	37.12
Pa	12,650	4.82	3.72	43.34	Cr	9, 620	9.68	2.22	37.35

Appendix 1. *Continued*

Element	T_c, K	p_c kbar	ρ_c $g \cdot cm^{-3}$	S_c $\frac{cal}{mol \cdot K}$	Element	T_c K	p_c kbar	ρ_c $g \cdot cm^{-3}$	S_c $\frac{cal}{mol \cdot K}$
Po	2,050	0.62	2.67	37.52	Ce	8, 860	3.03	2.03	40.51
Pt	14,330	8.70	5.02	41.8	Ca	4, 180	1.21	0.49	32.70
Pd	10,760	7.64	3.20	36.64	Cd	2, 790	1.60	2.74	33.20
Os	17,110	14.49	6.83	42.60	Be	8, 080	11.70	0.55	27.02
Nb	19,040	12.52	2.59	41.65	Ba	4, 100	0.81	1.15	37.06
Ni	10,330	9.12	2.19	36.50	Al	8, 000	4.47	0.64	33.52
Nd	7,920	2.65	2.05	41.33	Sb	2, 570	1.30	2.61	36.1
Mo	16,140	12.63	3.18	41.21	B	8, 200	9.57	0.69	28.9
Lu	7,060	2.84	2.97	39.41	Se	1, 010	0.37	1.60	31.8
Fr	1,810	0.12	0.65	39.50	S	730	0.30	0.72	27.7
Bi	4,200	1.26	2.66	40.2					

Appendix 2

Transport Cross-Sections of Electron Scattering by Inert Gas Atoms [388].

Transport cross-sections of electron scattering by inert gas atoms at different values of E, eV

Gas	0	0.01	0.02	0.03	0.05	0.07	0.1	0.15
He	–	5.21^{-16}	5.35^{-16}	5.46^{-16}	5.62^{-16}	5.74^{-16}	5.86^{-16}	6.04^{-16}
Ne	–	–	–	4.69^{-17}	5.36^{-17}	6.01^{-17}	7.01^{-17}	8.28^{-17}
Ar	8.05^{-16}	6.1^{-16}	3.74^{-16}	2.8^{-16}	1.84^{-16}	1.14^{-16}	4.5^{-17}	2.2^{-17}
Kr	3.07^{-15}	2.6^{-15}	1.97^{-15}	1.6^{-15}	1.14^{-15}	9.0^{-16}	6.8^{-16}	4.4^{-16}
Xe	1.78^{-14}	1.16^{-14}	8.0^{-15}	6.13^{-15}	3.95^{-15}	2.9^{-15}	2.1^{-15}	1.29^{-15}
Gas	**0.2**	**0.25**	**0.3**	**0.4**	**0.5**	**0.7**	**1.0**	**2.0**
He	6.16^{-16}	6.27^{-16}	6.35^{-16}	6.49^{-16}	6.59^{-16}	6.73^{-16}	6.85^{-16}	6.99^{-16}
Ne	9.3^{-17}	1.02^{-16}	1.09^{-16}	1.22^{-16}	1.32^{-16}	1.47^{-16}	1.62^{-16}	1.81^{-16}
Ar	1.77^{-17}	1.56^{-16}	1.09^{-16}	1.82^{-16}	2.83^{-17}	–	1.05^{-16}	2.48^{-16}
Kr	2.47^{-16}	1.5^{-16}	1.0^{-16}	6.02^{-17}	5.3^{-17}	5.0^{-17}	6.17^{-17}	2.1^{-16}
Xe	8.4^{-16}	5.35^{-16}	3.15^{-16}	1.75^{-16}	1.38^{-16}	1.35^{-16}	2.47^{-16}	8.25^{-16}
Gas	**3.0**	**4.0**	**5.0**	**6.0**	**7.0**	**8.0**	**10.0**	**–**
He	6.89^{-16}	5.6^{-16}	6.26^{-16}	6.01^{-16}	–	–	–	–
Ne	1.9^{-16}	1.98^{-16}	2.07^{-16}	2.14^{-16}	2.21^{-16}	–	–	–
Ar	4.07^{-16}	5.8^{-16}	–	8.7^{-16}	–	1.17^{-15}	1.38^{-15}	–
Kr	4.84^{-16}	–	1.02^{-15}	–	1.5^{-16}	–	1.93^{-16}	–
Xe	1.7^{-15}	2.48^{-15}	3.08^{-15}	–	–	3.37^{-15}	3.2^{-15}	–

Note: 5.2^{-16} is taken to be $5.2 \cdot 10^{-16}$

Appendix 3

The Electron Work Functions for Metals.

Metal	W, eV	Metal	W, eV	Metal	W, eV	Metal	W, eV
Li	2.38	Be	3.92	Cu	4.4	Pb	4.0
Na	2.35	Mg	3.64	Zn	4.24	Mo	4.3
K	2.22	Ca	2.80	Cd	4.1	W	4.5
Rb	2.16	Sr	2.35	Hg	4.25	U	3.3
Cs	1.81	Ba	2.49	Sn	4.38		

References

[1] L. D. Landau & E. M. Lifshits, *Statistical Physics, Part 1* (Nauka, Moscow, 1976)

[2] A. A. Vedenov, Thermodynamics of Plasma, in *Problems of Plasma Theory, Issue 1*, ed. M. A. Leontovich (Atomizdat, Moscow, 1963), pp. 273–285

[3] L. P. Kudrin, *Statistical Physics of Plasma* (Atomizdat, Moscow, 1974)

[4] A. G. Khrapak & I. T. Iakubov, *Electrons in Dense Gases and Plasmas* (Nauka, Moscow, 1981)

[5] N. I. Kluchnikov & S. A. Triger, *Zh. Eksp. Teor. Fiz.* (*Soviet Physics — JETP*) **52** (1967) 276

[6] B. M. Smirnov, *Introduction to Plasma Physics* (Nauka, Moscow, 1982)

[7] W. Ebeling, W. D. Kraeft & D. Kremp, *The Theory of Bound States and Equilibrium in Ionization of Plasma and Solid States.* Transl. from Ger. (Mir, Moscow, 1979)

[8] I. K. Kikoin, A. P. Senchenkov, E. B. Gelman *et al.*, *Zh. Eksp. Teor. Fiz.* (*Soviet Physics — JETP*) **49** (1965) 124

[9] D. A. Kirzhnits, Yu. E. Losovik & G. V. Shpatkovskaya, *Uspekhi Fiz. Nauk.* (*Soviet Physics — Uspekhy*) **117** (1975) 3

[10] V. K. Gryaznov, I. L. Iosilevskii, Yu. G. Krasnikov *et al.*, *Thermophysical Properties of Working Media in Nuclear Energy Devices with Gas-Phase Reactors*, ed. V. M. Ievlev (Atomizdat, Moscow, 1980)

[11] *Proc. Symp. Research of Uranium Plasmas and Their Technological Applications. Gainesvills*, ed. K. Thom & R. T. Schneider (NASA, Wash., 1971); *Proc. 2nd Symp. Uranium Plasmas. Atlanta.* (NASA, Wash., 1971)

[12] C. E. Ragan, M. G. Silbert & B. C. Diven, *J. Appl. Phys.* **48** (1977) 2860

[13] A. M. Prokhorov, S. I. Anisimov & P. P. Pashinin, *Uspekhi Fiz. Nauk.* (*Soviet Physics — Uspekhi*) **119** (1976) 401

[14] M. Lepouter, *Metal Ammoniac Solutions* (Acad. Press, New York, 1965)

[15] E. Shatzman, in *Strongly Coupled Plasmas*, ed. A. Kalman & P. Corini (Plenum Press, New York, 1978), pp. 409–425

[16] Ya. B. Zel'dovich & I. D. Novikov, *Relativistic Astrophysics* (Nauka, Moscow, 1971)

[17] V. I. Kucherenko, G. A. Pavlov, V. K. Gryaznov *et al.*, *Thermophysical Properties of Plasmas of Helium and Hydrogen Mixtures. Preprint* (Inst. Chem. Phys. USSR Acad. Sci., Chernogolovka, 1978)

[18] *Lasers and Thermonuclear Problems*, ed. B. B. Cadomtsev (Atomizdat, Moscow, 1973)

[19] R. M. More, in *Atomic and Molecular Physics of Controlled Thermonuclear Fusion*, ed. Ch. J. Joachein & D. E. Post (Plenum, New York, 1983), pp. 339–440

[20] V. M. Ievlev, *Izv. Acad. Nauk SSSR, Ser. Energy and Transport*, N 6 (1977) 24

[21] I. T. Iakubov & V. S. Vorobjev, *Astronautica Acta* **18** (1974) 79

[22] A. V. Nedospasov, *Uspekhi Fiz. Nauk* (*Soviet Physics — Uspekhi*) **123** (1977) 333

[23] F. J. Rogers, M. Ross, G. L. Haygein & L. K. Wong, *TBT Report UCID — 19557* (1980)

[24] L. M. Biberman, A. A. Likal'ter & I. T. Iakubov, *Teplofiz. Vis. Temp.* (*High Temperature*) **20** (1982) 565

[25] V. A. Alekseev, A. A. Vedenov, L. S. Krasitskaya *et al.*, *Zh. Eksp. Teor. Fiz., Pisma* (*Soviet Physics — JETP Letters*) **12** (1970) 501

[26] D. I. Zhukhovitskii, A. G. Khrapak & I. T. Iakubov, in *Chemistry of Plasmas, Issue 11*, ed. B. M. Smirnov (Energoatomizdat, Moscow, 1984), pp. 130–169

[27] V. V. Ivanov, I. S. Shvets & A. V. Ivanov, *Underwater Spark Discharges* (Naukova Dumka, Kiev, 1982)

[28] V. A. Alekseev & I. T. Iakubov, *Phys. Repts* **96** (1983) 1

[29] V. A. Alekseev & I. T. Iakubov, in *Handbook on Thermodynamic and Transport Properties of Alkali Metals*, ed. R. W. Ohse (Blackwell Sci. Publ., Oxford, 1986), pp. 703–734

[30] V. A. Alekseev, *Teplofiz. Vis. Temp.* (*High Temperature*) **6** (1968) 961

[31] I. K. Kikoin & A. P. Senchenkov, *Fiz. Met. Metalloved.* **24** (1967) 843

[32] F. Hensel & E. U. Frank, *Rev. Mod. Phys.* **44** (1968) 697

[33] V. A. Alekseev, V. G. Ovcharenko, Yu. F. Rizhkov *et al.*, *Zh. Eksp. Teor. Fiz., Pisma* (*Soviet Physics — JETP Letters*) **12** (1970) 306

[34] V. A. Alekseev, V. G. Ovcharenko & Yu. F. Rizhkov, *Uspekhi Fiz. Nauk* (*Soviet Physics — Uspekhi*) **120** (1976) 699

[35] A. A. Borzhievski, V. A. Sechenov & V. I. Horunzhenko, in *Proc. 18th Int. Confer. Ionization Phenomena in Gases, Vol. 1* (Swansea, 1987), pp. 250–251

[36] V. A. Alekseev, A. A. Vedenov, V. G. Ovcharenko *et al.*, *High Temperature — High Pressure* **7** (1975) 677

[37] H. Renkert, P. Hensel & E. U. Frank, *Ber. Bunsenges. Phys. Chem.* **75** (1971) 507

[38] H. Renkert, F. Hensel & E. U. Frank, *Phys. Lett. A* **30** (1969) 494

[39] P. P. Kulik, V. A. Ryabii & N. V. Ermokhin, *Non-ideal Plasma* (Energoatomizdat, Moscow, 1984)

[40] P. Pfeifer, W. F. Freyland & F. Hensel, *Phys. Lett. A* **43** (1973) 111

[41] I. M. Isakov & B. N. Lomakin, *Teplofiz. Vis. Temp.* (*High Temperature*) **17** (1979) 262

[42] V. A. Sechenov, E. E. Son & O. E. Shchekotov, *Teplofiz. Vis. Temp.* (*High Temperature*) **15** (1977) 413

[43] I. Ya. Dikhter & V. A. Zeigarnik, *Teplofiz. Vis. Temp.* (*High Temperature*) **15** (1977) 196

[44] S. G. Barol'skii, N. V. Ermokhin & P. P. Kulik, *Zh. Eksp. Teor. Fiz.* (*Soviet Physics — JETP*) **62** (1972) 176

[45] G. Franz, W. Freyland & F. Hensel, in *Proc. 4th Int. Conf. Liquid and Amorphous Metals* (Grenoble, 1981), pp. 44–46

[46] P. Pfeifer, W. Freyland & F. Hensel, *Ber. Bunsenges. Phys. Chem.* **83** (1979) 204

[47] W. Freyland & F. Hensel, *Ber. Bunsenges. Phys. Chem.* **76** (1972) 347

[48] B. N. Lomakin & A. D. Lopatin, *Teplofiz. Vis. Temp.* (*High Temperature*) **21** (1983) 163

[49] V. V. Gogoleva, V. Yu. Zitserman, A. Ya. Polishchuk & I. T. Iakubov, *Teplofiz. Vis. Temp.* (*High Temperature*) **22** (1984) 208

[50] N. V. Ermokhin, B. M. Kovalev & P. P. Kulik, *Teplofiz. Vis. Temp.* (*High Temperature*) **9** (1971) 665

[51] G. Rochling, in *Advances Energy Conversion* **3** (1963) 69

[52] H. Uchtmann & F. Hensel, *Phys. Lett. A* **53** (1975) 239

[53] L. J. Dukers & R. G. Ross, *Phys. Lett. A* **38** (1972) 291

[54] V. A. Alekseev, A. A. Vedenov, V. G. Ovcharenko *et al.*, *Zh. Eksp. Teor. Fiz., Pisma* (*Soviet Physics — JETP Letters*) **16** (1972) 73

[55] K. Tsuji, M. Yao & H. Endo, *J. Phys. Soc. Japan* **42** (1977) 1594

[56] F. E. Neale & N. E. Cusack, *J. Phys. F* **9** (1979) 85

[57] F. Hensel, *Ber. Bunsenges. Phys. Chem.* **76** (1971) 847

[58] F. Hensel, *Phys. Letter A* **31** (1970) 88

[59] H. Ikezi, K. Schwarzenegger, A. Simons *et al.*, *Phys. Rev. A* **18** (1978) 2494

[60] F. Hensel, in *Proc. 7th Int. Confer. Amorphous Liquid Semiconductors* (Edinburgh, 1977), pp. 58–70

[61] W. Hefner, R. W. Schmutzler & F. Hensel, *J. de Phys.* **41** (1980) 8

[62] O. Cheshnovsky, U. Even & J. Jortner, *Philos. Mag. B* **44** (1981) 1

[63] W. Hefner & F. Hensel, *Phys. Rev. Lett.* **48** (1982) 1026

[64] H. Uchtman, J. Popielawski & F. Hensel, *Ber. Bunsenges. Phys. Chem.* **85** (1981) 555

[65] Yu. S. Korshunov, A. P. Senchenkov, E. I. Asinovskii *et al.*, *Teplofiz. Vis. Temp.* (*High Temperature*) **8** (1970) 1288

[66] J. P. Stone, C. T. Ewing, J. R. Spann *et al.*, *J. Chem. Engng Data* **11** (1966) 320

[67] I. I. Novikov & V. V. Roshchupkin, *Izmerit. Tekhnica*, N 10 (1967) 27

[68] L. D. Voliak & A. K. Chelebaev, *Teplofiz. Vis. Temp.* (*High Temperature*) **14** (1976) 913

[69] I. K. Kikoin, A. P. Senchenkov, A. P. Naurzakov *et al.*, *Preprint IAE N 2310* (Kurchatov's Inst. Atomic Energy, Moscow, 1973)

[70] I. M. Korsunskii, *Zh. Eksp. Teor. Fiz.* (*Soviet Physics — JETP*) **89** (1985) 875

[71] A. K. Chelebaev, *Measurements of Equation of State Parameters for Dense Steams of Cesium. Theses* (Moscow, 1978)

[72] S. Jungst, B. Knuth & F. Hensel, *Phys. Rev. Lett.* **55** (1985) 2160

[73] *Physics of Simple Liquids*, ed. G. M. Themperly. Transl. from Engl. (Mir, Moscow, 1971)

[74] F. Noll, W. C. Pilgrim & R. Winter, *Z. Phys. Chemie* **156** (1988) 303

[75] R. Winter, F. Hensel, T. Bodensteiner & W. Gläser, *Z. Phys. Chemie* **156** (1988) 145

[76] I. N. Vasiliev & Yu. S. Trelin, *Teplofiz. Vis. Temp.* (*High Temperature*) **7** (1969) 1112

[77] K. Suzuki, M. Inutake & S. Fujiwaka, in *Metal — Dielectric Transitions* (L'vov State Univ., Moscow — L'vov, 1977), pp. 123–125

[78] I. Ya. Dikhter & V. A. Zeigarnik, *Review on Thermophysical Properties of Substances N 4* (High Energy Density Res. Center, Moscow, 1981), pp. 59–91

[79] V. V. Vorobiev, P. P. Kulik, A. V. Pallo *et al.*, in *Proc. 15th Int. Conf. Phenomena in Ionized Gases, Part 1* (Minsk, 1981), pp. 361–362

[80] N. V. Yermokhin, P. P. Kulik, V. A. Riabyi *et al.*, in *Proc. 15th Int. Conf. Phenomena in Ionized Gases, Part 1* (Minsk, 1981), pp. 363–364

[81] N. P. Kuznetsova & G. B. Lappo, *Teplofiz. Vis. Temp.* (*High Temperature*) **17** (1979) 37

[82] G. R. Gathers, J. W. Shaner & D. A. Young, *Phys. Rev. Lett.* **33** (1974) 70

[83] G. Gasers, D. Zh. Sheiner & R. Bri, *Prib. Nauchn. Issled.* **47** (1976) 71

[84] I. Ya. Dikhter & V. A. Zeigarnik, *Teplofiz. Vis. Temp.* (*High Temperature*) **13** (1975) 483

[85] I. T. Iakubov, *Beitr. Plasma Phys.* **17** (1977) 221

[86] G. R. Gathers, J. W. Shaner & W. M. Hodson, *High Temperatures — High Pressures* **11** (1979) 529

[87] G. R. Gathers, *Int. J. Thermophys.* **4** (1983) 209

[88] J. W. Shaner, R. S. Hixon & M. A. Winkler, in *Proc. Int. Symp. Physics of Ionized Gases — 1986*, ed. J. Puric & D. Belic (Beograd Univ., Beograd, 1986), pp. 331–346

[89] L. V. Al'tshuler, A. B. Bushman, M. V. Zhernokletov *et al.*, *Zh. Eksp. Teor. Fiz.* (*Soviet Physics — JETP*) **78** (1980) 741

[90] E. I. Asinovskii & V. A. Zheigarnik, *Teplofiz. Vis. Temp.* (*High Temperature*) **12** (1974) 1278

[91] V. M. Batenin & P. V. Minaev, *High Temperatures — High Pressures* **2** (1970) 597

[92] T. H. Peters, *Z. Phys.* **135** (1953) 573

[93] K. Günter, *Beitr. Plasma Phys.* **8** (1968) 383

[94] R. Radtke & K. Günter, *J. Phys. D* **9** (1976) 1131

[95] M. M. Popovic, S. S. Popovic & S. M. Vukovic, *Fizika* **6** (1974) 29

[96] A. Mokhtari, M. Skowronek & Y. Vitel, in *Proc. 14th Int. Symp. Physics of Ionized Gases*, ed. N. Konjevic (Sarajevo Univ., Sarajevo, 1988), pp. 373–376

[97] P. V. Mitin, in *Properties of Low Temperature Plasmas and Diagnostics Methods*, ed. M. F. Zhukov (Nauka, Novosibirsk, 1977), pp. 105–120

[98] S. I. Andreev & T. V. Gavrilova, *Teplofiz. Vis. Temp.* (*High Temperature*) **12** (1974) 1293

[99] S. I. Andreev & T. V. Gavrilova, *Teplofiz. Vis. Temp.* (*High Temperature*) **13** (1975) 176

[100] S. I. Andreev & T. V. Gavrilova, *Teplofiz. Vis. Temp.* (*High Temperature*) **13** (1975) 647

[101] E. A. Martin, *J. Appl. Phys.* **31** (1960) 225

[102] J. W. Robinson, *J. Appl. Phys.* **38** (1957) 210

[103] K. A. Naugol'nikh & N. A. Royn, *Electrical Discharges in Water* (Nauka, Moscow, 1971)

[104] N. N. Ogurtsova & I. V. Podmoshenskii, in *Low Temperature Plasmas, Proc. XX Congress IUPAC*. Transl. from Engl. (Mir, Moscow, 1967), pp. 432–440

[105] N. N. Ogurtsova, I. V. Podmoshenskii & V. L. Smirnov, *Teplofiz. Vis. Temp.* (*High Temperature*) **12** (1974) 650

[106] V. E. Fortov, *Uspekhi Fiz. Nauk* (*Soviet Physics — Uspekhi*) **138** (1982) 361

[107] L. V. Al'tshuler, *Uspekhi Fiz. Nauk* (*Soviet Physics — Uspekhi*) **85** (1965) 197

[108] S. B. Kormer, *Uspekhi Fiz. Nauk* (*Soviet Physics — Uspekhi*) **94** (1968) 641

[109] V. N. Minaev & A. G. Ivanov, *Uspekhi Fiz. Nauk* (*Soviet Physics — Uspekhi*) **119** (1976) 75

[110] L. Davison & R. A. Graham, *Phys. Repts* **5** (1979) 256

[111] G. E. Duvall, *Rev. Mod. Phys.* **49** (1977) 525

[112] J. W. Shaner & G. R. Gathers, in *High Pressure Science and Technology*, ed. K. D. Timerhous (Plenum, New York, 1979), pp. 847–867

[113] Ya. B. Zheldovich & Yu. P. Raizer, *Physics of Shock Waves and High Temperature Hydrodynamic Phenomena* (Nauka, Moscow, 1966)

[114] L. V. Al'tshuler, B. P. Moiseev & L. V. Popov, *Zh. Eksp. Teor. Fiz.* (*Soviet Physics — JETP*) **54** (1968) 785

[115] R. F. Trunin, M. A. Podurets, G. V. Simakov *et al.*, *Zh. Eksp. Teor. Fiz.* (*Soviet Physics — JETP*) **62** (1972) 785

[116] R. F. Trunin, M. A. Podurets, B. N. Moiseev *et al.*, *Zh. Eksp. Teor. Fiz.* (*Soviet Physics — JETP*) **56** (1969) 1172

[117] V. A. Simonenko, N. P. Voloshin, L. P. Volkov *et al.*, *Zh. Eksp. Teor. Fiz.* (*Soviet Physics — JETP*) **88** (1985) 1452

[118] E. N. Avrorin, B. K. Vodolaga, L. P. Volkov *et al.*, *Zh. Eksp. Teor. Fiz.* (*Soviet Physics — JETP*) **93** (1987) 613

[119] A. S. Vladimirov, N. P. Voloshin, V. A. Simonenko *et al.*, *Zh. Eksp. Teor. Fiz., Pisma* (*Soviet Physics — JETP Letters*) **39** (1984) 69

[120] S. I. Anisimov, A. M. Prokhorov & V. E. Fortov, *Uspekhi Fiz. Nauk* (*Soviet Physics — Uspekhi*) **134** (1983) 1000

[121] D. A. Kirzhnits, Yu. E. Losovik & G. V. Shpatkovskaya, *Uspekhi Fiz. Nauk* (*Soviet Physics — Uspekhi*) **119** (1976) 357

[122] B. N. Lomakin & V. E. Fortov, *Zh. Eksp. Teor. Fiz.* (*Soviet Physics — JETP*) **63** (1972) 42

[123] A. G. Kunavin, A. V. Kirillin & Yu. S. Korshunov, *Teplofiz. Vis. Temp.* (*High Temperature*) **12** (1974) 1302

[124] P. C. Lysne & D. R. Hardesty, *J. Chem. Phys.* **59** (1973) 6512

[125] V. E. Fortov, *Fiz. Gorenija i Vzriva*, N 8 (1972) 428

[126] V. E. Fortov & Yu. P. Krasnikov, *Zh. Eksp. Teor. Fiz.* (*Soviet Physics — JETP*) **59** (1970) 1945

[127] V. E. Fortov, *Zh. Prikl. Mekh. Tekhn. Fiz.* **6** (1972) 156; V. E. Fortov, in *Combustion and Flame* (Nauka, Moscow, 1972), pp. 56–87

[128] A. V. Bushman, B. N. Lomakin, V. A. Sechenov *et al.*, *Zh. Eksp. Teor. Fiz.* (*Soviet Physics — JETP*) **69** (1975) 1624

[129] Yu. N. Ryabinin, *Gases at High Densities and Temperatures* (Fizmatgiz, Moscow, 1959)

[130] V. A. Volkov, A. B. Karasev & B. K. Tkachenko, *Teplofiz. Vis. Temp.* (*High Temperature*) **16** (1978) 411

[131] B. K. Tkachenko, S. I. Titarov, A. B. Karasev *et al.*, *Fiz. Gorenija i Vzriva*, N 5 (1976) 763

[132] R. N. Cristian & F. L. Jarder, *J. Chem. Phys.* **33** (1955) 2042

[133] W. E. Peal, *J. Appl. Phys.* **28** (1957) 782

[134] M. A. Tsikulin & E. G. Popov, *Radiation Properties of Shock Waves in Gases* (Nauka, Moscow, 1977)

[135] V. K. Gryaznov, I. L. Iosilevskii & V. E. Fortov, *Zh. Prikl. Mekh. Tekhn. Fiz.*, N 3 (1973) 70

[136] V. E. Fortov, Yu. V. Ivanov, A. N. Dremin *et al.*, *Zh. Eksp. Teor. Fiz.* (*Soviet Physics — JETP*) **69** (1975) 1634

[137] V. E. Bespalov, V. K. Gryaznov, A. N. Dremin *et al.*, *Zh. Eksp. Teor. Fiz.* (*Soviet Physics — JETP*) **69** (1975) 2059

[138] Yu. V. Ivanov, A. N. Dremin, V. B. Mintsev *et al.*, *Zh. Eksp. Teor. Fiz.* (*Soviet Physics — JETP*) **71** (1976) 215

[139] V. B. Mintsev & V. E. Fortov, *Zh. Eksp. Teor. Fiz., Pisma* (*Soviet Physics — JETP Letters*) **30** (1979) 401

[140] I. T. Model, *Zh. Eksp. Teor. Fiz.* (*Soviet Physics — JETP*) **32** (1957) 714

[141] V. E. Bespalov, L. G. Djachkov, G. A. Kobsev *et al.*, *Teplofiz. Vis. Temp.* (*High Temperature*) **17** (1979) 266

[142] W. Ebeling, V. Fortov, A. Forster *et al.*, *Thermophysical Properties of Hot Dense Plasmas* (AVG, Leipzig, 1991)

[143] V. B. Mintsev, V. K. Gryaznov & V. E. Fortov, *Zh. Eksp. Teor. Fiz.* (*Soviet Physics — JETP*) **79** (1980) 116

[144] Yu. B. Zaporozhets, V. B. Mintsev & V. E. Fortov, *Zh. Eksp. Teor. Fiz., Pisma* (*Soviet Physics — JETP Letters*) **45** (1987) 204

[145] V. E. Fortov, A. A. Leont'ev, V. K. Gryaznov *et al.*, *Zh. Eksp. Teor. Fiz.* (*Soviet Physics — JETP*) **71** (1976) 225

[146] V. K. Gryaznov, M. V. Zhernokletov, V. N. Zubarev *et al.*, *Zh. Eksp. Teor. Fiz.* (*Soviet Physics — JETP*) **78** (1980) 573

[147] R. N. Keeler & B. J. Alder, *Physica* **21** (1980) 1658

[148] W. J. Nellis, F. H. Ree, M. Van Thiel & A. C. Mitchell, *Phys. Rev. Lett.* **48** (1988) 816

[149] M. Ross, W. J. Nellis & A. C. Mitchell, *Chem. Phys. Lett.* **68** (1979) 532

[150] W. J. Nellis, M. Ross, A. C. Mitchell *et al.*, *Phys. Rev. A: Gen. Phys.* **27** (1983) 608

[151] V. E. Bespalov, V. K. Gryaznov & V. E. Fortov, *Zh. Eksp. Teor. Fiz.* (*Soviet Physics — JETP*) **76** (1979) 140

[152] V. E. Bespalov, M. I. Kulish & V. E. Fortov, *Teplofiz. Vis. Temp.* (*High Temperature*) **24** (1986) 995

[153] H. Hess, *Contrib. Plasma Phys.* **29** (1989) 473

[154] V. E. Fortov, A. N. Dremin, A. A. Leontiev *et al.*, *Teplofiz. Vis. Temp.* (*High Temperature*) **13** (1975) 1072

[155] V. E. Fortov, A. N. Dremin & A. A. Leontiev, *Zh. Eksp. Teor. Fiz., Pisma* (*Soviet Physics — JETP Letters*) **20** (1974) 30

[156] A. A. Leontiev, V. E. Fortov & A. N. Dremin, in *Combustion and Flame* (Nauka, Moscow, 1977), pp. 515–521

[157] L. V. Al'tshuler, A. A. Bakanova, A. V. Bushman *et al.*, *Zh. Eksp. Teor. Fiz.* (*Soviet Physics — JETP*) **73** (1977) 1866

[158] A. A. Leontiev & V. E. Fortov, *Zh. Prikl. Mekh. Tekhn. Fiz.* N 3 (1974) 162

[159] V. E. Fortov & A. A. Leontiev, *Teplofiz. Vis. Temp.* (*High Temperature*) **14** (1976) 711

[160] A. V. Bushman & V. E. Fortov, *Uspekhi Fiz. Nauk* (*Soviet Physics — Uspekhi*) **140** (1983) 177

[161] S. B. Kormer, L. I. Funtikov, V. D. Urlin *et al.*, *Zh. Eksp. Teor. Fiz.* (*Soviet Physics — JETP*) **42** (1962) 686

[162] V. D. Urlin, S. B. Kormer & L. T. Popova, *Fiz. Tverd. Tela* **3** (1961) 2131

[163] Ph. de Beaumont & L. J. Leygonie, in *Proc. 5th Int. Symp. Detonation* (Pasadena, 1970), pp. 430–442

[164] K. Hornung & K. W. Michel, *J. Chem. Phys.* **56** (1972) 2072

[165] G. J. Kerley & J. Wise, in *Shock Waves in Condensed Matter*, ed. S. C. Shmidt & N. C. Holmes (Elsevier, New York, 1987), pp. 155–158

[166] J. R. Assay, T. G. Trucano & L. C. Chhabildas, *ibid.*, 159–162

[167] A. V. Bushman, B. L. Glushak & V. E. Fortov, *Zh. Eksp. Teor. Fiz., Pisma* (*Soviet Physics — JETP Letters*) **44** (1986) 375

[168] O. V. Badanov, V. E. Bespalov, V. A. Ternovoi *et al.*, *Teplofiz. Vis. Temp.* (*High Temperature*) **23** (1985) 976

[169] A. N. Zharkov, B. L. Glushak, V. E. Fortov *et al.*, *Zh. Eksp. Teor. Fiz.* (*Soviet Physics — JETP*) **96** (1989) 1301

[170] V. G. Ageev, A. V. Bushman, V. E. Fortov *et al.*, *Zh. Eksp. Teor. Fiz., Pisma* (*Soviet Physics — JETP Letters*) **48** (1988) 608

[171] M. V. Zhernokletov, V. M. Zhubarev, Yu. N. Sutulov *et al.*, *Zh. Prikl. Mekh. Tekhn. Fiz.* N 1 (1984) 119

[172] V. E. Fortov & A. N. Dremin, *Fiz. Gorenija i Vzriva* **9** (1973) 743

[173] C. Scidmore & E. Morris, in *Proc. Symp. Thermodynamics Nuclear Materials* (J.A.E.A., Vienna, 1962), pp. 173–216

[174] M. M. Rusakov, R. I. Ivanov, B. K. Shaidullin *et al.*, *Teplofiz. Vis. Temp.* (*High Temperature*) **15** (1977) 449

[175] Yu. A. Alekseev, V. P. Ratnikov & A. P. Ribakov, *Zh. Prikl. Mekh. Tekhn. Fiz.* N 2 (1971) 101

[176] A. V. Bushman, B. L. Glushak, V. K. Gryaznov *et al.*, in *Proc. 7th All-Union Symp. Combustion and Flame* (Inst. Chem. Phys. USSR Acad. Sci., Chernogolovka, 1986), pp. 123–126

[177] A. V. Bushman, I. K. Krasyuk, V. E. Fortov *et al.*, *Zh. Eksp. Teor. Fiz., Pisma* (*Soviet Physics — JETP Letters*) **39** (1984) 341

[178] E. I. Zababakhin, in *The Mechanics in USSR for 50 Years* (Nauka, Moscow, 1970), pp. 313

[179] A. V. Bushman & V. E. Fortov, in *Thermophysics, Vol. 1*, ed. V. Fortov & A. Sheindlin (Harwood. Academic Publishing, New York), pp. 162–181

[180] L. P. Volkov, N. P. Vladimirov, E. N. Avrorin *et al.*, *Zh. Eksp. Teor. Fiz., Pisma* (*Soviet Physics — JETP Letters*) **31** (1980) 623

[181] A. V. Bushman, G. I. Kanel, A. L. Ni & V. E. Fortov, *Thermophysics and Dynamics of Intense Pulsed Loading* (Int. Chem. Phys. USSR Acad. Sci., Chernogolovka, 1988); *Intense Dynamic Loading of Condensed Matter* (Taylor & Francis, 1993)

[182] E. N. Avrorin, B. K. Vodolaga, V. A. Simonenko *et al.*, *High Power Shock Waves and Extreme States of Matter* (High Temper. Inst. USSR Acad Sci., Moscow, 1990)

[183] L. P. Volkov, N. P. Volosh, R. A. Mangasarov *et al.*, *Zh. Eksp. Teor. Fiz., Pisma* (*Soviet Physics — JETP Letters*) **31** (1980) 546

[184] I. Sh. Model & A. T. Narozhnii, *Zh. Eksp. Teor. Fiz., Pisma* (*Soviet Physics — JETP Letters*) **41** (1985) 270

[185] S. I. Anisimov, A. M. Prokhorov & V. E. Fortov, *Uspekhi Fiz. Nauk* (*Soviet Physics — Uspekhi*) **142** (1984) 395

[186] J. Duderstadt & G. Modes, *Inertial Confinement Synthesis*. Transl. from Engl. (Energoatomizdat, Moscow, 1985)

[187] A. F. Akerman, B. A. Demidov, V. E. Fortov *et al.*, *Application of Strong Current Relativistic Electron Beams in Dynamic Physics of High Pressures and Temperatures* (Inst. Chem. Phys. USSR Acad. Sci., Chernogolovka, 1986)

[188] A. V. Bushman, O. Yu. Vorobiev, V. E. Fortov *et al.*, *Numerical Simulation of High Power Ion Beam Effects on Metal Targets* (Inst. Chem. Phys. USSR Acad. Sci., Chernogolovka, 1987)

[189] R. J. Trainor, M. C. Graboske & K. S. Long, *Rep. VCRL-52565, 80257, 82141*

[190] R. M. Hore, H. J. Schwar & H. Hoval, in *Laser Interaction and Related Plasma Phenomena, Vol. 5* (Plenum, New York, 1981), pp. 253–277

[191] R. Sigel, *Phys. Rev.* **33** (1974) 1020

[192] L. Vessel & J. Solem, *Phys. Rev. Lett.* **40** (1978) 1391

[193] R. J. Trainor, J. W. Shaner, J. M. Auerbach & N. C. Holmes, *Phys. Rev. Lett.* **42** (1979) 1154

[194] R. J. Trainor, N. C. Holmes & R. A. Anderson, in *Shock Waves in Condensed Matter-81*, ed. W. J. Nellis, L. Seaman & R. A. Grolham (North Holland, New York, 1982), pp. 145
[195] N. C. Holmes & R. J. Trainor, *Phys. Rev. A* **25** (1982) 160
[196] P. D. Gold Stone, R. F. Benjamin & R. B. Schaltz, *Phys. Lett.* **38** (1979) 761
[197] M. D. Rosen & D. W. Phillipson, *Shock Waves in Condensed Matter-83*, ed. J. R. Assay, R. A. Graham & G. K. Straub (Elsevier, New York, 1984), pp. 323
[198] Yu. A. Bondarenko, I. N. Burdonskii, V. V. Gavrilov *et al.*, *Zh. Eksp. Teor. Fiz.* (*Soviet Physics — JETP*) **81** (1981) 170
[199] J. S. Rennet, *Appl. Phys. Lett.* **38** (1981) 226
[200] A. Ng, D. Parfeniuk & A. DaSilva, *Phys. Rev. Lett.* **54** (1985) 2604
[201] A. Ng, D. Parfeniuk & A. DaSilva, *Phys. Rev. Lett.* **56** (1986) 1595
[202] T. A. Hall & A. Djaamet, *Phys. Rev. Lett.* **60** (1988) 2004
[203] J. P. Roman & F. Cottete, *Physica* **139** (1986) 545
[204] B. A. Demidov, A. I. Martinov, L. I. Rudakov *et al.*, *Zh. Eksp. Teor. Fiz.* (*Soviet Physics — JETP*) **80** (1981) 738
[205] A. F. Akkerman, A. V. Bushman, V. E. Fortov *et al.*, *Zh. Eksp. Teor. Fiz.* (*Soviet Physics — JETP*) **91** (1986) 1762
[206] F. C. Perry & H. M. Winder, *J. Appl. Phys.* **47** (1976) 127
[207] S. A. Bogolyubskii & B. L. Gerasimov, *Zh. Eksp. Teor. Fiz., Pisma* (*Soviet Physics — JETP Letters*) **24** (1976) 202
[208] B. M. Manzon, *Uspekhi Fiz. Nauk* (*Soviet Physics — Uspekhi*) **134** (1981) 611
[209] S. C. Rashleigh & R. A. Marshall, *J. Appl. Phys.* **49** (1978) 2540
[210] M. M. Kondratenko, E. F. Lebedev, V. E. Fortov *et al.*, *Teplofiz. Vis. Temp.* (*High Temperature*) **26** (1988) 159
[211] K. F. Froschner & R. S. Lee, *Shock Waves in Condensed Matter-83*, ed. J. R. Assay, R. A. Graham & G. K. Straub (Elsevier, New York, 1984), pp. 87
[212] A. A. Vedenov & A. I. Larkin, *Zh. Eksp. Teor. Fiz.* (*Soviet Physics — JETP*) **36** (1959) 1133
[213] V. P. Kopishev, *Zh. Eksp. Teor. Fiz.* (*Soviet Physics — JETP*) **55** (1968) 1304
[214] A. A. Likal'ter, *Zh. Eksp. Teor. Fiz.* (*Soviet Physics — JETP*) **56** (1969) 240
[215] A. I. Larkin, *Zh. Eksp. Teor. Fiz.* (*Soviet Physics — JETP*) **38** (1960) 1896
[216] W. D. Kraeft, D. Kremp, W. Ebeling & G. Röpke, *Quantum Statistics of Charged Particle Systems* (Akad. Verlag, Berlin, 1985)
[217] V. A. Alekseev, V. E. Fortov & I. T. Iakubov, in *Proc. 15th Int. Conf. Phenomena in Ionized Gases* (Minsk, 1981), pp. 73–85
[218] N. V. Iermokhin, P. P. Kulik, V. A. Ryabyi *et al.*, *J. Phys.* **39** (1978) C1-200
[219] W. Freyland, *Phys. Rev. B* **20** (1979) 5104
[220] I. T. Iakubov & A. A. Likal'ter, *Contr. Plasma Physics* **27** (1987) 479
[221] F. Hensel, *Ber. Bunsenges. Phys. Chem.* **80** (1976) 786
[222] A. A. Vedenov, in *Int. Conf. Quiscent Plasma. Abstracts* (Fraskatti, 1968), pp. 107–108
[223] V. A. Alekseev & A. A. Vedenov, *Uspekhi Fiz. Nauk* (*Soviet Physics — Uspekhi*) **102** (1970) 665

[224] B. M. Smirnov, *Dokl. Akad. Nauk SSSR* (*Soviet Physics — Doklady*) **195** (1970) 75
[225] J. Popielawski & J. Gryko, *J. Chem. Phys.* **66** (1977) 2257
[226] A. G. Khrapak & I. T. Iakubov, *Uspekhi Fiz. Nauk* (*Soviet Physics — Uspekhi*) **129** (1979) 45
[227] A. N. Lagar'kov & I. T. Iakubov, *Chemistry of Plasma, Issue 7*, ed. B. M. Smirnov (Atomizdat, Moscow, 1980), pp. 75–109
[228] R. Morrow & J. D. Grags, *J. Phys. D* **6** (1973) 1274
[229] A. A. Likal'ter, *Teplofiz. Vis. Temp.* (*High Temperature*) **16** (1978) 1219
[230] J. L. Magee & K. Funabashi, *Radiat. Res.* **10** (1959) 622
[231] P. A. Montano, G. K. Shenoy, E. E. Alp *et al.*, *Phys. Rev. Lett.* **56** (1986) 2076
[232] Ya. B. Zel'dovich & L. D. Landau, *Acta Phys. — Chim. USSR* **18** (1943) 194
[233] G. Franz, W. Freyland & F. Hensel, *J. de Phys.* **41** (1980) 70
[234] N. Mott & E. Devis, *Electron Processes in Non-Crystal Substances.* Transl. from Engl. (Mir, Moscow, 1974)
[235] Ya. I. Frenkel, *Kinetic Theory of Liquids* (Nauka, Leningrad, 1975)
[236] A. G. Khrapak & I. T. Yakubov, *Zh. Eksp. Teor. Fiz.* (*Soviet Physics — JETP*) **59** (1971) 945
[237] A. N. Starostin, *Some Problems of Non-ideal Plasma Theory. Theses* (Moscow, 1971)
[238] R. Redmer & G. Röpke, *Physica* **130** (1985) 523
[239] W. Richert, S. A. Insepov & W. Ebeling, *Ann. Phys.* **41** (1984) 139
[240] G. E. Norman & A. N. Starostin, *Teplofiz. Vis. Temp.* (*High Temperature*) **8** (1970) 413
[241] B. M. Smirnov, *Atomic Collision and Elementary Processes* (Atomizdat, Moscow, 1968)
[242] A. Ya. Polishchuk, *Teplofiz. Vis. Temp.* (*High Temperature*) **23** (1985) 10
[243] Yu. K. Gus'kov, R. V. Savvov & V. A. Slobodniuk, *Zh. Tekhn. Fiz.* (*Soviet Physics — JTP*) **48** (1978) 277
[244] G. Messi & E. Barchop, *Electron and Ion Collisions.* Transl. from Engl. (Inostr. Literatura, Moscow, 1958)
[245] E. Karule, *J. Phys. B* **5** (1972) 2051
[246] D. W. Norcross, *Phys. Rev. Lett.* **32** (1974) 192
[247] T. Hill, *Statistical Mechanics.* Transl. from Engl. (Inostr. Literatura, Moscow, 1960)
[248] B. M. Smirnov, *Claster Ions* (Nauka, Moscow, 1983)
[249] P. J. Forster, R. E. Leckenby & E. J. Robbins, *J. Phys. B* **2** (1969) 478
[250] A. Herrmann, S. Schumacher & L. Woste, *J. Chem. Phys.* **68** (1978) 2327
[251] M. H. Kappes, M. Schär, P. Radi & S. Schumacher, *J. Chem. Phys.* **84** (1986) 1863
[252] K. Rademann, B. Kaiser, U. Even & F. Hensel, *Phys. Rev. Lett.* **59** (1987) 2319
[253] Yu. P. Korchevoi & V. N. Makarchuk, *Ukr. Fiz. Zh.* **24** (1979) 799
[254] A. G. Khrapak, *Teplofiz. Vis. Temp.* (*High Temperature*) **17** (1979) 1147
[255] I. T. Iakubov, *Dokl. Akad. Nauk SSSR* (*Soviet Physics — Doklady*) **247** (1979) 841

[256] A. N. Lagarkov & A. K. Sarychev, *J. Phys.* **40** (1979) C7-691
[257] J. P. Hernandez, G. Schonherr, W. Gotzlaff & F. Hensel, *J. Phys. Chem.* **17** (1984) 4421
[258] *Electrons in Fluids*, ed. J. Jortner & N. R. Kestner (Springer, Berlin, 1974)
[259] I. T. Iakubov & A. G. Khrapak, *Rep. Progr. Phys.* **45** (1982) 697
[260] I. T. Iakubov, A. G. Khrapak, V. V. Pogosov & S. A. Trigger, *Solid-State Commun.* **60** (1986) 377
[261] V. V. Pogosov & A. G. Khrapak, *Teplofiz. Vis. Temp.* (*High Temperature*) **26** (1988) 209
[262] D. I. Zhukhovitskii, *Teplofiz. Vis. Temp.* (*High Temperature*) **27** (1989) 15
[263] N. B. Vargaftic, L. D. Voliak & V. G. Stepanov, *Handbook of Thermodynamic and Transport Properties of Alkali Metals*, ed. R. W. Ohse (Blackwell Scientific Publications, Oxford, 1985), pp. 641–666
[264] A. A. Borzhievskii, V. A. Sechenov & V. I. Horunzhenko, *Teplofiz. Vis. Temp.* (*High Temperature*) **26** (1988) 722
[265] A. A. Likal'ter, *Teplofiz. Vis. Temp.* (*High Temperature*) **20** (1982) 1076
[266] V. S. Vorobiev & A. A. Likal'ter, *The Chemistry of Plasma, Issue 15*, ed. B. M. Smirnov (Energoatomizdat, Moscow, 1988), pp. 163–207
[267] A. A. Likal'ter, *Teplofiz. Vis. Temp.* (*High Temperature*) **23** (1985) 465
[268] D. A. Young & B. J. Alder, *Phys. Rev. A* **3** (1971) 363
[269] A. V. Bushman, A. L. Ni & V. E. Fortov, in *Equations of State for Extreme Conditions* (IPTM Siber. Branch USSR Acad. Sci., Novosibirsk, 1981), pp. 178
[270] W. E. Freyland & F. Hensel, *Ber. Bunsenges. Phys. Chem.* **76** (1972) 16
[271] I. G. Dillon, P. A. Nelson & B. S. Swanson, *J. Chem. Phys.* **44** (1966) 4229
[272] S. Brush, H. L. Sahlin & E. Teller, *J. Chem. Phys.* **45** (1966) 2102
[273] M. Baus & J. P. Hansen, *Phys. Repts.* **59** (1980) 228
[274] H. E. DeWitt, *Strongly Coupled Plasmas*, ed. G. Kalman (Plenum, New York, 1978), pp. 83
[275] S. Ichimaru, *Rev. Mod. Phys.* **54** (1982) 1017
[276] J. P. Hansen, *Phys. Rev. A: Gen. Phys.* **8** (1973) 3110
[277] E. L. Pollock & J. P. Hansen, *Phys. Rev. A: Gen. Phys.* **8** (1973) 3096
[278] H. E. DeWitt & W. B. Hubbard, *Astrophys. J.* **205** (1976) 295
[279] S. Galam & J. P. Hansen, *Phys. Rev. A: Gen. Phys.* **14** (1976) 816
[280] K. C. Ng, *J. Chem. Phys.* **61** (1974) 2680
[281] W. L. Slattery, G. D. Doolen & H. E. DeWitt, *Phys. Rev. A: Gen. Phys.* **21** (1980) 2087
[282] E. H. Lieb & H. J. Narnhofer, *J. Stat. Phys.* **12** (1975) 291
[283] R. C. Gann, Sudip Chakravatry & G. V. Chester, *Phys. Rev.* **20** (1979) 326
[284] D. S. Fisher, B. I. Halperin & P. M. Platzman, *Phys. Rev. Lett.* **42** (1979) 798
[285] S. Ichimaru, *Strongly Coupled Plasmas*, ed. G. Kalman (Plenum, New York, 1978), pp. 189
[286] V. M. Zamalin, G. E. Norman & V. S. Filinov, *Monte Carlo Method in Statistical Thermodynamics* (Nauka, Moscow, 1977)
[287] N. P. Kovalenko & I. Z. Fisher, *Uspekhi Fiz. Nauk* (*Soviet Physics — Uspekhi*) **108** (1972) 209

[288] J. Rashbruk, *Physics of Simple Liquids*, ed. G. Temperly & J. Roulingson. Transl. from Engl. (Mir, Moscow, 1971)

[289] Y. Rosenfeld & N. W. Ashcrott, *Phys. Rev. A* **20** (1979) 1208

[290] J. F. Springer, M. A. Pokrant & F. A. Stevens, *J. Chem. Phys.* **58** (1973) 4865

[291] H. E. DeWitt, *Phys. Rev.* **14** (1976) 1290

[292] V. D. Gorobchenko & E. G. Maksimov, *Uspekhi Fiz. Nauk* (*Soviet Physics — Uspekhi*) **130** (1980) 65

[293] W. Ebeling, A. Förster, V. E. Fortov *et al.*, *Thermophysical Properties of Hot Dense Plasma* (Teubner Verlagsgeselschaft, Stuttgart, 1991)

[294] H. E. DeWitt & Rosenfeld, *Phys. Lett.* **75** (1979) 79

[295] E. P. Wigner, *Phys. Rev.* **46** (1934) 1002

[296] I. M. Zidil'kovskii, *Uspekhi Fiz. Nauk* (*Soviet Physics — Uspekhi*) **152** (1987) 583

[297] S. L. Gilbert, J. J. Bollinger & D. J. Wineland, *Phys. Rev. Lett.* **60** (1988) 2022

[298] T. A. Hall, A. Djaoui & R. W. Eason, *Phys. Rev. Lett.* **60** (1988) 2034

[299] A. W. Overhauser, *Phys. Rev.* **167** (1968) 691; *Adv. Phys.* **27** (1978) 343

[300] C. Deutsch, Y. Furutani & M. M. Gombert, *Phys. Repts.* **69** (1981) 1358

[301] F. H. Stillinger & D. Lovett, *J. Chem. Phys.* **49** (1968) 1991

[302] B. Jankovic, *J. Stat. Phys.* **17** (1977) 357

[303] E. E. Salpeter, *Ausr. J. Phys.* **7** (1954) 373

[304] H. C. Graboske, H. E. DeWitt, A. S. Grosmann & M. S. Cooper, *Astrophys. J.* **181** (1973) 457

[305] R. Abe, *Progr. Theor. Phys.* **22** (1959) 21

[306] F. J. Rogers & H. E. DeWitt, *Phys. Rev.* **8** (1973) 1061

[307] Ph. Choguard, in *Strongly Coupled Plasmas*, ed. G. Kalman (Plenum Press, New York, 1978), pp. 83–98

[308] V. K. Gryaznov & I. L. Iosilevskii, in *Thermophysical Properties of Low Temperature Plasmas*, ed. V. M. Ievlev (Nauka, Moscow, 1976), pp. 25–37

[309] V. K. Gryaznov & I. L. Iosilevskii, in *Numerical Methods of Solid-State Mechanics, Issue 4* (Siber. Branch of USSR Acad. Sci., Novosibirsk, 1973), pp. 166–184

[310] E. G. Cohen & T. J. Murphy, *Phys. Fluids* **12** (1969) 1404

[311] A. J. Gutman, B. W. Ninham & C. J. Tompson, *Phys. Lett.* **26** (1968) 180

[312] D. D. Garley, *Phys. Rev. A: Gen. Phys.* **10** (1974) 863

[313] C. W. Hirt, *Phys. Fluids* **10** (1967) 565

[314] M. J. Gillan, *J. Phys. C: Solid State Phys.* **7** (1974) 1

[315] J. L. Lebowitz & J. K. Percus, *Phys. Rev.* **144** (1966) 251

[316] J. P. Hansen, G. M. Torrie & P. Viellefosse, *Phys. Rev. A: Gen. Phys.* **16** (1977) 2153

[317] D. J. Stevenson, *Phys. Rev. B* **12** (1975) 3999

[318] E. L. Pollock & B. J. Alder, *Phys. Rev. A* **15** (1977) 1263

[319] G. E. Norman & A. N. Starostin, *Teplofiz. Vis. Temp.* (*High Temperature*) **8** (1970) 413

[320] V. Ebeling, V. Kraeft & D. Kremp, *The Theory of Bound States and Ionization Equilibrium in Plasmas and Solid States*. Transl. from Engl. (Mir, Moscow, 1979)

[321] M. Gell'Mann & K. A. Bruckner, *Phys. Rev.* **106** (1957) 364

[322] N. C. Graboske, Jr. D. J. Harwood & H. E. DeWitt, *Phys. Rev. A* **13** (1971) 1419

[323] H. E. DeWitt, *J. Math Phys.* **3** (1962) 1216

[324] H. C. Graboske, D. J. Harwood & F. J. Rogers, *Phys. Rev.* **186** (1969) 210

[325] V. B. Zelener, G. E. Norman & V. S. Filinov, *The Theory of Pertubations and Pseudo-Potential in Statistical Thermodynamics* (Nauka, Moscow, 1981)

[326] W. Ebeling & R. Sanding, *Ann. Phys.* **13** (1964) 219

[327] A. A. Barker, *J. Chem. Phys.* **55** (1971) 4

[328] W. Ebeling & R. Sanding, *Ann. Phys.* **28** (1973) 289

[329] I. L. Iosilevskii, *Teplofiz. Vis. Temp.* (*High Temperature*) **18** (1980) 447

[330] I. L. Iosilevskii, *Teplofiz. Vis. Temp.* (*High Temperature*) **23** (1985) 1041

[331] V. H. Armstrong, in *Progr. High Temperature Physics and Chemistry, Vol. 1* (Pergamon Press, New York, 1967), pp. 231–248

[332] V. E. Fortov, B. N. Lomakin & Yu. G. Krasnikov, *Teplofiz. Vis. Temp.* (*High Temperature*) **9** (1971) 869

[333] A. I. Larkin, *Zh. Eksp. Teor. Fiz.* (*Soviet Physics — JETP*) **38** (1960) 1896

[334] A. Yu. Savukinas & A. P. Fidjunas, *Litovsk. Fiz. Sborn.* **14** (1974) 73

[335] V. K. Gryaznov, A. G. Ivanov *et al.*, *Zh. Strukt. Khim.* **30** (1989) 132

[336] N. F. Carnahan & K. E. Starling, *J. Chem. Phys.* **51** (1969) 635

[337] V. P. Kopishev, *Preprint* N 59 (IPM, Moscow, 1978)

[338] R. Latter, *Phys. Rev.* **99** (1955) 1854; *J. Chem. Phys.* **24** (1959) 280

[339] N. N. Kalitkin, *Zh. Eksp. Teor. Fiz.* (*Soviet Physics — JETP*) **38** (1960) 1534

[340] H. H. Kalitkin, L. V. Kuzmina & V. S. Rogov, *Preprint* N 21 (IPM, Moscow, 1972)

[341] L. V. Al'tshuler, N. N. Kalitkin & L. V. Kuzmina, *Zh. Eksp. Teor. Fiz.* (*Soviet Physics — JETP*) **72** (1977) 317

[342] N. N. Kalitkin & L. V. Kuzmina, *Preprint* N 35 (IPM, Moscow, 1975); *Preprint* N 14 (IPM, Moscow, 1979)

[343] I. L. Iosilevskii & V. K. Gryaznov, *Teplofiz. Vis. Temp.* (*High Temperature*) **19** (1981) 1121

[344] *The Theory of Non-Homogeneous Electron Gas*, ed. S. Lundkvist & H. March. Transl. from Engl. (Mir, Moscow, 1987)

[345] A. K. Rajagopal, *Adv. Chem. Phys.* **41** (1980) 59

[346] N. I. Kluchnikov & I. A. Lubimova, *Reviews on Thermophysical Properties of Substances*, N 4(60) (Inst. High Temperatures USSR Acad. Sci., Moscow, 1987), pp. 1–104

[347] D. A. Liberman, *Phys. Rev. B* **20** (1979) 4981

[348] G. V. Sin'ko, *Teplofiz. Vis. Temp.* (*High Temperature*) **21** (1983) 1041

[349] A. F. Nikiforov, V. G. Novikov & V. B. Uvarov, *Vopr. Atomn. Nauki i Tekhniki, Ser. Methods and Programs*, N 4 (6) (1979) 16

[350] V. G. Novikov, *Preprint* N 133 (IPM, Moscow, 1985)

[351] M. W. C. Dharma-Wardana & F. Perrot, *Phys. Rev. A* **26** (1982) 2096

[352] R. G. Ross & D. A. Greenwood, *Progr. Mater. Sci.* **14** (1971) 173

[353] I. Ya. Dikhter & V. A. Zeigarnik, *Teplofiz. Vis. Temp.* (*High Temperature*) **15** (1977) 471
[354] W. J. Nellis, M. Van-Thiel & A. C. Mitchell, *Phys. Rev. Lett.* **48** (1979) 916
[355] B. N. Lomakin, V. E. Fortov & O. E. Shchekotov, *Teplofiz. Vis. Temp.* (*High Temperature*) **8** (1970) 154, 620, 869
[356] R. N. Keeler, M. Van-Thiel & B. J. Alder, *Physica* (*Utrecht*) **31** (1965) 1437
[357] M. Ross & A. K. McMahan, *Phys. Rev. B* **21** (1980) 1658
[358] M. Ross, W. Nellis & A. C. Mitchell, *Law. Liv. Nat. Lab. Rept UCRL-82316* (LLNL, Livermore, 1979)
[359] L. P. Volkov, N. P. Volosh & G. V. Sin'ko, *Zh. Eksp. Teor. Fiz., Pisma* (*Soviet Physics — JETP Letters*) **31** (1980) 546
[360] V. K. Gryaznov, I. L. Iosilevskii & V. E. Fortov, *Zh. Eksp. Teor. Fiz., Pisma* (*Soviet Physics — JETP Letters*) **8** (1982) 1378
[361] D. A. Liberman, *J. Quantum. Spectrosc. Rad. Transfer* **27** (1982) 335
[362] S. B. Kormer, A. I. Funtikov & V. D. Urlin, *Zh. Eksp. Teor. Fiz.* (*Soviet Physics — JETP*) **42** (1962) 686
[363] G. E. Norman, *Notes on Physics and Chemistry of Low Temperature Plasmas*, ed. L. S. Pollak (Nauka, Moscow, 1971)
[364] E. G. Maksimov & O. V. Dolgov, *Uspekhi Fiz. Nauk.* (*Soviet Physics — Uspekhi*) **135** (1981) 260
[365] G. Carmi, *J. Math. Phys.* **9** (1968) 2120
[366] D. Ceperrey, *Phys. Rev.* **18** (1978) 3126
[367] V. A. Ivanov, J. N. Makarenko & S. M. Stishov, *Phys. Lett. A* **47** (1974) 75
[368] D. A. Kirzhnits, *Uspekhi Fiz. Nauk.* (*Soviet Physics — Uspekhi*) **119** (1976) 375
[369] N. F. Mott, *Phil. Mag.* **16** (1967) 49
[370] W. Ebeling, A. Forster, W. Richert & H. Hess, *Physica A* **50** (1988) 159
[371] V. E. Fortov, *Teplofiz. Vis. Temp.* (*High Temperature*) **10** (1972) 168
[372] M. M. Martinyuk, *Fiz. Gor. i Vzriva* **13** (1977) 213
[373] J. W. Shaner & G. R. Gathers, in *High Pressure Science and Technology*, ed. K. D. Timmerhous & M. S. Barber (Plenum, New York, 1979), pp. 847
[374] M. M. Rusakov, *Teplofiz. Vis. Temp.* (*High Temperature*) **13** (1975) 20; M. M. Rusakov, R. I. Ivanov, B. K. Shaidullin *et al.*, *ibid.* **15** (1977) 449
[375] M. A. Coock, *J. Appl. Phys.* **29** (1958) 1612; *ibid.* **26** (1955) 426; *J. Chem. Phys.* **24** (1955) 60; *Proc. Roy. Soc. A* **508** (1961)
[376] M. A. Coock, *The Science of High Explosives* (Publ. Comp., Reinhold, New York, 1958)
[377] M. A. Coock, *J. Appl. Phys.* **52** (1981) 1881
[378] V. E. Fortov, S. I. Musiankov, V. V. Yakushev *et al.*, *Teplofiz. Vis. Temp.* (*High Temperature*) **12** (1974) 957
[379] W. C. Davis & A. W. Campbell, *J. Appl. Phys.* **31** (1960) 1225; A. Kendrew & L. Whitbread, in *Proc. 3rd Int. Symp. Detonation* (1960)
[380] N. M. Kuznetsov, *Tables of Thermodynamical Functions and Shock Wave Adiabats for Air at High Temperatures* (Mashinostr., Moscow, 1945)
[381] E. B. Royce, *Phys. Rev.* **164** (1967) 929

[382] R. G. Arkhipov, *Zh. Eksp. Teor. Fiz.* (*Soviet Physics — JETP*) **49** (1965) 1601; *Fiz. Tverd. Tela* **4** (1962) 1077

[383] L. V. Al'tshuler & A. A. Bakonova, *Uspekhi Fiz. Nauk.* (*Soviet Physics — Uspekhi*) **96** (1968) 193

[384] S. G. Brush, H. E. DeWitt & H. E. Trulio, *Phys. Rev.* **8** (1973) 1061

[385] D. A. Young, *Shock Compression of Condensed Matter-1989*, ed. S. C. Schmidt *et al.* (North Holland, New York, 1990), p. 45

[386] R. Reichlin & K. E. Brister, *Phys. Rev. Lett.* **62** (1989) 669

[387] E. M. Lifshits & L. P. Pitaevskii, *Physical Kinetics* (Nauka, Moscow, 1979)

[388] L. J. Kieffer, *Atom. Data* **2** (1971) 293

[389] L. Spitzer, *Physics of Completely Ionized Plasmas.* Transl. from Engl. (Inostr. Liter., Moscow, 1963)

[390] L. I. Podlubnii & V. S. Rostovskii, *Physics of Plasma* **6** (1980) 1339

[391] V. M. Atrazhev & I. T. Iakubov, *Teplofiz. Vis. Temp.* (*High Temperature*) **18** (1980) 1292

[392] I. T. Iakubov & A. Ya. Polishchuk, *J. Phys. B* **15** (1982) 4029

[393] D. Pains & F. Nozieres, *The Theory of Quantum Liquids.* Transl. from Engl. (Mir, Moscow, 1967)

[394] L. I. Podlubnii, V. S. Rostovskii & V. S. Filinov, *Teplofiz. Vis. Temp.* (*High Temperature*) **26** (1988) 218

[395] I. M. Lifshits, *Uspekhi Fiz. Nauk.* (*Soviet Physics — Uspekhi*) **83** (1964) 617

[396] J. Zeiman, *Disorder Models.* Transl. from Engl. (Mir, Moscow, 1982)

[397] B. I. Shklovskii & A. L. Efros, *Electron Properties of Alloyed Semiconductors* (Nauka, Moscow, 1979)

[398] A. N. Lagar'kov & A. K. Sarichev, *Zh. Eksp. Teor. Fiz.* (*Soviet Physics — JETP*) **68** (1975) 641

[399] A. N. Lagar'kov & A. K. Sarichev, *Teplofiz. Vis. Temp.* (*High Temperature*) **16** (1979) 903

[400] A. G. Khrapak & I. T. Iakubov, *Teplofiz. Vis. Temp.* (*High Temperature*) **9** (1971) 1139

[401] R. Redmer, H. Reinholz & G. Ropke, *Private Commun.* (1990)

[402] L. A. Bolshov, A. N. Starostin, *Proc. 10th Int. Conf. Phenomena of Ionized Gases* (Oxford, 1971), pp. 268

[403] A. N. Lagar'kov & A. K. Sarichev, *Teplofiz. Vis. Temp.* (*High Temperature*) **17** (1979) 429

[404] R. Schmutsler & F. Hensel, *Ber. Bunsenges. Phys. Chem.* **76** (1973) 531

[405] F. E. Neale & N. E. Cusack, *J. Phys. F* **9** (1979) 85

[406] Yu. L. Klimontovich, *The Kinetic Theory of Non-ideal Gas and Plasma* (Nauka, Moscow, 1975)

[407] J. Hubbard, *Proc. Roy. Soc.* **261** (1961) 371

[408] H. A. Gould & H. E. DeWitt, *Phys. Rev.* **155** (1967) 68

[409] R. H. Williams & H. E. DeWitt, *Phys. Fluids* **12** (1969) 2326

[410] B. B. Kadomtsev, *Zh. Eksp. Teor. Fiz.* (*Soviet Physics — JETP*) **33** (1957) 151

[411] W. Ebeling & G. Röpke, *Ann. der Phys.* **B36** (1979) 429

[412] G. Röpke, *Phys. Rev. A* **38** (1988) 3001
[413] M. G. Kivelson & D. F. DuBois, *Phys. Fluids* **7** (1964) 1578
[414] S. Ichimaru & S. Tanaka, *Phys. Rev. A* **32** (1985) 1790
[415] F. J. Rogers, H. E. DeWitt & D. B. Boercker, *Phys. Lett. A* **82** (1981) 331
[416] E. I. Asinovsky, A. V. Kirillin, E. P. Pakhomov & V. I. Shabashov, in *Proc. IEEE* **59** (1971) 592
[417] R. S. Devoto, *Phys. Fluids* **16** (1973) 616
[418] A. L. Khomkin, *Teplofiz. Vis. Temp.* (*High Temperature*) **12** (1974) 879
[419] J. Kopainsky, *Z. Phys.* **B248** (1971) 417
[420] J. Bues, H. J. Patt & J. Richter, *Z. Angew. Phys.* **B22** (1967) 345
[421] J. Hackhman & J. Ulenbush, *Proc. 10th Int. Conf. Phenomena in Ionized Gases*, ed. R. N. Franklin (Donald Parsons, Oxford, 1971), p. 260
[422] U. Bauder, R. Devote & D. Mukherjee, *Phys. Fluids* **16** (1973) 2143
[423] C. Goldbach, G. Nollez, S. Popovic & M. Popovic, *Z. Naturforsch. A* **33** (1968) 11
[424] A. A. Bakeev & R. E. Rovinskii, *Teplofiz. Vis. Temp.* (*High Temperature*) **8** (1970) 1121
[425] M. M. Popovic, Y. Vitel & A. A. Mihajlov, in *Strongly Coupled Plasma Physics*, ed. S. Ichimaru (Elsevier, New York, 1990), pp. 561–569
[426] K. Günter, S. Lang & R. Radtke, *J. Phys. D* **16** (1983) 1235
[427] V. S. Vorobiev & A. L. Khomkin, *Teplofiz. Vis. Temp.* (*High Temperature*) **15** (1977) 188
[428] V. E. Fortov & I. T. Iakubov, *Physics of Non-ideal Plasmas* (Inst. Chem. Phys. USSR Acad. Sci., Chernogolovka, 1984)
[429] A. L. Khomkin, *Teplofiz. Vis. Temp.* (*High Temperature*) **16** (1978) 37
[430] A. Mokhgtari, M. Skowronek & Y. Vitel, in *Proc. 9th ESCMPIG* (Lisbon, 1988)
[431] Yu. V. Ivanov, V. B. Mintsev, V. E. Fortov & A. N. Dremin, *Zh. Tekhn. Fiz., Pisma* (*Soviet Physics — JTP Letters*) **2** (1976) 97
[432] V. A. Sechenov, E. E. Son & O. E. Shchekotov, *Zh. Tekhn. Fiz., Pisma* (*Soviet Physics — JTP Letters*) **1** (1975) 891
[433] B. K. Tkachenko, S. I. Titarov, A. V. Karasev *et al.*, *Fiz. Gorenija i Vzriva*, **12** (1976) 763
[434] R. N. Keeler, M. Van Thiel & B. J. Alder, *Physica* **31** (1965) 1437
[435] K. Günter, M. M. Popovic, S. S. Popovic & R. Radtke, *J. Phys. D* **9** (1976) 1139
[436] S. I. Andreev, *Optika i Spektrosk.* **38** (1975) 432
[437] S. G. Barol'skii, H. V. Ermokhin, P. P. Kulic *et al.*, *Teplofiz. Vis. Temp.* (*High Temperature*) **15** (1977) 193
[438] G. A. Pavlov & V. I. Kucherenko, *Teplofiz. Vis. Temp.* (*High Temperature*) **15** (1977) 409
[439] H. M. Milchberg & R. R. Freeman, *Phys. Fluids B* **2** (1990) 1395
[440] V. K. Gryaznov, Yu. V. Ivanov, A. N. Starostin & V. E. Fortov, *Teplofiz. Vis. Temp.* (*High Temperature*) **14** (1976) 643
[441] F. E. Höhne, R. Redmer, G. Röpke & H. Wegener, *Physica A* **128** (1984) 643
[442] M. Schlanges & D. Kremp & H. Kewer, *Ann. der Phys.* **41** (1984) 54

[443] Y. T. Lee & R. M. More, *Phys. Fluids* **27** (1984) 1273
[444] J. P. Hansen & J. R. McDonald, *Phys. Rev. A* **23** (1981) 2041
[445] J. P. Hansen, J. R. McDonald & E. L. Pollok, *Phys. Rev. A* **11** (1975) 1026
[446] I. T. Iakubov, in *Proc. 20th Int. Conf. Phenomena in Ionized Gases, Vol. 4* (Piza, 1991), pp. 881–882
[447] A. A. Likal'ter, *Teplofiz. Vis. Temp.* (*High Temperature*) **25** (1987) 424
[448] V. S. Vorobiev, *Teplofiz. Vis. Temp.* (*High Temperature*) **25** (1987) 323
[449] A. K. Sarichev, *Teplofiz. Vis. Temp.* (*High Temperature*) **23** (1985) 228
[450] Yu. K. Kurilenkov & A. A. Valuev, *Beitr. Plasma Phys.* **24** (1984) 161
[451] N. V. Ermokhin, B. M. Kovalev & P. P. Kulik, *Teplofiz. Vis. Temp.* (*High Temperature*) **15** (1977) 695
[452] N. N. Kalitkin & I. V. Kuz'mina, *Fiz. Plasmi* **2** (1976) 858
[453] S. A. Maev, *Zh. Tekhn. Fiz.* (*Soviet Physics — JTP*) **40** (1970) 567
[454] L. M. Biberman & G. E. Norman, *Uspekhi Fiz. Nauk.* (*Soviet Physics — Uspekhi*) **91** (1967) 193
[455] I. V. Avilova, L. M. Biberman, V. S. Vorobiev *et al.*, *Optical Properties of Hot Air*, ed. L. M. Biberman (Nauka, Moscow, 1970)
[456] Yu. P. Raizer, *The Physics of Gas Discharges* (Nauka, Moscow, 1987)
[457] W. Wiese, M. Smith & B. Glenon, *Atomic Transitions Probabilities* (NBS, N3RDS, Washington, 1966)
[458] Ya. B. Zel'dovich & Yu. P. Raizer, *Physics of Shock Waves and High Temperature Hydrodynamic Phenomena* (Nauka, Moscow, 1966)
[459] L. G. D'yachkov & G. A. Kobzev, in *Chemistry of Plasma, Issue 8*, ed. B. M. Smirnov (Atomizdat, Moscow, 1981), pp. 122–155
[460] H. Griem, *Plasma Spectroscopy.* Transl. from. Engl. (Atomizdat, Moscow, 1969)
[461] H. Griem, *Extension of Spectral Lines in Plasmas.* Transl. from. Engl. (Atomizdat, Moscow, 1976)
[462] M. Baranger, in *Atomic and Molecular Processes*, ed. D. Beits. Transl. from Engl. (Mir, Moscow, 1964), pp. 429–477
[463] V. S. Lisitsa, *Uspekhi Fiz. Nauk.* (*Soviet Physics — Uspekhi*) **122** (1977) 449
[464] P. Klein & D. Meiners, *JQSRT* **17** (1977) 197
[465] G. A. Kobsev, *Optical Properties of Air Plasmas, Preprint IVT N 1-134* (Inst. High. Temper., Moscow, 1984)
[466] G. D. Tsarikis & K. Eidman, *JQSRT* **38** (1987) 353
[467] Bach-Truong, J. Richou, A. Lesage & M. H. Miller, *Phys. Rev. A* **24** (1981) 2550
[468] K. Guenter, H. Hess & R. Radtke, in *Proc. 17th Int. Conf. Phenomena in Ionized Gases, Invited Lectures* (Budapest, 1985), pp. 68–82
[469] M. Kettlitz, R. Radtke, R. Spanke & L. Hitzschke, *JQSRT* **34** (1985) 275
[470] V. A. Sechenov & V. N. Novikov, in *Proc. 18th Int. Conf. Phenomena in Ionized Gases, Vol. 1*, (Swansea, 1987), pp. 242–243
[471] B. H. Armstrong, *JQSRT* **4** (1964) 207
[472] V. S. Vorobiev, *Teplofiz. Vis. Temp.* (*High Temperature*) **16** (1978) 464
[473] V. M. Batenin & A. V. Minaev, *The Chemistry of Plasmas, Issue 2*, ed. B. M. Smirnov (Atomizdat, Moscow, 1975), pp. 199–221

[474] V. G. Sevast'yanenko, *Preprints N 30 and N 32* (ITPM Siber. Branch USSR Acad. Sci., Novosibirsk, 1980)

[475] G. Ekker, *The Theory of Completely Ionized Plasmas.* Transl. from Engl. (Mir, Moscow, 1974)

[476] R. J. Tighe & C. F. Hooper, Jr., *Phys. Rev. A* **15** (1977) 1773

[477] C. A. Iglesias, J. L. Lebowitx & G. MacGowan, *Phys. Rev. A* **28** (1983) 1667

[478] M. W. C. Dharma-Wardana & F. Perrot, *Phys. Rev. A* **33** (1986) 3303

[479] Yu. K. Kurilenkov & V. S. Filinov, *Teplofiz. Vis. Temp.* (*High Temperature*) **14** (1976) 657

[480] Yu. K. Kurilenkov & V. S. Filinov, *Teplofiz. Vis. Temp.* (*High Temperature*) **18** (1980) 433

[481] D. W. Jones, K. Musiol & W. L. Wiese, *Spectral Line Shapes, Vol. 2* (Walter de Cruyter, New York, 1983)

[482] L. Boder & R. S. Tankin, *JQSRT* **10** (1970) 991

[483] H. Gündel, *Beitr. Plasma Phys.* **1** (1971) 1

[484] E. E. Antonov & V. I. Popovich, in *6th All-Union Conf. Physics of Low Temperature Plasmas, Abstracts, Vol. 2* (Lening. Inst. Nuclear Phys., Lenigrad), pp. 272–275

[485] G. A. Kobzev, Yu. K. Kurilenkov & G. E. Norman, *Teplofiz. Vis. Temp.* (*High Temperature*) **15** (1977) 193

[486] F. E. Höhe & R. Zimmermann, *J. Phys. B* **15** (1982) 2551

[487] L. G. Diachkov, *Optica i Spectrosk.* **61** (1986) 688

[488] L. G. Diachkov, G. A. Kobsev & P. M. Pankratov, *Optica i Spectrosk.* **63** (1987) 250

[489] I. B. Levinson & A. A. Nikitin, in *Handbook on Theoretical Calculations of Atom Spectra Line Intensities* (Nauka, Leningrad, 1962), p. 48

[490] W. L. Wiese, D. S. Kellecher & D. R. Paquette, *Phys. Rev. A* **6** (1972) 1132

[491] R. Radtke & K. Günter, *Beitr. Plasma Phys.* **26** (1986) 129

[492] R. Radtke, F. Serick, R. Spanke & R. Zimmerman, *J. Phys. D* **21** (1988) 535

[493] S. I. Andreev & V. E. Gavrilov, *Zh. Prikl. Spektrosk.* **13** (1970) 988

[494] S. I. Andreev, *Zh. Tekhn. Fiz.* (*Soviet Physics — JTP*) **45** (1975) 1010

[495] V. M. Batenin & P. V. Minaev, *Teplofiz. Vis. Temp.* (*High Temperature*) **15** (1977) 647

[496] V. E. Gavrilov, T. V. Gavrilova & T. N. Fedorova, *Optica i Spectrosc.* **61** (1986) 951

[497] V. A. Sechenov, *Fizika Plasmi* **7** (1981) 1172

[498] A. F. Nikiforov, V. G. Novikov, N. Yu. Orlov & V. B. Uvarov, *Preprint N172* (IPM USSR Acad. Sci., Moscow, 1979)

[499] V. E. Fortov, V. B. Mintsev & Yu. B. Zaporogets, in *Proc. 3rd Int. Workshop on Non-ideal Plasmas, Biesenthal*, ed. R. Radtke & H. Hess (Central Inst. Electron. Phys., Berlin, 1984) pp. 26–29

[500] V. M. Atrazhev & I. T. Iakubov, in *Proc. 19th Int. Conf. Phenomena in Ionized Gases, Vol. 1*, ed. J. M. Labat (Beograd Univ., Beograd, 1989), pp. 62–63

[501] Kim Longhvan, R. H. Pratt & H. H. Tseng, *Phys. Rev. A* **32** (1985) 1693

[502] H. Totsuju, *Phys. Rev. A* **32** (1985) 3005

[503] A. A. Artem'ev & I. T. Iakubov, in *Proc. 14th Int. Symp. Physics of Ionized Gases*, ed. N. Konjevic, L. Tanovic & N. Tanovic (Sarajevo Univ., Sarajevo, 1988), pp. 365–368

[504] M. A. Berkovsky, D. Djordjevic, Yu. K. Kurilenkov & M. M. Popovic, *ibid.*, pp. 361–364

[505] R. Kawakami, K. Mima, H. Totsuju *et al.*, *Phys. Rev. A* **38** (1988) 3618

[506] R. N. Bhatt & T. M. Rice, *Phys. Rev. B* **20** (1979) 466

[507] J. Popielawski, H. Uchtmann & F. Hensel, *Ber. Bunsenges. Phys. Chem.* **83** (1979) 123

[508] H. Overhoff, H. Uchtmann & F. Hensel, *J. Phys. F* **6** (1976) 526

[509] H. Uchtmann, F. Hensel & H. Overhoff, *Philos. Mag. B* **42** (1980) 583

[510] A. N. Lagar'kov & A. K. Sarichev, *Teplofiz. Vis. Temp.* (*High Temperature*) **16** (1978) 903

[511] H. W. Verleur, *J. Opt. Soc. Amer.* **58** (1968) 1356

[512] H. Uchtmann, U. Bruisius, M. Yao *et al.*, *Z. Phys. Chem.* **156** (1988) 151

[513] W. Hefner, R. W. Schmutzler & F. Hensel, *J. de Phys.* **41** (1980) 58

[514] M. Yao, H. Uchtmann & F. Hensel, in *Proc. 3rd Int. Workshop Non-ideal Plasma* (GDR Acad. Sci., Berlin, 1984), pp. 13–16

[515] I. T. Iakubov & A. G. Khrapak, *Soviet Technol. Rev. B, Therm. Phys. Rev.* **2** (1989) 269

[516] S. Sou, *Hydrodynamics of Multi-phase Systems.* Transl. from Engl. (Mir, Moscow, 1971)

[517] M. S. Sodha & S. Guha, *Adv. Plasma Phys.* **4** (1971) 219

[518] D. I. Zhukhovitskii, A. G. Khrapak & I. T. Iakubov, in *Proc. 6th All-Union Conf. Physics of Low Temperature Plasmas* (Leningr. State Univ., Leningrad, 1983), pp. 57–68

[519] D. I. Zhukhovitskii, A. G. Khrapak & I. T. Iakubov, *Teplofiz. Vis. Temp.* (*High Temperature*) **22** (1984) 833

[520] A. A. Arshinov & A. K. Musin, *Dokl. Akad. Nauk SSSR* (*Soviet Physics — Doklady*) **118** (1958) 461

[521] A. A. Arshinov & A. K. Musin, *Dokl. Akad. Nauk SSSR* (*Soviet Physics — Doklady*) **120** (1958) 747

[522] A. A. Arshinov & A. K. Musin, *Radiotekhn. i Electron.* **7** (1962) 890

[523] F. J. Smith, *J. Chem. Phys.* **28** (1958) 746

[524] F. Dennery, in *Proc. Int. Symp. MHD Transformation of Energy, Part 1.* Transl. from Engl., ed. V. A. Popov (USSR Acad. Sci., Moscow, 1966), pp. 104–114

[525] E. P. Zimin, Z. G. Mikhnevich & V. A. Popov, in *Proc. Int. Symp. Electricity from MHD, Vol. 3* (IAEA, Vienna, 1966), pp. 97–106

[526] N. Einbinder, *J. Chem. Phys.* **26** (1957) 948

[527] E. G. Gibson, *Phys. Fluids* **9** (1966) 2389

[528] L. P. Kudrin, *Statistical Physics of Plasmas* (Atomizdat, Moscow, 1974)

[529] E. V. Samuilov, N. V. Voskresenskaya & Ya. P. Rabiner, in *Thermophysical Properties of Gases* (Nauka, Moscow, 1973), pp. 153–160

[530] D. I. Zhukhovitskii & I. T. Iakubov, *Teplofiz. Vis. Temp.* (*High Temperature*) **23** (1985) 842

[531] E. V. Samuilov & A. V. Gorbatov, *Teplofiz. Vis. Temp.* (*High Temperature*) **17** (1979) 1058

[532] A. A. Artemiev & I. T. Iakubov, *Teplofiz. Vis. Temp.* (*High Temperature*) **28** (1990) 1064

[533] D. I. Zhukhovitskii, *Teplofiz. Vis. Temp.* (*High Temperature*) **23** (1985) 1050

[534] L. D. Landau & E. M. Lifshits, *The Theory of the Field* (Nauka, Moscow, 1973)

[535] M. Ya. Amus'ya & V. N. Tsitovich, *Radiation of High Energy Particles, Preprint N 100* (Phys. Inst. USSR Acad. Sci., Moscow, 1988)

[536] L. D. Landau & E. M. Lifshits, *Electrodynamics of Solid Media* (Fizmatgiz, Moscow, 1959)

[537] K. Ohtake, I. Ikezaki, K. Urebe & Y. Mori, in *Proc. 8th Int. Conf. MHD Transformation of Energy* (Inst. High Temper. USSR Acad. Sci., Moscow, 1983), pp. 24–27

[538] G. S. Dragan, A. A. Malgota, S. K. Protas *et al.*, in *Mater. Meet. The Methods of Preparation of Propellants and Construction of Combustion Chambers for Coal MHDES* (USSR Acad. Sci., Moscow, 1984), pp. 191–194

[539] E. P. Zimin & V. A. Popov, in *Proc. Int. Symp. Electricity from MHD-1968, Vol. 4* (IAEA, Warshava, Vena, 1968), pp. 2221–2237

[540] I. T. Iakubov, A. G. Khrapak & D. I. Zhukhovitskii, in *Proc. Int. Symp. Physics of Ionized Gases*, ed. M. Y. Kupera (Beograd Univ., Beograd, 1986), pp. 443–446

[541] G. K. Goncharov, V. I. Koraban', A. V. Kolesnik & I. M. Radyuk, *Kvant. Electron.* **15** (1988) 2575

[542] O. F. Petrov, *Preprint N 3-254* (Inst. High Temperatures USSR Acad. Sci., Moscow, 1988)

[543] A. F. Volkov & Sh. M. Kogan, *Uspekhi Fiz. Nauk.* (*Soviet Physics — Uspekhi*) **96** (1968) 633

[544] G. A. Kobzev & Yu. K. Kurilenkov, *Teplofiz. Vis. Temp.* (*High Temperature*) **16** (1978) 458

[545] I. M. Rutkevich & O. A. Sinkevich, *Teplofiz. Vis. Temp.* (*High Temperature*) **18** (1980) 27

[546] I. M. Rutkevich, *Izv. Akad. Nauk SSSR*, N 4 (1977) 3

[547] I. M. Rutkevich & P. M. Tokar', *Teplofiz. Vis. Temp.* (*High Temperature*) **19** (1981) 1228

[548] V. M. Atrazhev & I. T. Iakubov, *Teplofiz. Vis. Temp.* (*High Temperature*) **18** (1980) 16

[549] E. P. Velikhov, A. S. Kovalev & A. T. Rakhimov, *Physical Phenomena in Gas-Discharge Plasmas* (Nauka, Moscow, 1987)

[550] Yu. P. Raizer, *Zh. Eksp. Teor. Fiz.* (*Soviet Physics — JETP*) **58** (1970) 2127

[551] S. G. Barol'skii, M. B. Ermokhin, P. P. Kulik *et al.*, *Teplofiz. Vis. Temp.* (*High Temperature*) **14** (1976) 702

[552] Ya. B. Zel'dovich, *Zh. Fiz. Khim.* **22** (1948) 27

[553] A. V. Gurevich & L. P. Pitaevskii, *Zh. Eksp. Teor. Fiz.* (*Soviet Physics — JETP*) **46** (1964) 1281
[554] L. M. Biberman, V. S. Vorobiev & I. T. Iakubov, *Dokl. Akad. Nauk SSSR* (*Soviet Physics — Doklady*) **296** (1987) 576
[555] L. D. Landau & E. M. Lifshits, *The Mechanics* (Fizmatgiz, Moscow, 1958)
[556] K. Kilimann & W. Ebeling, *Z. Naturforsch.* **45a** (1990) 613
[557] M. Schlanges, Th. Bornath & D. Kremp, *Phys. Rev.* **38A** (1988) 2174
[558] G. A. Pavlov, V. E. Fortov & A. A. Ovchinnikov, *Dokl. Akad. Nauk SSSR* (*Soviet Physics — Doklady*) **251** (1980) 1085
[559] S. A. Bastandjiyan, A. G. Merzhanov & S. I. Khudyaev, *Dokl. Akad. Nauk SSSR* (*Soviet Physics — Doklady*) **163** (1965) 133
[560] S. V. Vladimirov, *Equations of Mathematical Physics* (Nauka, Moscow, 1967)
[561] G. Korn & T. Korn, *The Handbook on Mathematics for Scientific Workers* (Nauka, Moscow, 1968)
[562] J. Horie, *J. Phys. Chem. Solids* **28** (1967) 1569
[563] L. Davison & R. A. Graham, *Phys. Rept.* **55** (1979) 256
[564] S. M. Stishov, *Uspekhi Fiz. Nauk.* (*Soviet Physics — Uspekhi*) **96** (1968) 467
[565] V. D. Urlin, *Zh. Eksp. Teor. Fiz.* (*Soviet Physics — JETP*) **49** (1965) 485
[566] S. B. Kormer, *Uspekhi Fiz. Nauk.* (*Soviet Physics — Uspekhi*) **94** (1968) 641
[567] Ya. B. Zel'dovich, *The Theory of Shock Waves and Gas Dynamics* (USSR Acad. Sci., Moscow — Leningrad, 1946)
[568] G. D. Kahl & D. S. Mylin, *Phys. Fluids* **12** (1969) 2283
[569] G. Ya. Gelin, *Dokl. Akad. Nauk SSSR* (*Soviet Physics — Doklady*) **119** (1958) 1106; **120** (1958) 730
[570] A. D. Sidorenko, *Dokl. Akad. Nauk SSSR* (*Soviet Physics — Doklady*) **178** (1968) 818
[571] I. V. Zharkov & V. A. Kalinin, *Solids State Equations at High Pressures and Temperatures* (Nauka, Moscow, 1968)
[572] E. I. Zababakhin & V. A. Simonenko, *Zh. Eksp. Teor. Fiz.* (*Soviet Physics — JETP*) **52** (1967) 1317
[573] Z. Kurant & K. Fridrikhs, *Supersound Flows and Shock Waves* (GITTL, Moscow, 1950)
[574] N. M. Kuznetsov, *Zh. Eksp. Teor. Fiz.* (*Soviet Physics — JETP*) **88** (1985) 470
[575] A. G. Ivanov, S. A. Novikov & Yu. I. Garasover, *Fiz. Tverd. Tela* **4** (1962) 249
[576] S. S. Kutateladze, A. A. Borisov & V. E. Nakoryakov, *Dokl. Akad. Nauk SSSR* (*Soviet Physics — Doklady*) **252** (1980) 595
[577] F. D. Bennet, G. D. Kahil & E. H. Wedemeyer, *Exploding Wires, Vol. 3*, ed. W. C. Chace (Plenum, New York, 1964), p. 65; *ibid.*, *Vol. 4* (1968), p. 1
[578] G. W. Paxton, *Phys. Fluids* **10** (1967) 2360
[579] S. P. D'yakov, *Zh. Eksp. Teor. Fiz.* (*Soviet Physics — JETP*) **27** (1954) 288
[580] S. V. Iordanskii, *Prikl. Matem. Mekhan.* **21** (1957) 465
[581] N. S. Freeman, *Proc. Roy. Soc. A* **228** (1955) 341
[582] P. M. Zaidel, *Prikl. Matem. Mekhan.* **24** (1960) 219

[583] V. M. Kantorovich, *Zh. Eksp. Teor. Fiz.* (*Soviet Physics — JETP*) **33** (1957) 1525
[584] J. J. Erpenbek, *Phys. Fluids* **5** (1962) 1181
[585] G. W. Swan & G. R. Fowles, *Phys. Fluids* **18** (1975) 28
[586] G. R. Fowles, *Phys. Fluids* **19** (1976) 227
[587] A. D. Sakharov, P. M. Zaidel, V. N. Mineev *et al.*, *Dokl. Akad. Nauk SSSR* (*Soviet Physics — Doklady*) **159** (1964) 1019
[588] V. N. Mineev & E. V. Savinov, *Zh. Eksp. Teor. Fiz.* (*Soviet Physics — JETP*) **52** (1967) 629
[589] A. S. Plashenov, *Fiz. Gorenija i Vriva* **4** (1968) 95
[590] T. N. Fortova, A. N. Dremin & V. E. Fortov, *Tshisl. Met. Mekhan. Sploshn. Sredy* **4** (1973) 143
[591] N. M. Kuznetsov, *Uspekhi Fiz. Nauk.* (*Soviet Physics — Uspekhi*) **159** (1989) 493
[592] V. E. Fortov, *Zh. Tekhn. Fiz.* (*Soviet Physics — JTP*) **42** (1972) 333
[593] R. W. Griffith, R. J. Sanderman & H. G. Hornung, *J. Phys. D: Appl. Phys.* **9** (1976) 1681
[594] G. K. Tukamaev, *Zh. Tekhn. Fiz., Pisma* (*Soviet Physics — JTP Letters*) **6** (1980) 1239
[595] V. E. Fortov, A. L. Ni & S. A. Sugak, in *Proc. 8th Int. Colloq. Gasgynamics of Explosions and Reactive Systems* (Minsk, 1981), pp. 138–140
[596] A. L. Ni, S. A. Sugak & V. E. Fortov, *Teplofiz. Vis. Temp.* (*High Temperature*) **24** (1986) 564
[597] J. S. Thomsen, *Amer. J. Phys.* **38** (1970) 560
[598] J. Kestin, *Amer. J. Phys.* **29** (1961) 329
[599] J. S. Thomsen & I. J. Hartka, *Amer. J. Phys.* **30** (1962) 26; S. M. Curry & G. K. Henrey, *Amer. J. Phys.* **36** (1968) 838
[600] J. S. Thomsen, *Phys. Fluids* **11** (1968) 1338
[601] M. Ross, *Shock Waves in Condensed Matter*, ed. S. C. Shmidt & N. C. Holmes (North Holland, Amsterdam, 1988), p. 87
[602] W. J. Nellis, *ibid.*, p. 43